IUTAM SYMPOSIUM ON NONLINEAR ANALYSIS OF FRACTURE

SOLID MECHANICS AND ITS APPLICATIONS

Volume 49

Aims and Scope of the Series

The fundamental questions arising in mechanics are: *Why?*, *How?*, and *How much?* The aim of this series is to provide lucid accounts written by authoritative researchers giving vision and insight in answering these questions on the subject of mechanics as it relates to solids.

The scope of the series covers the entire spectrum of solid mechanics. Thus it includes the foundation of mechanics; variational formulations; computational mechanics; statics, kinematics and dynamics of rigid and elastic bodies; vibrations of solids and structures; dynamical systems and chaos; the theories of elasticity, plasticity and viscoelasticity; composite materials; rods, beams, shells and membranes; structural control and stability; soils, rocks and geomechanics; fracture; tribology; experimental mechanics; biomechanics and machine design.

The median level of presentation is the first year graduate student. Some texts are monographs defining the current state of the field; others are accessible to final year undergraduates; but essentially the emphasis is on readability and clarity.

For a list of related mechanics titles, see final pages.

IUTAM Symposium on

Nonlinear Analysis of Fracture

Proceedings of the IUTAM Symposium
held in Cambridge, U.K.,
3–7 September 1995

Edited by

J. R. WILLIS

Department of Applied Mathematics and Theoretical Physics,
University of Cambridge, U.K.

SPRINGER-SCIENCE+BUSINESS MEDIA, B.V.

A C.I.P. Catalogue record for this book is available from the Library of Congress

ISBN 978-94-010-6379-1 ISBN 978-94-011-5642-4 (eBook)
DOI 10.1007/978-94-011-5642-4

Printed on acid-free paper

CONTENTS

PREFACE

This volume constitutes the Proceedings of the IUTAM Symposium on 'Nonlinear Analysis of Fracture', held in Cambridge from 3rd to 7th September 1995. Its objective was to assess and place on record the current state of understanding of this important class of phenomena, from the standpoints of mathematics, materials science, physics and engineering. All fracture phenomena are nonlinear; the reason for inclusion of this qualification in the title was to reflect the intention that emphasis should be placed on distinctive aspects of nonlinearity, not only with regard to material constitutive behaviour but also with regard to insights gained, particularly from the mathematics and physics communities, during the recent dramatic advances in understanding of nonlinear systems in general. The expertise represented in the Symposium was accordingly very wide, and many of the world's greatest authorities in their respective fields participated.

The Symposium remained focussed on issues of practical significance for fracture phenomena, with concentration on aspects that are still imperfectly understood. The most significant unifying issue in this regard is that of scale: this theme was addressed from several perspectives. One important aspect is the problem of passing information on one scale up or down, as an input for analysis at another scale. Although this is not always the case, it may be that the microscopic process of fracture is understood in some particular class of materials. The problem then becomes one of constructing an appropriate model at the macroscopic scale, that retains the essential features of the microscopic process but avoids unmanageable complexity. Dually, considering the problem from the macroscopic end, it is important to assess which particular aspects of the macroscopic stress field interact directly with the fracture process. In the simplest cases, the process is driven by the crack tip singularity in the macroscopic field; then, at least some problems relating to scale disappear. The focus of interest of course is in the régime where this 'singularity dominance' is not realised. 'Fracture process zones', and their extent, vary with the material under consideration: the underlying scales are self-evidently different for composite materials than for metals, for example, most immediately because of their very different microstructural features and length scales. Particularly for brittle materials, local statistical variations can be a major source of sensitivity to scale and variability of performance.

Dynamical problems present a major challenge: crack stability is still not completely resolved, even for materials for which elastic stress analysis is appropriate. The phenomena are of interest both on laboratory and terrestrial scales; indeed, much of the recent progress has been achieved by those whose primary concern is with geophysics.

All of the aspects mentioned above require sound physical modelling coupled with analysis. In some cases the analysis may be of the classical kind, while for others the natural approach is numerical simulation. In all cases, it is important to recognise any simple unifying features, such as may follow from recognition of similarity or scaling. These Proceedings address all of these different strands, and provide a reasonable reflection of understanding as it exists at present.

The Symposium consisted of forty-three lectures, all of which were invited and accorded equal weight in the programme. In addition, two poster sessions allowed a further twenty-three presentations. Only the content of the lectures is reflected in this volume, except that a full record of the programme features as an Appendix. A few of the lectures are not represented, mainly because of prior commitments to publish elsewhere.

The International Scientific Committee responsible for the Symposium comprised the following:

Prof. J.R. Willis (UK) – Chairman

Prof. M.F. Ashby (UK)	Prof. D. Gross (Germany)
Prof. G.I. Barenblatt (UK/Russia)	Prof. J.W. Hancock (UK)
Prof. P. Duxbury (USA)	Prof. R. Madariaga (France)
Prof. L.B. Freund (USA)	Prof. V. Tvergaard (Denmark)

The Committee gratefully acknowledges financial support for the Symposium from the International Union of Theoretical and Applied Mechanics, the United States Office of Naval Research, AEA Technology, the Royal Society, and the International Science Foundation.

The smooth running of the Symposium owes much to the unstinting efforts of Tom Gosling, Anne-Marie Harte, John Huber and Alex Korsunsky, and it would not have happened at all without a great deal of work before, during and after, by Lin Hardiman, to whom particular thanks are due.

SCALING IN NONLINEAR FRACTURE MECHANICS

Z. P. BAŽANT
Departments of Civil Engrg. and Materials Science
Northwestern University, Evanston, IL 60208 USA

Abstract. The paper[1] presents a review of recent results on the problem of size effect (or the scaling problem) in nonlinear fracture mechanics of quasibrittle materials and on the validity of recent claims that the observed size effect may be caused by the fractal nature of crack surfaces. The problem of scaling is approached through dimensional analysis and asymptotic matching. Large-size and small-size asymptotic expansions of the size effect on the nominal strength of structures are presented, considering not only specimens with large notches (or traction-free cracks) but also structures with no notches. Simple size effect formulas matching the required asymptotic properties are given. Regarding the fractal nature of crack surfaces, it is concluded that it cannot be the cause of the observed size effect.

1. Introduction

Scaling is a salient aspect of all physical theories. Nevertheless, little attention has been paid to the problem of scaling or size effect in solid mechanics. Up to the middle 1980's, observations of the size effect on the nominal strength of a structure have generally been explained by Weibull-type theory of random strength. However, recent in-depth analysis (Bažant and Xi, 1991) has shown that this Weibull-type theory does not capture the essential cause of size effect for quasibrittle materials such as rocks, toughened ceramics, concretes, mortars, brittle fiber composites, ice (especially sea ice), wood particle board and paper, in which the fracture process zone is not small compared to structural dimensions and large stable crack growth occurs prior to failure. The dominant source of size effect in these materials is not statistical but consists in the release of stored energy from the structure engendered by a large fracture.

By approximate analysis of energy release from the structure, a simple

[1]Supported partly by NSF grant MSS-911447-6 to Northwestern University and partly by ACBM Center at Northwestern University.

J. R. Willis (ed.), IUTAM Symposium on Nonlinear Analysis of Fracture, 1–12.

size effect law (Bažant 1983, 1984) has been derived for quasibrittle fracture. This law subsequently received extensive justifications, based on: (1) comparisons with tests of notched fracture specimens of concretes, mortars, rocks, ceramics, fiber composites (Bažant and Pfeiffer, 1987; Bažant and Kazemi, 1991, 1992; Bažant, Gettu and Kazemi, 1990; Gettu, Bažant and Karr, 1991, Bažant, Ožbolt and Eligehausen, 1994; Bažant, Daniel and Li, 1995) as well as unnotched reinforced concrete structures, (2) similitude in energy release and dimensional analysis, (3) comparison with discrete element (random particle) numerical modeling of fracture (e.g. Jirásek and Bažant, 1995), (4) derivation as a deterministic limit of a nonlocal generalization of Weibull statistical theory of strength (Bažant and Xi, 1991), and (5) comparison with finite element solutions based on nonlocal model of damage (Bažant, Ožbolt and Eligehausen, 1994). The simple size effect law has been shown useful for evaluation of material fracture characteristics from tests. Important contributions to the study of size effects in quasibrittle fracture have also been made by Carpinteri (1986), Planas and Elices (1988a,b, 1989, 1993), van Mier (1986), and others.

Recently, the fractal nature of crack surfaces in quasibrittle materials (Mandelbrot et al. 1984; Mecholsky and Mackin, 1988; Mosolov and Borodich, 1992; Borodich, 1992; Xie, 1993; etc.) has been studied intensively. It has been proposed that the crack surface fractality might be an alternative source of the observed size effect (Carpinteri 1994; Carpinteri et al. 1993, 1995; Lange et al., 1993, and Saouma et al., 1990, 1994).

This paper outlines a generalized asymptotic theory of scaling of quasibrittle fracture and also explores the possible role of the crack surface fractality in the size effect.

2. Large-Size Asymptotic Expansion of Size Effect

For the sake of brevity, the analysis will be made in general for fractal cracks and the nonfractal case will then simply be obtained as a limit case. Consider a crack representing a fractal curve (Fig. 1a) whose length is defined as $a_\delta = \delta_0(a/\delta_0)^{d_f}$ where d_f = fractal dimension of the crack curve (≥ 1) and δ_0 = lower limit of fractality implied by material microstructure, which may be regarded as the length of a ruler by which the crack length is measured (Mandelbrot et al., 1984). Unlike the case of classical, nonfractal fracture mechanics, the energy $\mathcal{W}_f$ dissipated per unit length of a fractal crack cannot be a material constant because the length of a fractal curve is infinite. Rather, it must be defined as

$$\mathcal{W}_f/b = G_{fl}\, a^{d_f} \tag{1}$$

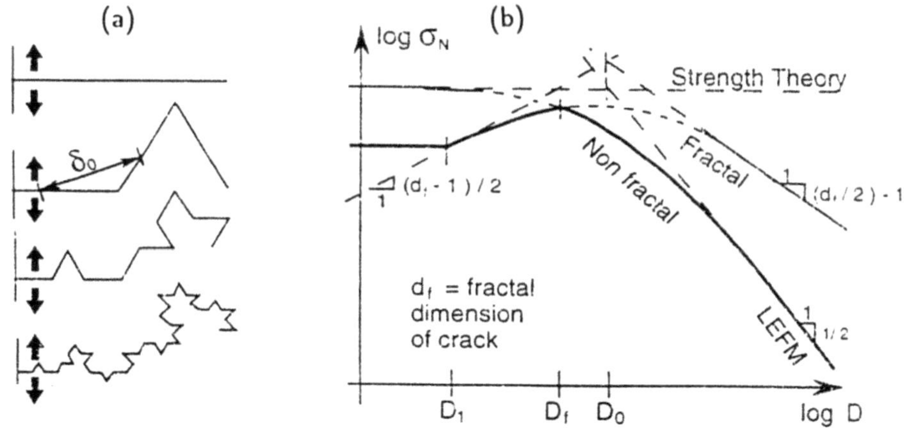

Figure 1: (a) Von Koch curves as examples of fractal crack at progressive refinement. (b) Size effect curves obtained for geometrically similar specimens with nonfractal and fractal cracks and finite size of fracture process zone (possible transition to horizontal line for nonfractal behavior is shown for $D < D_1$)

where b = thickness of the structure (considered to be two-dimensional), and G_{fl} = fractal fracture energy, of dimension Jm^{-d_f-1}. A nonfractal crack is the special case for $d_f = 1$, and then $G_{fl} = G_f$, representing the standard fracture energy, of dimension Jm^{-2}).

The rate of macroscopic energy dissipation $\mathcal{G}_{cr}$ with respect to the 'smooth' (projected, Euclidean) crack length a is:

$$\mathcal{G}_{cr} = \frac{1}{b}\frac{\partial W_f}{\partial a} = G_{fl} d_f a^{d_f-1} \tag{2}$$

(e.g., Borodich, 1992; Mosolov and Borodich, 1992). To characterize the size effect in geometrically similar structures of different sizes D, we introduce the usual nominal stress $\sigma_N = P/bD$ where D = characteristic size (dimension) of the structure, and P = applied load. If $P = P_{max}$ = maximum load, σ_N is called the nominal strength.

The problem will be analyzed under the following three hypotheses: (1) Within a certain range of sufficiently small scales, the failure is caused by propagation of a single fractal crack. (2) The fractal fracture energy, G_{fl} is a material constant correctly defining energy dissipation. (3) The material may (but need not) exhibit a material length, c_f.

The material length, c_f, may be regarded as the size (smooth, or projected) of the fractal fracture process zone in an infinitely large specimen (in which the structure geometry effects on the process zone disappear). The special case $c_f = 0$ represents the fractal generalization of linear elastic fracture mechanics (LEFM). Alternatively, if we imagine the fracture process zone to be described by smeared cracking or continuum damage mechanics, we may define $c_f = (G_{fl}/W_d)^{1/(2-d_f)}$ in which W_d = energy dissipated per unit volume of the continuum representing in a smeared way the fracture process zone (area under the complete stress-strain curve with strain softening). As still another alternative, in view of nonlinear fracture mechanics such as the cohesive crack model, we may define $c_f = (EG_{fl}/f_t^2)^{1/(2-d_f)}$ in which E = Young's modulus and f_t = material tensile strength.

We have two basic variables, a and c_f, both of the dimension of length. Let us now introduce two dimensionless variables: $\alpha = a/D$ and $\theta = c_f/D$. In view of Buckingham's theorem of dimensional analysis, the complementary energy Π^* of the structure with a fractal crack may be expressed in the form:

$$\Pi^* = \frac{\sigma_N^2}{E} bD^2 f(\alpha, \theta) \tag{3}$$

in which f is a dimensionless continuous function characterizing structure geometry.

The energy balance during crack propagation (first law of thermodynamics) must be satisfied by nonfractal as well as fractal cracks. The energy release from the structure as a whole is a global characteristic of the state of the structure and must be calculated on the basis of the smooth (projected, Euclidean) crack length a rather than the fractal curve length a_δ, i.e.

$$\frac{\partial \Pi^*}{\partial a} = \frac{\partial W_f}{\partial a}. \tag{4}$$

Substituting (3) and differentiating, we obtain an equation (see Bažant, 1995a,b) containing the derivative $g(\alpha, \theta) = \partial f(\alpha, \theta)\partial\alpha$, which represents the dimensionless energy release rate. The derivative of (3) must be calculated at constant load (or constant σ_N) because, as known from fracture mechanics, the energy release rate of a crack is the derivative of the complementary energy at constant load, i.e. $\partial\sigma_N/\partial a = 0$. In this manner (Bažant, 1995a,b) one obtains the equation $\sigma_N = \sqrt{E\mathcal{G}_{cr}/Dg(\alpha_0, \theta)}$ where α_0 = relative crack length α at maximum load. Because $g(\alpha_0, \theta)$ ought to be a smooth function, we may expand it in a Taylor series about the point $(\alpha, \theta) \equiv (\alpha_0, 0)$. This leads to the result (Bažant, 1995a,b,c):

$$\sigma_N = \sqrt{\frac{E\mathcal{G}_{cr}}{D}} \left[g(\alpha_0, 0) + g_1(\alpha_0, 0)\frac{c_f}{D} + \frac{1}{2!}g_2(\alpha_0, 0)\left(\frac{c_f}{D}\right)^2 + ... \right]^{-1/2} \tag{5}$$

in which $g_1(\alpha_0, 0) = \partial g(\alpha_0, \theta)/\partial\theta$, $g_2(\alpha_0, 0) = \partial^2 g(\alpha_0, \theta)/\partial\theta^2, ...,$ all evaluated at $\theta = 0$. This equation represents the large-size asymptotic series expansion of the size effect. To obtain a simplified approximation, one may truncate the asymptotic series after the linear term, i.e.

$$\sigma_N = Bf_t' D^{(d_f - 1)/2} \left(1 + \frac{D}{D_0}\right)^{-1/2} \tag{6}$$

in which D_0 and B are certain constants depending on both material and structure properties, expressed in terms of function $g(\alpha_0, 0)$ and its derivative. For the nonfractal case, $d_f \to 1$, this reduces to the size effect law deduced by Bažant (1983, 1984, 1993), which reads $\sigma_N = Bf_t'/\sqrt{1+\beta}$, $\beta = D/D_0$, in which β is called the brittleness number (Bažant and Pfeiffer, 1987).

If only geometrically similar fracture test specimens are considered, α_0 is constant (independent of D), and so is D_0. For brittle failures of geometrically similar quasibrittle structures without notches, it is often observed that the crack lengths at maximum load are approximately geometrically similar. For concrete structures, the geometric similarity of cracks at maximum load has been experimentally demonstrated for diagonal shear of beams, punching of slabs, torsion, anchor pullout or bar pullout, and bar splice failure, and is also supported by finite element solutions (e.g. ACI, 1992; Bažant et al. 1994) and discrete element (random particle) simulations (Jirásek and Bažant, 1995), albeit for only a limited size range of D. Thus, k, c_0, D_0, σ_N^0 and Bf_t' are all constant. In these typical cases, (6) describes the dependence of σ_N on size D only, that is, the size effect. Fig. 1b shows the size effect plot of $\log \sigma_N$ versus $\log D$ at constant α_0. Two size effect curves are seen: (1) the fractal curve and (2) the nonfractal curve (for the latter, the possibility of termination of fractality at the left end is considered in the plot).

The curve of fractal scaling obtained in Fig. 1b disagrees with the bulk of experimental evidence (for concrete, see e.g. the review in Bažant et al. 1994); for carbon fiber epoxy composites used in aerospace industry, see (Bažant, Daniel and Li, 1995). It follows that crack fractality cannot be the cause of the observed size effect.

What aspect of the fracture process causes the crack fractality to have no significant effect on scaling of failure? The fracture front in quasibrittle materials does not consist of a single crack, but a wide band of microcracks, which all must form and dissipate energy before the fracture can propagate. Only very few of the microcracks and slip planes eventually coalesce into a single continuous crack, which forms the final crack surface with fractal characteristics. Thus, even though the final crack surface may be to a large extent fractal, the fractality cannot be relevant for the fracture process zone advance. Most of the energy is dissipated in the fracture process zone by microcracks (as well as plastic-frictional slips) that do not become part of the final crack surface and thus can have nothing to do with the fractality of the final crack surface.

3. Generalizations and Ramifications of Asymptotic Analysis

Material length c_f can be defined as the LEFM-effective length of the fracture process zone, measured in the direction of propagation in a specimen of infinite size. In that case, $\theta = c_f/D = (a - a_0)/D = \alpha - \alpha_0$, and so $g(\alpha, \theta)$ reduces to the LEFM function of one variable, $g(\alpha)$. Thus Eq. (6) yields (Bažant, 1995a,b,c):

$$D_0 = c_f \frac{g'(\alpha_0)}{g(\alpha_0)}, \qquad Bf_t' = \sqrt{\frac{EG_f}{c_f g'(\alpha_0)}}, \qquad \sigma_N^0 = \sqrt{\frac{EG_{fl} d_f \alpha_0^{d_f-1}}{c_f g'(\alpha_0)}} \tag{7}$$

and so Eq. (6) takes the form:

$$\sigma_N = \sqrt{\frac{EG_{fl} d_f \alpha_0^{d_f-1}}{g'(\alpha_0)c_f + g(\alpha_0)D}} \tag{8}$$

The advantage of this equation is that its parameters are directly the material fracture parameters. For $d_f = 1$, Eq. (8) reduces to the form of size effect law derived in a different manner by Bažant and Kazemi (1990, 1991) (also Eq. 12.2.11 in Bažant and Cedolin, 1991). Fitting this equation to size effect data, which can be done easily by rearranging the equation to a linear regression plot, one can determine G_f or G_{fl} and c_f. This serves as the basis of the size effect method for measuring the material fracture parameters, which has been adopted by RILEM as an international standard for concrete.

More generally, one may introduce general dimensionless variables $\xi = \theta^r = (c_f/D)^r, h(\alpha_0, \xi) = [g(\alpha_0, \theta]^r$, with any $r > 0$. Then, expanding the function $h(\alpha_0, \xi)$ in a Taylor series with respect to ξ, one obtains by a similar procedure as before a more general large-size asymptotic series expansion (whose nonfractal special case was derived by Bažant in 1985 (see ACI, 1992):

$$\sigma_N = \sigma_P \left[\beta^r + 1 + \kappa_1\beta^{-r} + \kappa_2\beta^{-2r} + \kappa_3\beta^{-3r} + ...\right]^{-1/2r} \tag{9}$$

in which $\beta = D/D_0$ and $\kappa_1, \kappa_2, ...$ are certain constants. However, based on experiments as well some limit properties, it seems that $r = 1$ is the appropriate value for most cases.

The large-size asymptotic expansion (9) diverges for $D \to 0$. For small sizes, one needs a small-size asymptotic expansion. The previous energy release rate equation $(\sigma_N^2/E)Dg(\alpha, \vartheta) = \mathcal{G}_{cr}$ is not meaningful for the small size limit because the zone of distributed cracking is relatively large. In that case, the material failure must be characterized by W_f rather than G_f. In that case, the energy balance equation (first law) for $\partial\sigma_N/\partial a = 0$ (second law) must be written in the form $\sigma_N^2[\psi(\alpha,\eta)]^r/E = W_f$ where $\psi(\alpha,\eta)$ = dimensionless function of dimensionless variables $\alpha = a/D$ and $\eta = (D/c_f)^r = \vartheta^{-r}$ (variable ϑ is now unsuitable because $\vartheta \to \infty$ for $D \to 0$), and exponent $r > 0$ is introduced for the sake of generality, as before. Because, for very small D, there is a diffuse failure zone, a must now be interpreted as the characteristic size of the failure zone, e.g., the length of cracking band. The same procedure as before now leads to the result (Bažant, 1995a,c):

$$\sigma_N = \sigma_P \left[1 + \beta^r + b_2\beta^{2r} + b_3\beta^{3r} + ...\right]^{-1/2r} \tag{10}$$

in which $\beta = D/D_0$ and $\sigma_P, D_0, b_2, b_3, ...$ are certain constants depending on both material and structure properties and can be expressed in terms of function $\psi(\alpha_0, 0)$ and its derivatives. Eq. (10) represents the small-size asymptotic series expansion.

An important common characteristic of the large-size and small-size asymptotic series expansions in Eqs. (9) and (10) is that they have the first two terms in common. Therefore, if either series is truncated after the second term, it reduces to the same generalized size effect law derived by Bažant in 1985 (see ACI, 1992):

$$\sigma_N = \sigma_P(1 + \beta^r)^{-1/2r} \tag{11}$$

Because this law is anchored to the asymptotic cases on both sides and shares with both expansions the first two terms, it may be regarded as a matched asymptotic (e.g. Bender and Orszag, 1978), that is, an intermediate approximation of uniform applicability for any size. The value $r = 1$ appears, for several reasons, most appropriate.

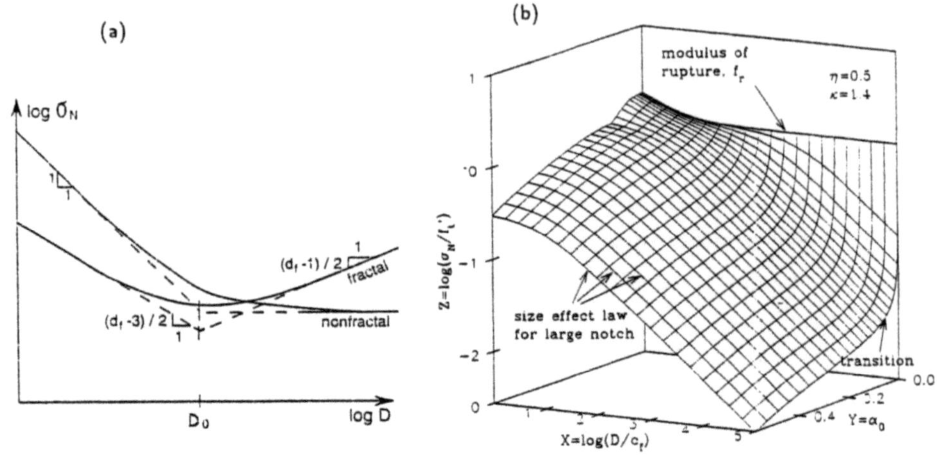

Figure 2: (a) Size effect curves obtained for unnotched specimens, nonfractal and fractal. (b) The surface of universal size effect law for notched as well as unnotched fracture specimens

A different approach is needed for unnotched quasibrittle structures that reach the maximum load when the crack initiates from a smooth surface, as exemplified by the standardized bending test of modulus of rupture f_r of a plain concrete beam. Applying the size effect law in Eq. (6 for the case $\alpha_0 \to 0$ is impossible because $g(\alpha_0, 0)$ vanishes as $\alpha_0 \to 0$. To deal with this case, one must truncate the large-size asymptotic series expansion only after the third term. Then, considering that $r = 1$ and $g(\alpha_0, 0) = 0$, restricting attention to the nonfractal case only, and using a similar procedure as that which led to Eq. (8), one obtains after some further asymptotic approximations (Bažant, 1995a,c) the following size effect law (Fig. 2a) for failures at crack initiation from a smooth surface:

$$\sigma_N = B f_r^\infty \left(1 + \frac{D_b}{D}\right) = f_r^\infty \left[1 - 0.0634 g''(0) \frac{\bar{c}_f}{D}\right] \tag{12}$$

(the first part of this equation was derived by Bažant and Li (1995) in a different manner). Here f_r^∞ is the modulus of rupture for an infinitely large beam (but not so large that Weibull statistical size effect would become significant), and B is a dimensionless parameter. This equation can be arranged as a linear regression plot of σ_N versus $1/D$, which is again helpful for easy identification of the constants from tests.

Asymptotic matching of the three asymptotic expansions, namely: (1) the large-size expansion for large α_0, (2) the large-size expansion for vanishing α_0, and (3) the small-size expansion for large α_0, leads (Bažant, 1995a,c) to the following approximated universal size effect law (Fig. 2b) valid for failures at both large cracks and crack initiation from a smooth surface:

$$\sigma_N = \sigma_0 \left(1 + \frac{D}{D_0}\right)^{-1/2} \left\{1 + \frac{1}{s}\left[\left(\eta + \frac{D}{D_b}\right)\left(1 + \frac{D}{D_0}\right)\right]^{-1}\right\}^s \tag{13}$$

in which σ_0, D_0, D_b and $\bar{c}_f$ are constants expressed in terms of $g(\alpha_0)$ and its first and second derivatives and of EG_f, and η and κ are additional empirical constants.

4. Summary and Conclusion

In quasibrittle structures, the size effect can be generally characterized on the basis of asymptotic series expansions and asymptotic matching. Whereas for normal sizes the scaling problem is extremely difficult, it becomes much simpler both for very large sizes (LEFM) and for very small sizes (plasticity). Asymptotic matching is an effective way to obtain a simplified description of the size effect in the normal, intermediate range of sizes. The size effect at crack initiation from a smooth surface can also be described the basis of the asymptotic energy release analysis, and a universal size effect law comprising both types of size effect can be formulated The fractal morphology of crack surfaces in quasibrittle materials does not appear to play a significant role in fracture propagation and the size effect.

Appendix. Is Weibull-Type Size Effect Important for Quasibrittle Failure?

It is proper to explain at least briefly why strength randomness is not considered in the present analysis of size effect. The main reason is the redistribution of stresses caused by stable fracture growth prior to maximum load and localization of damage into a fracture process zone. If the Weibull probability integral is applied to the redistributed stress field, which has high stress peaks near the crack tip, the dominant contribution to the integral comes from the fracture process zone. The important point is that the size of this zone is nearly independent of structure size D. The contribution from the rest of the structure is nearly vanishing, which corresponds to the fact that the fracture cannot occur outside the process zone. Because, in specimens of different sizes, this zone has about the same size, the Weibull-type size effect must, therefore, disappear. In other words, the fracture is probabilistic, but only the random properties of the material in a zone of the same size decide the failure, even though the structures have different sizes.

A generalized version of Weibull-type theory, in which the material failure probability depends not on the local stress but on the average strain of a characteristic volume of the material, has been shown to yield the approximate size effect formula (Bažant and Xi, 1991):

$$\sigma_N = \frac{Bf'_t}{\sqrt{\beta^{2n/m} + \beta}} \tag{14}$$

in which m = Weibull modulus (exponent of Weibull distribution of random strength), which is typically about 12 for concrete, and $n = 1$, 2 or 3 for one-, two- and three-dimensional similarity. Typically, for $n = 2$ or 3, $2n/m \ll 1$, for concrete. Then, for $m \to \infty$, which is the deterministic limit, this formula approaches the size effect law in (6). Also, for $D \to 0$, this formula asymptotically approaches the classical Weibull size effect law, and for large sizes and any m, this formula asymptotically approaches Eq. (6). It has been shown that the difference between these two formulas for concrete structures is significant only for extremely small sizes, which are below the applicability of continuum modeling.

References

ACI Comm. 446 (1992). "Fracture mechanics of concrete: Concepts, models and determination of material properties," State-of-Art Report of Am. Concrete Institute (ACI), *Fracture Mechanics of Concrete Structures*, Z.P. Bažant, Ed., Elsevier, London, 1–140.

Bažant, Z.P. (1983). "Fracture in concrete and reinforced concrete", *Mechanics of Geomaterials: Rocks, Concretes, Soils*, Pre-prints, IUTAM Prager Symposium held at Northwestern University, eds Z.P. Bažant, Evanston, IL, 281–317.

Bažant, Z. P. (1984). "Size effect in blunt fracture: Concrete, rock, metal." *J. of Engng. Mechanics*, ASCE,**110**, 518–535.

Bažant, Z.P. (1993). "Scaling Laws in Mechanics of Failure." *J. of Engrg. Mech.*, ASCE, **119**, 1828–1844.

Bažant, Z.P. (1995a). "Scaling theories for quasibrittle fracture: Recent advances and new directions." in *Fracture Mechanics of Concrete Structures* (Proc., 2nd Int. Conf. on Fracture Mech. of Concrete and Concrete Strucutres (FraMCoS-2), held at ETH, Zürich), ed. by F.H. Wittmann,

Aedificatio Publishers, Freiburg, Germany, 515–534.

Bažant, Z.P. (1995b). "Scaling of quasibrittle fracture and the fractal question." *ASME J. of Materials and Technology* **117**, 361–367 (Materials Division Special 75th Anniversary Issue).

Bažant, Z.P. (1995c). "Scaling of quasibrittle fracture: II. The fractal hypothesis, its critique and Weibull connection." *Int. Journal of Fracture*, submitted to.

Bažant, Z.P., and Cedolin, L. (1991). *Stability of Structures: Elastic, Inelastic, Fracture and Damage Theories* (textbook and reference volume), Oxford University Press, New York.

Bažant, Z.P., Daniel, I., and Li, Z. (1995). "Size effect and fracture characteristics of fiber-composite laminates", *Report*, Dept. of Civil Engng., Northwestern University, Evanston, Illinois; also *ASME JEMT*, in press.

Bažant, Z.P., and Kazemi, M. T. (1990). "Size effect in fracture of ceramics and its use to determine fracture energy and effective process zone length." *J. of American Ceramic Society* **73**, 1841–1853.

Bažant, Z.P., and Kazemi, M.T. (1991). "Size effect on diagonal shear failure of beams without stirrups." *ACI Structural J.* **88**, 268–276.

Bažant, Z.P., and Li, Z. (1995). "Modulus of rupture: size effect due to fracture initiation in boundary layer." *J. of Struct. Engrg. ASCE*, **121**, 739-746.

Bažant, Z.P., Lin, F.-B., and Lippmann, H. (1993). "Fracture energy release and size effect in borehole breakout. *Int. Journal for Numerical and Analytical Methods in Geomechanics*, **17**, 1-14.

Bažant, Z.P., Ožbolt, J., and Eligehausen, R. (1994). "Fracture size effect: review of evidence for concrete structures." *J. of Struct. Engrg., ASCE*, **120**, 2377–2398.

Bažant, Z. P., and Pfeiffer, P. A. (1987). "Determination of fracture energy from size effect and brittleness number." *ACI Materials Jour.*, **84**, 463–480.

Bažant, Z.P., and Xi, Y. (1991). "Statistical size effect in quasi-brittle structures: II. Nonlocal theory." ASCE *J. of Engineering Mechanics*, **117**, 2623–2640.

Bender, M.C., and Orszag, S.A. (1978). *Advanced mathematical methods for scientists and engineers.* McGraw Hill, New York (chapters 9–11).

Borodich, F. (1992). "Fracture energy of fractal crack, propagation in concrete and rock" (in Russian). *Doklady Akademii Nauk* **325**, 1138–1141.

Carpinteri, A. (1986) *Mechanical Damage and Crack Growth in Concrete.* Martinus Nijhoff Publishers, Doordrecht.

Carpinteri, A., Chiaia, B., and Ferro, G. (1993). "Multifractal scaling law for the nominal strength variation of concrete structures", in *Size effect in concrete structures* (Proc., Japan Concrete Institute Intern. Workshop held in Sendai, Japan, Nov. 1995), eds M. Mihashi, H. Okamura and Z.P. Bažant, E. & F.N. Spon, London–New York, 193–206.

Carpinteri, A. (1994). "Fractal nature of material microstructure and size effects on apparent mechanical properties." *Mechanics of Materials* **18**, 89–101.

Cahn, R. (1989). "Fractal dimension and fracture." *Nature* **338**, 201–202.

Gettu, R., Bažant, and Karr, M. E. (1990). "Fracture properties and brittleness of high-strength concrete", *ACI Materials Journal* **87**, 608-618.

Jirásek, M., and Bažant (1995). "Macroscopic fracture characteristics of random particle systems." *Intern. J. of Fracture*, **69**, 201-228.

Lange, D.A., Jennings, H.M., and Shah, S.P. (1993). "Relationship between fracture surface roughness and fracture behavior of cement paste and mortar." *J. of Am. Ceramic Soc.* **76**, 589–597.

Mandelbrot, B.B., Passoja, D.E., and Paullay, A. (1984). "Fractal character of fracture surfaces of metals." *Nature* **308**, 721–722.

Mecholsky, J.J., and Mackin, T.J. (1988). "Fractal analysis of fracture in ocala chert." *J. Mat. Sci. Letters* **7**, 1145–1147.

Mosolov, A.B., and Borodich, F.M. (1992). "Fractal fracture of brittle bodies under compression" (in Russian). *Doklady Akademii Nauk* **324**, 546–549.

Planas, J., and Elices, M. (1988a). "Size effect in concrete structures: mathematical approximations and experimental validation." *Cracking and Damage, Strain Localization and Size Effect*, Proc. of France-U.S. Workshop, Cachan, France, eds J. Mazars and Z.P. Bažant, pp. 462-476.

Planas, J., and Elices, M. (1988b). "Conceptual and experimental problems in the determination of the fracture energy of concrete." Proc., *Int. Workshop on Fracture Toughness and Fracture Energy, Test Methods for Concrete and Rock*, Tohoku Univ., Sendai, Japan, pp. 203-212.

Planas, J., and Elices, M. (1989), "Size effct in concrete structures: mathematical approximations and experimental validation", in *Cracking and Damage*, ed. by J. Mazars and Z.P. Bažant, Elsevier, London, 462–476.

Planas, J., and Elices, M. (1993). "Drying shrinkage effect on the modulus of rupture." *Creep and Shrinkage in Concrete Structures* (Proc., ConCreep 5, Barcelona), eds Z.P. Bažant and I. Carol, E. & F.N. Spon, London, 357-368.

Saouma, V.C., Barton, C., and Gamal-el-Din, N. (1990). "Fractal characterization of concrete crack surfaces." *Engrg. Fracture Mechanics* **35**.

Saouma, V.C., and Barton, C.C. (1994). "Fractals, fracture and size effect in concrete." *J. of Engrg. Mechanics ASCE* **120**, 835–854.

Xie, Heping (1993). *Fractals in Rock Mechanics.* Balkema, Rotterdam.

SCALING EFFECTS IN THE FRACTURE OF QUASI-BRITTLE MATERIALS AND STRUCTURES

B. L. KARIHALOO[†]
School of Civil and Mining Engineering
The University of Sydney
NSW 2006
Australia

Abstract - Quasi-brittle materials are characterised by tension softening behaviour after the attainment of their tensile strength. The micromechanical description of this behaviour introduces an internal length scale, called the characteristic length of the material. Structures made from quasi-brittle materials are known to exhibit a strong size effect. Several cohesive crack models have been proposed to explain this effect on the basis of dimensional considerations through the so-called brittleness number.

In this paper an inverse procedure for a cohesive crack model is exploited to provide explicit definitions of characteristic length and brittleness number from physical rather than dimensional considerations. The physical cause for the size effect is also identified.

1. Introduction

Materials, such as concrete, rocks and some ceramics, that have traditionally being regarded as brittle do in fact exhibit a moderately strain-hardening behaviour prior to the attainment of the ultimate tensile strength (region AB in Fig. 1a), and an increase in deformation with decreasing tensile carrying capacity, i.e. by tension softening (region BD in Fig. 1a).

The pre-peak strain hardening behaviour of quasi-brittle materials is due to the formation of microcracks along the interfaces between the matrix and other phases and of their deflection into the matrix. This behaviour has been successfully explained using the concepts of damage mechanics (see, e.g. Karihaloo, 1995).

[†]Now at Institute of Mechanical Engineering, Aalborg University, DK 9220 Aalborg East, Denmark

J. R. Willis (ed.), IUTAM Symposium on Nonlinear Analysis of Fracture, 13–22.

The tension softening response is due to the localization of damage in the form of a macroflaw along the eventual failure plane whose catastrophic advance is prevented by a bridging mechanism which breaks its continuity. The micromechanical description of the tension softening behaviour introduces an internal length scale, called the characteristic length of the material.

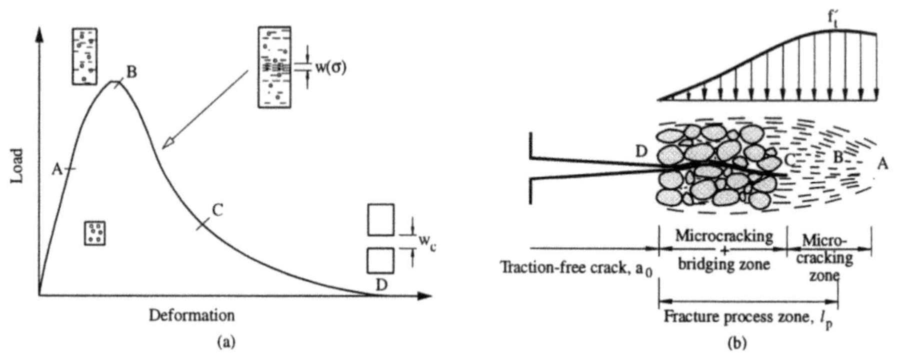

Figure 1. Typical tensile load-deflection response of a pre-cracked quasi-brittle specimen (a), and the fracture process zone ahead of the real traction-free crack (b).

Structures made from quasi-brittle materials are known to exhibit a strong size effect whereby small structures appear to fail in a ductile manner but large structures in a brittle and often catastrophic manner. Several cohesive crack models have been proposed to explain this ductile to brittle transition on the basis of dimensional considerations through the so-called brittleness number. We shall briefly review the various brittleness measures in the next Section. We shall also describe briefly an inverse procedure for a cohesive crack to provide explicit definitions of characteristic length and brittleness measures from physical rather than dimensional considerations.

2. Nonlinear Fracture Theories for Quasi-Brittle Materials

The primary reason why the Griffith linear elastic fracture theory is inapplicable to quasi-brittle materials is the formation of an extensive fracture process zone (FPZ) ahead of a pre-existing notch/crack. The material in this zone progressively softens due to microcracking and other bridging mechanisms. This is schematically illustrated in Fig. 1b on the example of a notched specimen subjected to a tensile load.

A fracture theory capable of describing material softening will necessarily be a nonlinear one, but because of the size of the FPZ, it will differ from the nonlinear fracture theory for ductile materials. In a quasi-brittle material the FPZ practically occupies the entire zone of nonlinear deformation (Fig. 2)

The first nonlinear theory of fracture for quasi-brittle materials was proposed by Hillerborg et al. (1976). It includes the tension softening FPZ through a 'fictitious' crack ahead of a pre-existing real crack whose faces are acted upon by certain closing stresses in the spirit of Barenblatt, Dugdale-BCS models (Fig. 3). The term 'fictitious' is used to underline the fact that this portion of the crack cannot be continuous with full separation of faces. Unlike the Barenblatt model, the size of FPZ may not be small. In consequence, a knowledge of the distribution of closing stresses $\sigma(w)$ is now essential. Thus, the fracture of a quasi-brittle material requires at least two material parameters. In the nonlinear theory of Hillerborg et al. (1976), these are the shape of tension softening relation $\sigma(w)$ and the area under this curve G_F (Fig. 3b), besides $f'(t)$.

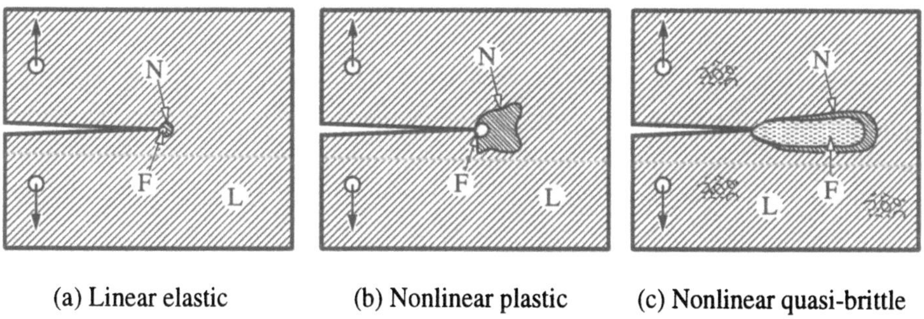

Figure 2. Distinguishing features of fracture in (a) a linear elastic, (b) a ductile, (c) a quasi-brittle material. L=linear, N=nonlinear, F=FPZ (after ACI Report 446.1, 1989).

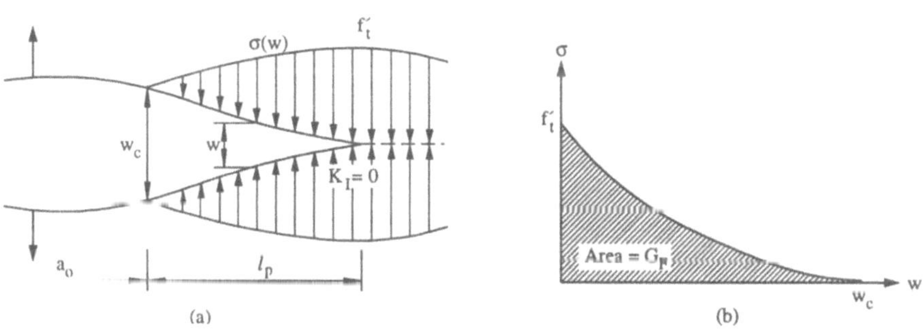

Figure 3. A real traction-free crack a_0 terminating in a fictitious crack with residual stress transmission capacity $\sigma(w)$ whose faces close smoothly near its tip ($K_I = 0$).

Another nonlinear theory of fracture proposed by Bažant (1976) treats the tension softening in a "smeared" manner through a strain-softening constitutive relation. To relate the inelastic strain ε to w and G_F of Hillerborg et al. (1976), it is now necessary to introduce a gauge length over which the microcracks in the FPZ are assumed to be distributed (Fig. 4)

3. Size Effect and Brittleness Number

The size of FPZ is commensurate with that of a small cracked structure made of a quasi-brittle material so that its response will be ductile. On the other hand, the FPZ occupies but a small fraction of the structural volume in an "infinitely" large structure, so that its response will be brittle. If we therefore define the fracture energy, G_f and size of FPZ, c_f with respect to an infinitely large structure ($W \to \infty$), then in order to determine G_f (akin to the modulus of cohesion in the Barenblatt cohesive model) and c_f from laboratory specimens of moderate size, Bažant (1984) proposed the following scaling law based on dimensional considerations and smeared crack model (Fig. 4)

$$(\sigma_N)_u = [E' G_f/(g'(\alpha_0)\, c_f)]^{\frac{1}{2}}\ [1 + g(\alpha_0)\, W/(g'(\alpha_0)\, c_f)]^{-\frac{1}{2}} \tag{1}$$

where $(\sigma_N)_u$ is the nominal stress at maximum load P_u. In (1) $\alpha_0 = a_0/W$, a prime on $g'(\alpha_0)$ denotes differentiation with respect to α_0 and $g(\alpha_0)$ is a function of the relative notch depth α_0 only. It is related to the geometry function $Y(\alpha_0)$ appearing in the stress intensity factor K_I.

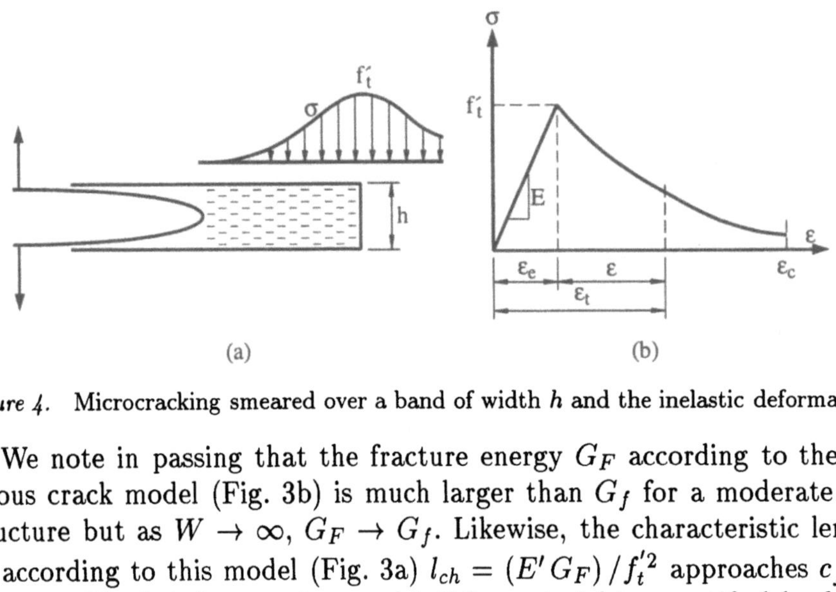

Figure 4. Microcracking smeared over a band of width h and the inelastic deformation.

We note in passing that the fracture energy G_F according to the fictitious crack model (Fig. 3b) is much larger than G_f for a moderate size structure but as $W \to \infty$, $G_F \to G_f$. Likewise, the characteristic length l_{ch} according to this model (Fig. 3a) $l_{ch} = (E' G_F)/f_t'^2$ approaches c_f, as $W \to \infty$. The brittleness of a quasi-brittle material is quantified by l_{ch} or c_f; the smaller the value of l_{ch} or c_f, the more brittle the material.

As the strength of a structure is defined by the force per unit area, i.e. by energy per unit volume, while its toughness is energy per unit area, the brittleness of the structure includes its size from dimensional considerations

$$\beta = l_{ch}/W \qquad \text{or} \qquad \beta = c_f/W. \tag{2}$$

Another measure of brittleness can also be introduced, if E' and f_t' in l_{ch} have been independently measured, namely the energy brittleness number

(Carpinteri, 1986) $s_e = G_F/\left(f_t' W\right)$. In the next Section we shall study the physical cause of the size effect in quasi-brittle materials.

4. Physical Cause for Size Effect

Leicester (1969, 1973) seems to have been the first to investigate the effect of size on the strength of structures made out of metals, timber and concrete. He identified two fundamental causes of size effect in real structures as (i) the material heterogeneity, and (ii) the occurrence of notches or other discontinuities to the flow of stress. The size effect due to heterogeneity in material strength is explained by Weibull statistics but the size effect caused by the occurrence of notches, etc needs some explanation. Leicester argued that, as the LEFM predicts the occurrence of infinite stresses at the apex of notch root, the failure criterion for real materials must be stated in terms of the elastic stresses on some region R_0 encircling the notch root. Then provided (i) *the structural member is sufficiently large* so that at failure R_0 is contained with the range of applicability of asymptotic stress field at notch root, and (ii) *the details of the notch root are always the same* (i.e. geometric similarity of structures), the failure stress is given by $(\sigma_N)_u = A/W^s$.

The above two assumptions are in complete accord with the two hypotheses on which the size effect rule (1) is based. In fact, R_0 can be identified with c_f. For structures of moderate size, the requirement that $R_0 << W$ cannot be met. In consequence, it is insufficient to consider just the singular term in the asymptotic stress field in mode I, because now the terms that were unimportant due to the smallness of R_0 can no longer be ignored.

To understand the role played by the non-singular terms in the asymptotic solution, let us consider the configuration of the Griffith problem (a plane containing a crack $2a$) and write

$$\sigma_{yy}| \equiv \sigma_0(r) \approx K_I\,(1 + r/\delta)/\sqrt{2\pi\, r} \tag{3}$$

where $\delta = 4a/3$.

We shall consider the fictitious crack of Fig. 3 to investigate the effect of the second term in (3) on the result. We shall now investigate how σ and w vary individually in the FPZ. If we know these variations, then elimination of the distance from them will give us the relation $\sigma(w)$. For this, we will decompose the problem into two sub-problems and use superposition. The first sub problem is the elastic one with the FPZ. The stress intensity factor at the tip of the traction free crack O is equal to K_I and the stress distribution ahead of O is given by (3). In the second sub-problem, the stress and opening displacement are $[\sigma(s) - \sigma_0(l_p - s)]$ and $w(s)$, respectively. The stress intensity factor $k(s)$ at the crack tip and the opening displacement

$g(s,t)$ at the location t due to unit normal forces at s are known from handbooks

$$k(s) = -\sqrt{2/(\pi s)}, \qquad g(s,t) = (4/(\pi E')\,\log\left|(\sqrt{s}-\sqrt{t})/(\sqrt{s}+\sqrt{t})\right|. \quad (4)$$

$k(s)$ and $g(s,t)$ are the usual influence or Green's functions. The corresponding Green's functions for a finite crack depend on the crack length.

Superposition and smooth crack closure condition give

$$\int_0^{l_p} g(s,t)[\sigma(s)-\sigma_0(l_p-s)]ds = w(t), \; \int_0^{l_p} k(s)[\sigma(s)-\sigma_0(l_p-s)]ds = 0. \quad (5)$$

Substituting (3) and (4) into (5) and evaluating some elementary integrals, we get

$$\frac{4}{\pi E'}\int_0^{l_p} \log\left|\frac{\sqrt{s}-\sqrt{t}}{\sqrt{s}+\sqrt{t}}\right| \sigma(s)\,ds + \frac{8\,K_I\sqrt{t}}{E'\sqrt{2\pi}}\left(1+\frac{l_p}{2\delta}-\frac{t}{3\delta}\right) = w(t) \quad (6)$$

$$K_I\left(1+\frac{l_p}{2\delta}\right) - \sqrt{\frac{2}{\pi}}\int_0^{l_p}\frac{\sigma(s)}{\sqrt{s}} = 0. \quad (7)$$

These integral equations have to be solved numerically for a given value of K_I to obtain $\sigma(s)$ and l_p. Alternatively, l_p can be prescribed, and the equations solved for $\sigma(s)$ and K_I. Of course, in both cases it is necessary to know the tension softening relation $\sigma(w)$. Horii et al. (1987) have solved (6) and (7) numerically for a linear approximation to $\sigma(w)$. However, here we shall demonstrate an inverse procedure which allows us to solve these equations analytically.

In this procedure, we approximate $w(s)$, instead of $\sigma(w)$, in such a way as to solve (6) and (7) analytically for $\sigma(s)$ and l_p (for prescribed K_I). We then eliminate s between the assumed $w(s)$ and the calculated $\sigma(s)$ to establish $\sigma(w)$; hence the inverse nature of the procedure, first proposed by Smith (1974) (see also Smith, 1994).

Let

$$w(s) = a_n\, s^{n+\frac{3}{2}} \quad (8)$$

where n is positive integer or zero ($n \geq 0$) and a_n a constant to be determined. s varies between 0 and l_p, so that (8) identically meets the requirement that $w(0) = 0$ at the fictitious crack tip. Also $dw/ds = 0$ at $s = 0$, thereby satisfying the smooth closure condition.

Substitution of (8) into eqns (6) and (7) gives, after simplification,

$$K_I = \frac{E'}{2\sqrt{2}} a_n l_p^{n+1}\,\frac{\Gamma(n+\frac{5}{2})}{\Gamma(n+2)} = \frac{E'}{2\sqrt{2}}\frac{w_c}{l_p}\frac{\Gamma(n+\frac{5}{2})}{\Gamma(n+2)} \quad (9)$$

$$\frac{\sigma(y)}{f_t'} = \frac{\Gamma(n+1)}{\Gamma(n+\frac{1}{2})}\sqrt{1-y}\left\{\frac{J_n(y)}{\sqrt{\pi}} + 2\frac{l_p}{\delta}\frac{\Gamma(n+\frac{3}{2})}{\Gamma(n+2)}\right\}\left\{1+2\frac{l_p}{\delta}\frac{(n+\frac{1}{2})}{(n+1)}\right\}^{-1} \quad (10)$$

where $y = s/l_p$. From (8), we also have $w(y)/w_c = y^{n+3/2}$. Elimination of y between this and (10) gives the stress displacement relation $\sigma(w)$ in the FPZ corresponding to the assumed $w(s)$ (8). The length of FPZ, l_p can also be calculated, and is

$$l_p = \left[\frac{4\sqrt{\pi}\, f_t'}{E'\, w_c\, (n+\frac{3}{2})}\frac{\Gamma(n+1)}{\Gamma(n+\frac{1}{2})} - \frac{2}{\delta}\frac{(n+\frac{1}{2})}{(n+1)}\right]^{-1} . \tag{11}$$

The simplest example is one in which only one term ($n = 0$) is retained in (8). In this case, the stress-displacement law in the FPZ is

$$\sigma/f_t' = \left[1 - (w/w_c)^{\frac{2}{3}}\right]^{\frac{1}{2}} . \tag{12}$$

It is not always possible to eliminate y. This would be the case if we chose three terms ($n = 0, 1, 2$) in (8). It would then be necessary to perform the elimination numerically. When the numerical results have been fitted by polynomial approximation, the following relation results ($\gamma = w/w_c$)

$$\frac{\sigma}{f_t'} = 1 - 9.2431\gamma^2 + 33.8259\gamma^3 - 59.4248\gamma^4 + 49.3000\gamma^5 - 15.4722\gamma^6. \tag{13}$$

Irrespective of the number of terms chosen in (8), (9) and (11) give

$$l_p = l_{p\infty}(1-\lambda)^{-1}, \qquad K_I = K_{I\infty}(1-\lambda)^{1/2} \tag{14}$$

where $\lambda = l_{p\infty}/\delta$, and

$$l_{p\infty} = A_1\, E'\, w_c/f_t', \quad K_{I\infty} = \sqrt{A_2\, E'\, w_c\, f_t'}. \tag{15}$$

$l_{p\infty}$ is the length of FPZ when $\delta = \infty$ or $\lambda = 0$, i.e. when the second term in (3) is ignored. The constants A_1 and A_2 depend on the shape of the $\sigma(w)$ curve. For (12), $A_1 = 3/8$, $A_2 = 3\pi/16$, and for (13), $A_1 = 28/39$, $A_2 = 0.4969$.

We can also evaluate the energy expended in the FPZ, denoted W_d as well as the specific fracture energy G_F

$$W_d = \int_0^{l_p}\int_0^{w} \sigma(w)\, dw\, ds, \qquad G_F = \int_0^{w_c} \sigma(w)\, dw. \tag{16}$$

The integrals in (16) can be evaluated analytically for the $\sigma(w)$ relation (12), but for the relation (13), numerical integration is necessary

$$W_d = A_3\, f_t'\, w_c^2/(1-\lambda), \qquad G_F = A_4\, f_t'\, w_c \tag{17}$$

where $A_3 = (9\pi)/256$, $A_4 = (3\pi)/16$ for relation (12), and $A = 0.0521$, $A_4 = 0.4969$ for relation (13).

We shall now adopt an approximate procedure for finite bodies suggested by Horii et al. (1987), to extend the above results. In this procedure the variation of stress intensity factor corresponding to the stress distribution (3) is approximated by

$$K_I(r) = -\int_0^r k(s)\,\sigma_0(r-s)\,ds = K_I\,[1 + r/(2\delta)]. \tag{18}$$

$K_I(r)$ is used to identify the parameter λ for a body of finite size, say a three-point notched bend beam or a crack line wedge loaded specimen (CLWL) as follows. The ratio $K_I(r)/K_I = Y\left(\frac{a+r}{W}\right)/Y\left(\frac{a}{W}\right)$ is set up using the appropriate geometry function $Y\left(\frac{a}{W}\right)$. For a three-point bend specimen $Y\left(\frac{a}{W}\right)$ is given by

$$Y(\alpha) = A_0 + A_1\,\alpha + A_2\,\alpha^2 + A_3\,\alpha^3 + A_4\,\alpha^4 \tag{19}$$

where $\alpha = a/W$, and the constants $A_0, \cdots, A_4$ depend only on the span to depth ratio S/W. For $S/W = 4$, $A_0 = 1.93$, $A_1 = -3.07$, $A_2 = 14.53$, $A_3 = -25.11$, and $A_4 = 25.80$. For a CLWL the geometry function is

$$Y(\alpha) = (2+\alpha)(0.886+4.64\,\alpha-13.32\,\alpha^2+14.72\,\alpha^3-5.64\,\alpha^4)\,(1-\alpha)^{-\frac{3}{2}}. \tag{20}$$

Next the ratio is expanded in Taylor's series about $\frac{a}{W}$ and only the first two terms retained

$$K_I(r)/K_I \approx 1 + [Y'(\alpha)/Y(\alpha)]\,(r/W) \tag{21}$$

where $Y'(\alpha) = dY(\alpha)/d\alpha$.

A comparison of (21) with (18) immediately gives

$$\lambda = l_{p\infty}/\delta = 2[Y'(\alpha)/Y(\alpha)]\,(l_{p\infty}/W) \tag{22}$$

where $l_{p\infty}$ can be chosen corresponding to the semi-infinite crack geometry (15) provided the stress distribution (3) has been used in its calculation.

As an example, let us consider a three-point notched bend beam with $S/W = 4$. We find, after simplification

$$\frac{Y'(\alpha)}{Y(\alpha)} = -1.59 - 12.53\alpha + 7.13\alpha^2 + 72.89\alpha^3 - 122.01\alpha^4 = -1.59 - F(\alpha) \tag{23}$$

where $F(\alpha)$ stands for the terms involving α. Likewise, for a CLWL specimen we find

$$\frac{Y'(\alpha)}{Y(\alpha)} = 7.237 - 56.244\alpha + 431.3\alpha^2 - 3225.6\alpha^3 + 23326\alpha^4 = 7.237 - F(\alpha). \tag{24}$$

From (15) and (22) we can write

$$K_I(r)/K_{I\infty} = [1 - 2\{Y'(\alpha)/Y(\alpha)\}\,(l_{p\infty}/W)\,]^{\frac{1}{2}} \tag{25}$$

where $l_{p\infty}$ and $K_{I\infty}$ are given by (15). Substitution of (23) or (24) into (25) immediately gives the ratio of $K_I/K_{I\infty}$ for structures of one and the same geometry. Relation (25) therefore also gives the ratio of the corresponding failure stresses

$$(\sigma_N)_u/(\sigma_N)_{u\infty} = [1 - 2A_5(l_{p\infty}/W) + 2F(\alpha)\,(l_{p\infty}/W)\,]^{\frac{1}{2}} \tag{26}$$

where $A_5 = -1.59$ for three-point bend beam and $A_2 = 7.237$ for CLWL.

5. Discussion and Conclusions

Let us rewrite some of the equations derived above in a form suitable for explaining many features of the fracture of quasi-brittle structures, and not only the size effect.

First, we eliminate w_c from (15) using (17) to get

$$l_{p\infty} = (A_1/A_4)\left(E'G_F/f_t'^2\right). \tag{27}$$

Next, we introduce a parameter $\gamma(\alpha) = A_4\left[A_5 + F(\alpha)\right]/A_1$. $F(\alpha)$ clearly reflects the influence of the geometry and initial notch depth of the specimen. Substitution of (27) and $\gamma(\alpha)$ into (26) gives

$$(\sigma_N)_u/(\sigma_N)_{u\infty} = [1 + \gamma(\alpha)l_{ch}/W]^{\frac{1}{2}}. \tag{28}$$

Despite the approximations made in arriving at (28), we have revealed that the origin of the size effect is in the non-singular stress distribution ahead of the notch/crack, represented by δ (or λ). In the process, we have also identified the physical basis of the many brittleness measures that had their roots in dimensional analysis and geometric similitude.

As we have already identified l_{ch} with $l_{p\infty}$ (27), the physical basis of the structural brittleness measure namely l_{ch}/W (2) becomes immediately apparent, and by inference also that of energy brittleness number s_e. The structural brittleness number β of Bažant & Kazemi (1990), defined by (2) can also be immediately identified with $W/[l_{p\infty}\gamma(\alpha)]$ of (28).

It is a common feature of all cohesive crack models, including that of the FCM, that $G_F(\equiv J_c)$ is uniquely determined by the area under the tension softening diagram. In general, G_F is given by (17) where only the constant A_4 depends on the shape of this diagram. Consequently, if the shape of $\sigma(w)$ has been accurately determined and the uniaxial tensile strength has been independently measured, then G_F is known exactly, and one can use (17) to calculate $w_c = J_c/(A_4 f_t')$.

Let us now assume that the fracture of a test specimen occurs when $K_I = K_{Ic}^e$ at $P = P_{max}$. From (14) and (15) it follows that

$$(K_{Ic}^e)^2 = A_2\, E'\, w_c\, f_t'\,(1-\lambda) \tag{29}$$

which may be rewritten in terms of $G_F(\equiv J_c)$ using (17)

$$(K_{Ic}^e)^2 = (A_2/A_4)\, E'\, J_c\,(1-\lambda). \tag{30}$$

As λ is negative for the three-point notched beam geometry and positive for CLWL geometry, and as (29) is based on an effective LEFM concept, i.e. $(K_{Ic}^e)^2 = EJ_c$, it would appear that the relation (30) is violated. What we have actually proved is that if K_I attains the critical value K_{Ic}^e at the maximum load P_{max}, then at this instant the crack opening displacement

w must be different from its critical value w_c, say equal to w^*, so that the following relation is exactly met

$$A_2 \, w^* \, f_t' \, (1 - \lambda) = J_c. \tag{31}$$

$w^* < w_c$ for a three-point beam geometry, and $w^* > w_c$ for CLWL geometry.

We have proved from physical principles what is commonly observed in a test on three-point notched beam (or CLWL) specimen, namely that the peak load is attained before (or after) the stress-free crack begins to grow.

6. References

ACI Committee 446 Report 446-1R (1989) *Fracture Mechanics of Concrete: Concepts, Models and Determination of Material Properties*, American Concrete Institute, Detroit.

Bažant, Z. P. (1976) Instability, ductility and size effect in strain-softening concrete, *ASCE J. Engng. Mech.* **102**, 331–344.

Bažant, Z. P. (1984) Size effect in blunt fracture: concrete, rock and metal, *ASCE J. Engng. Mech.* **110**, 518–535.

Bažant, Z. P. and Kazemi, M. T. (1990) Determination of fracture energy, process-zone length and brittleness number from size effect, with application to rock and concrete, *Int. J. Fracture* **44**, 111-131.

Carpinteri, A. (1986) *Mechanical Damage and Crack Growth in Concrete*, Martinus Nijhoff, Dordrecht.

Hillerborg, A., Modéer, M. and Patersson, P. E. (1976) Analysis of crack formation and crack growth in concrete by means of fracture mechanics and finite elements, *Cement Concr. Res.* **6**, 773–782.

Horii, H., Hasegawa, A. and Nishino, F. (1987) Process zone model and influencing factors in fracture of concrete, in S. P. Shah and S. E. Swartz (eds.), *Fracture of Concrete and Rock*, Springer-Verlag, New York, pp. 205–219.

Karihaloo, B. L. (1995) *Fracture Mechanics and Structural Concrete*, Longman, U.K.

Leicester, R. H. (1969) The size effect of notches, in *Proc. Second Australasian Conf. Mech. Struct. Mater.*, Melbourne, pp. 4.1–4.20.

Leicester, R. H. (1973) Effect of size on the strength of structures, *Div. Bldg. Res. Paper 71*, CSIRO, Melbourne, pp. 1–13.

Smith, E. (1974) The structure in the vicinity of a crack tip: a general theory based on the cohesive zone model, *J. Engng. Fracture Mech.* **6**, 213–222.

Smith, E. (1994) The elastically equivalent softening zone size for an elastic-softening material: I. power law softening behaviour, II. a simple piece-wise softening law, *Mech. Mater.* **17**, 363–378.

CONSTRAINT ESTIMATION FOR EDGE CRACKED BEND BARS

A. D. KARSTENSEN, A. NEKKAL AND J. W. HANCOCK
Department of Mechanical Engineering
University of Glasgow
Glasgow G12 8QQ
Scotland

1. Introduction

The stress field close to a crack tip in an elastic isotropic solid can be expressed as an asymptotic expansion following Williams (1957):

$$\sigma_{ij}(r,\theta) = A_{ij}(\theta)r^{-\frac{1}{2}} + B_{ij}(\theta) + C_{ij}(\theta)r^{\frac{1}{2}} + \cdots \tag{1}$$

(r,θ) are polar co-ordinates centred at the crack tip, and σ_{ij} are the cartesian components of the stress tensor. Focussing interest on the non-zero terms at the crack tip, the elastic field can be expressed in the form:

$$\sigma_{ij}(r,\theta) = \frac{K_I}{\sqrt{2\pi r}} f_{ij}(\theta) + T\delta_{1,j}\delta_{i,1} \tag{2}$$

The first term embodies the stress intensity factor K, while the second term, denoted T, is a uniform stress $\sigma_{11} = T$, acting parallel to the crack flanks. The non-singular T-stress has now been tabulated for a wide range of geometries, in which the results are either expressed in terms of a stress concentration factor $\frac{T}{\sigma}$ or as a biaxiality parameter β:

$$\beta = \frac{T\sqrt{\pi a}}{K} \tag{3}$$

Results for some important crack geometries have been reviewed by Karstensen and Hancock (1994) and Sherry et al. (1994).

J. R. Willis (ed.), IUTAM Symposium on Nonlinear Analysis of Fracture, 23–32.

The role ot the non-singular T-stress on the development of elastic-plastic crack tip fields has been discussed by Larsson and Carlsson (1973), Harlin and Willis (1988, 1990) and Bilby et al. (1986). Betegón and Hancock, (1991) demonstrated that all geometries which initially develop constrained flow fields feature positive values of T, while geometries which exhibit unconstrained flow field feature negative values of T. In bending, single edge cracked bars with $a/W \geq 0.3$ exhibit positive values of T and develop fully constrained fields.

Shallow cracked bars develop a negative T stress and lose crack tip constraint. To predict the level of crack tip constraint in fully plastic specimens Betegón and Hancock, (1991) suggested that the full field solutions could be related to the modified boundary layer formulation at the same value of T. Modified boundary layer formulations are formal representation of contained yielding in which the plastic zone is contained within an outer elastic field defined by K and T. In both contained yielding and full plasticity, constraint loss arises from a largely hydrostatic term which is independent of the distance $\frac{r\sigma_0}{J}$ from the crack tip. This results in parallel stress profile at low levels of deformation when the loss of constraint only arises from T. O'Dowd and Shih (1991a,b) generalised these results by writing the stress field the form:

$$\frac{\sigma_{ij}}{\sigma_0} = \left(\frac{J}{\alpha\varepsilon_0\sigma_0 I_n r}\right)^{\frac{1}{(n+1)}} \tilde{\sigma}_{ij}(\theta, n) + Q\left((\frac{r\sigma_0}{J})^q \hat{\sigma}_{ij}(\theta, n)\right) + \ ... \tag{4}$$

The first term in (4) is the HRR field. The amplitude of the second term is denoted Q. The angular functions $\tilde{\sigma}_{ij}$ and the integration constants I_n have been tabulated by Shih (1983). Q controls the level of constraint in the stress field. It is argued that the exponent q can be approximated to zero, leading to a distance independent second order term. Numerical solutions using modified boundary layer formulations however suggest that the fields are better expressed in terms of the small scale yielding (T=0) field:

$$\sigma_{ij} = \sigma_{ij}^{SSY} + Q\sigma_0\delta_{1j} \tag{5}$$

The small scale yielding (T=0) field comprises the HRR field plus some minor but not insignificant higher order terms. Although modified boundary layer formulations are formal representations of contained yielding, Betegón and Hancock (1991) and Al-Ani and Hancock (1991) attempted to correlate modified boundary layer formulations with full field solutions of a wide range of geometries into full plasticity. In order to correlate fully plastic solutions with small scale yielding solutions using modified boundary conditions, T was calculated in the same manner from the applied load or equivalently from the elastic component of J, in both cases. Although T

can simply be determined from remote loading parameters, Q can currently only be determined by detailed local finite element analysis. The present work is motivated towards developing schemes to estimate Q from remote loading parameters, and present some results from a more comprehensive study by Karstensen, Nekkal and Hancock (1995).

2. Numerical Solutions

2.1. MATERIAL RESPONSE

Finite element solutions were obtained within the framework of small strain deformation as described by Hibbitt, Karlsson and Sorenson (1992). The material response was linear elastic at stress level less than the yield stress, σ_0. In uniaxial tension the material response can be described by Hooke's law:

$$\sigma = E\varepsilon \qquad (\sigma \leq \sigma_0) \tag{6}$$

where E is Young's modulus. Poisson's ratio, ν, was set to 0.3. Yield and associated plastic flow was modelled by incremental plasticity under the Prandtl-Reuss flow rules. The plastic response was approximated to a Ramberg-Osgood stress-strain relation which in uniaxial tension can be described by:

$$\frac{\varepsilon}{\varepsilon_0} = \frac{\sigma}{\sigma_0} + \alpha(\frac{\sigma}{\sigma_0})^n \tag{7}$$

Karstensen, Nekkal and Hancock (1995) have preformed numerical calculations with a wide range of hardening rates, under both bending and tensile loading. In the present work data for n=13, ε_0=0.001 and $\alpha = \frac{3}{7}$ are presented.

2.2. FULL FIELD SOLUTIONS

Numerical solutions have been obtained for a range of plane strain edge cracked bend bars shown schematically in figure 1. Symmetry allowed half the bar to be modelled. The crack length to width ratios which have been examined are $\frac{a}{W}$=0.1, 0.2, 0.3, 0.4, 0.5, 0.6, 0.7,0.8 and 0.9. The height $2H$ of the bars was 6 times the width.

3. Constraint Estimation

In order to illustrate the nature of the crack tip fields, numerical results for a shallow edge cracked bar (a/W=0.1, n=13) are shown in figure 2. The hoop stress directly ahead of the crack is given as a function of $\frac{r\sigma_0}{J}$ for

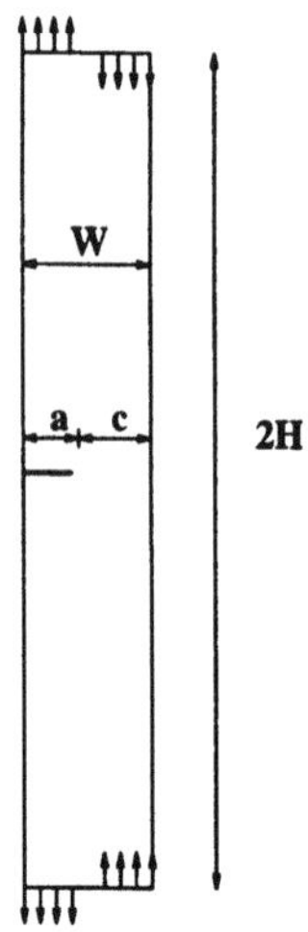

Figure 1. Geometry of the edge cracked bend bars.

increasing levels of deformation. The stress profiles are compared with the small scale yielding field which applies at very small levels of deformation when plasticity is a small perturbation of the elastic field. Initially the crack tip field can be expressed as the small scale yielding field plus a distance independent term. This can be seen at deformation levels of $\frac{a\sigma_0}{J} = 417$ and 53 when the difference between the small scale yielding field and the full field solution is independent of distance beyond $\frac{r\sigma_0}{J} = 10$. At higher levels of deformation $\frac{a\sigma_0}{J} = 1.8$ the difference between the full field solution and the small scale yielding field becomes distance dependent as the global bending field is encountered. This effect arises because the bar is subjected to a bending moment and the ligament remote from the tip is in compression.

In order to examine the nature of the constraint loss, Q has been decomposed into two terms :

$$Q = Q_T + Q_P \tag{8}$$

Q_T is determined from the modified boundary layer formulation as a function of T and is independent of the distance $\frac{r\sigma_0}{J}$ but dependent on the strain hardening rate. The residual term Q_P can be regarded as the difference between the total loss of constraint given by Q and the loss of constraint given by a negative T.

4. Q_P Estimation for Single Edge Cracked Bars in Bending

The nature of the residual constraint term Q_P is now investigated. The distance dependency of Q_P is shown in figure 3, by plotting Q_P as a function of level of deformation expressed in terms of the applied load, P normalised by the limit load P_{Limit}, at distances $\frac{r\sigma_0}{J}$=1,2 and 5 from the crack tip. The

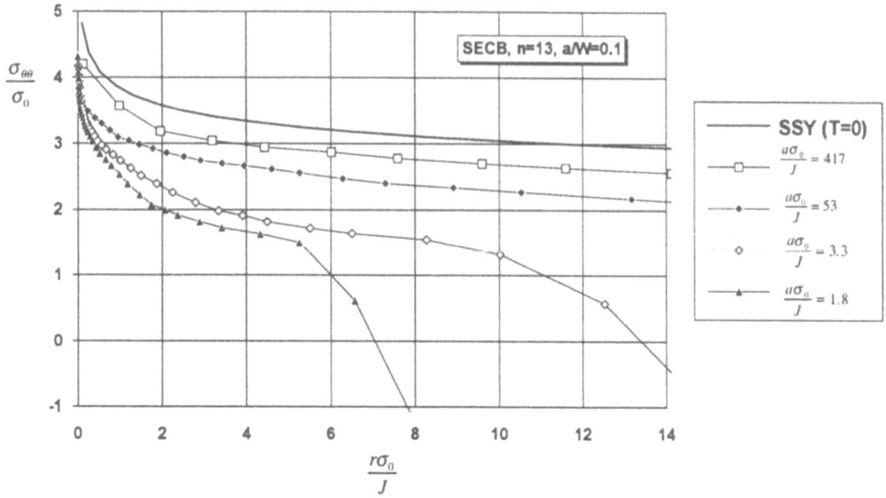

Figure 2. The hoop stress directly ahead of a crack SECB, a/W=0.1, n=13 at several levels of deformation

limit loads were determined numerically from the non-hardening analysis, but agree closely with the expression given by Miller (1987).

It is significant that the data for all geometries fall on the same curve, whose shape only depends on the strain hardening rate and the distance $\frac{r\sigma_0}{J}$ from the crack tip. Karstensen, Nekkal and Hancock (1995) have described the form of this relation by:

$$Q_P = k_2(n)\left(\frac{r\sigma_0}{J}\right)\left(\frac{P}{P_{Limit}}\right)^{n+1} \tag{9}$$

$k_2(n)$ is a tabulated proportionality constant, dependent on the strain hardening rate, but independent of geometry (a/W ratio or the ligament size c). This function is shown plotted for n=13 at a distance $\frac{r\sigma_0}{J} = 2$ in figure 4.

The relation between the plastic component of J and the load can be expressed in the form:

$$J_P = \alpha\sigma_0\varepsilon_0 c h_1(\frac{a}{W}, n)(\frac{P}{P_{Limit}})^{n+1} \tag{10}$$

$h_1(a/W, n)$ is a function of the $\frac{a}{W}$ ratio and the strain hardening, tabulated by Kumar, German and Shih (1980). Equations (9) and (10) suggest that Q_P is linearly dependent on J_P. Figure 5 thus shows Q_P as a function of

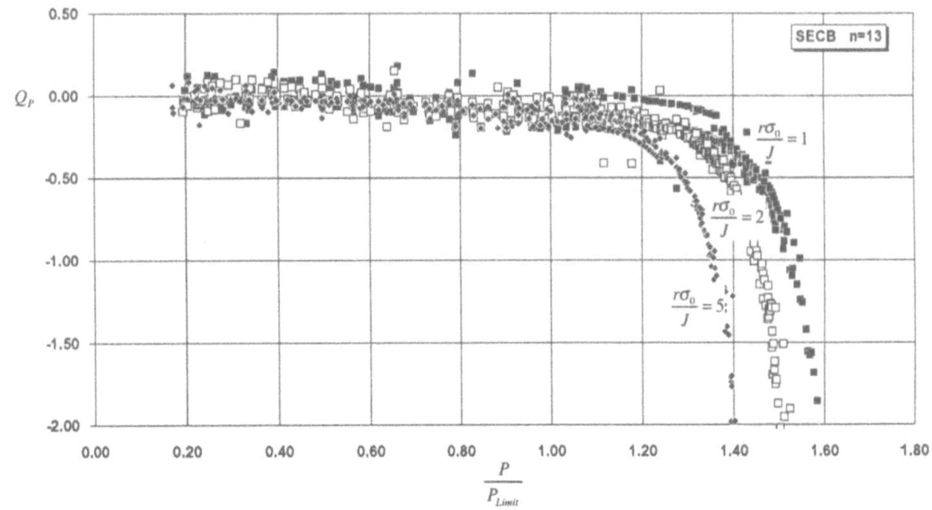

Figure 3. Q_P as a function of $\frac{P}{P_{Limit}}$ for Single edge bend bars at distances $\frac{r\sigma_0}{J}$=1, 2 and 5 from the crack tip.

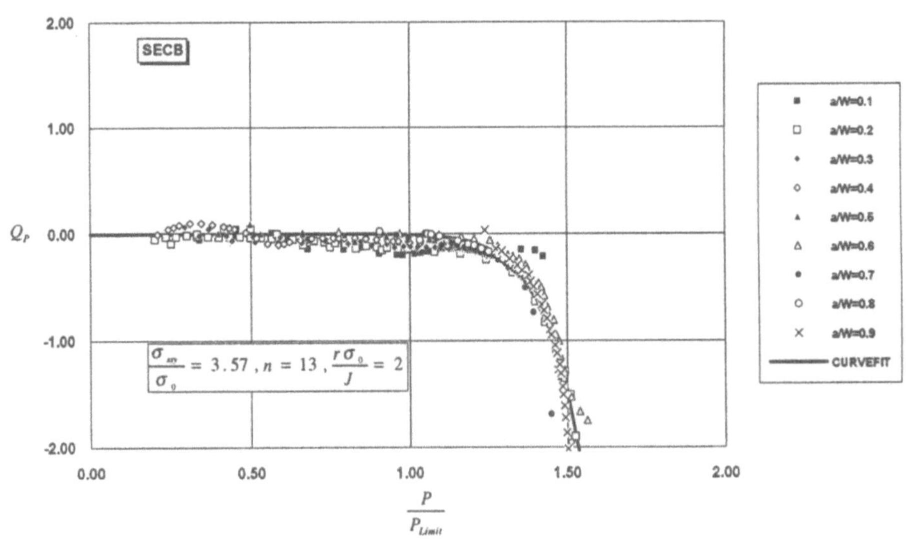

Figure 4. Q_P as a function of $\frac{P}{P_{Limit}}$ for single edge bend bars at a distance $\frac{r\sigma_0}{J}$ = 2 from the crack tip, for all a/W ratios and n=13

$\frac{J_P}{c\sigma_0}$ for n=13 at distances $\frac{r\sigma_0}{J}$=1, 2 and 5 from the tip of the complete range of edge cracked bars (a/W=0.1 to 0.9).

$$Q_{P_{(\frac{r\sigma_0}{J})}} = k_1(n)\left(\frac{r\sigma_0}{J}\right)\left(\frac{J_P}{c\sigma_0}\right) = k_1(n)(\frac{r}{c})\left(\frac{J_P}{J}\right) \tag{11}$$

It can be seen that $k_1(n)$ is insensitive to the geometry (a/W), as all the

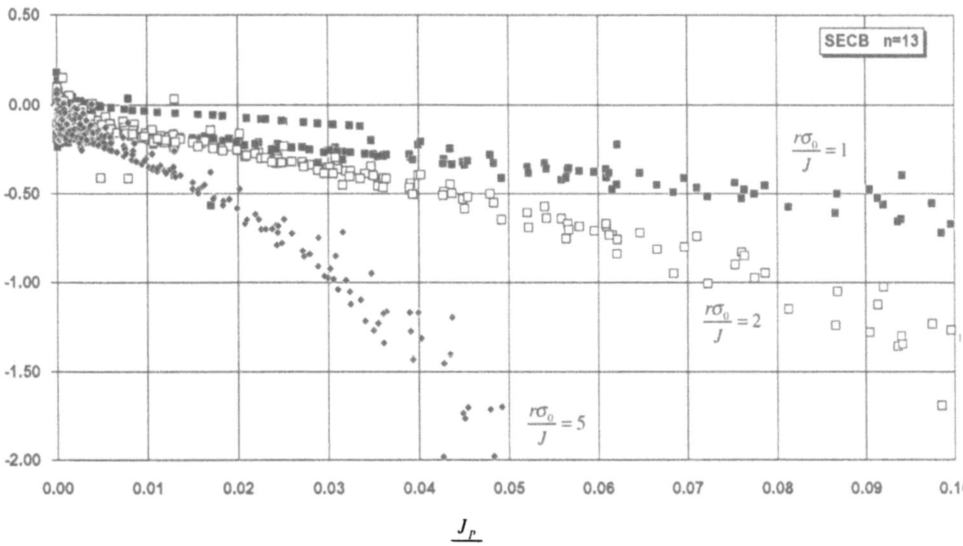

Figure 5. Q_P as a function of $\frac{J_P}{c\sigma_0}$ for single edge cracked bars for all a/W at distances $\frac{r\sigma_0}{J}$=1,2,5 from the crack tip.

curves fall on the same straight line. At moderate distances from the tip Q_P is linearly dependent on distance and may be thought of as a constraint gradient. Using the notation of Shih and O'Dowd (1992)

$$Q' = \frac{\partial Q}{\partial \frac{r\sigma_0}{J}} = \frac{\partial Q_P}{\partial \frac{r\sigma_0}{J}} = k_2(n)(\frac{P}{P_{Limit}})^{n+1} = k_1(\frac{J_P}{c\sigma_0}) \tag{12}$$

Complete expressions for the stress field may now be assembled from equation (9). Firstly the results are assembled in a form which enables the stress field to be determined from the applied load.

$$\frac{\sigma_{\theta\theta}}{\sigma_0} = \frac{\sigma_{SSY}}{\sigma_0} + f(\frac{T}{\sigma_0}, n) + k_2(n)\left(\frac{r\sigma_0}{J}\right)\left(\frac{P}{P_{Limit}}\right)^{n+1} \tag{13}$$

$$Q_{(\frac{r\sigma_0}{J})} = f\left(\frac{\beta}{\sigma_0}\sqrt{\frac{J_E}{E'\pi a}}, n\right) + k_2\left(\frac{r\sigma_0}{J}\right)\left(\frac{P}{P_{Limit}}\right)^{n+1} \tag{14}$$

Secondly the results are expressed in a form which enables the stress field to be determined from the elastic and plastic components of J.

$$\frac{\sigma_{\theta\theta}}{\sigma_0} = \frac{\sigma_{\mathrm{SSY}}}{\sigma_0} + f\left(\frac{\beta}{\sigma_0}\sqrt{\frac{J_E}{E'\pi a}}, n\right) + k_1(n)\left(\frac{r\sigma_0}{J}\right)\left(\frac{J_P}{c\sigma_0}\right) \tag{15}$$

$$Q_{(\frac{r\sigma_0}{J})} = f\left(\frac{\beta}{\sigma_0}\sqrt{\frac{J_E}{E'\pi a}}, n\right) + k_1(n)\left(\frac{r\sigma_0}{J}\right)\left(\frac{J_P}{c\sigma_0}\right) \tag{16}$$

Values of the constants k_1 and k_2 have been tabulated for both bending and tension for a range of strain hardening rates in Karstensen, Nekkal and Hancock (1995).

5. Discussion

The difference between the contained yielding prediction of Q based on T and the crack tip field in full field solutions has been denoted Q_P through equation (8). The existence of a valid Q field beyond the predictions based on T requires the existence of a distance independent Q_P term. Figures 3 and 5 all clearly show that at levels of deformation at which significant deviations occur from the modified boundary layer formulation, Q_P increases with distance from the tip. For force loaded edge cracked bars, the distance independent Q term is accounted for by T, and that the deviation from J-T characterisation at finite distances from the tip arises from the global bending field. This difference cannot be described by a valid distance independent term, and it is therefore necessary to conclude that Q does not significantly extend the two parameter characterisation of force loaded edge cracked bars in tension and bending beyond the limits of J-T characterisation. It is also interesting to note that the form of the expressions suggests that at the crack tip the global bending term disappears leaving Q only dependent on T.

In order to ensure a valid Q field, it is impractical to require that Q' is zero, and it is appropriate to allow a finite but restricted constraint gradient, Q'. In this context equation (13) provides a convenient way of determining $Q_{(\frac{r\sigma_0}{J})}$ from the applied load in a form which parallels the J estimation schemes advanced by Kumar, German and Shih (1980). In deeply cracked bend bars the fully plastic field initially develops high constraint, and is well described by J through the HRR field, within the limitations of J dominance. Further deformation leads to a loss of crack tip constraint and a loss of J-dominance which has been clearly identified to arise from the global bending field.

Deviations from the fully constrained field of deeply cracked bars arises only from the distance dependent term Q_P. A knowledge of J and k_2 thus characterises the crack tip field. This provides a two parameter characterisation of deeply cracked geometries in which loss of constraint arises from the global bending field, and extends crack tip characterisation beyond the limits of J-dominance.

6. Conclusions

The development of crack tip constraint has been systematically examined for edge cracked bars subject to bending. The initial loss of crack tip constraint is controlled by the sign of the non-singular T stress which is associated with fields which can be described by the small scale yielding field plus a distance independent term (Q). Within contained yielding crack tip characterisation can rigorously be achieved by T or equivalently Q. J-T characterisation does however extend in practice well beyond the formal limits of contained yielding if a notional value of T is calculated from the elastic component of J.

At high levels of deformation both J-T and J-Q characterisation break down simultaneously due to the global bending field impinging on the crack tip. This results in a distance dependent term. In this context Q has been decomposed into a distance independent term Q_T which is formally related to T and a distance dependent term which is related to the global bending field. This has been expressed in terms of far field parameters such as the applied load and J. This constitutes a two parameter characterisation, and associated fracture criterion for deeply cracked bend bars, beyond the limits of J-dominance.

Shallow cracked bars show a very limited region of single parameter characterisation, as constraint loss originates from the compressive T stress associated with the elastic field. Characterisation can be extended by the use of a two parameter approach using J and T (or equivalently J-Q). These fields eventually break down simultaneously due to the global bending field impinging on the crack tip. At these deformation levels characterisation has been achieved by the use of three parameters, J, Q_T and Q' or Q_P, this provides a complete crack tip constraint estimation scheme for edge cracked bars in bending.

7. Acknowledgments

Annette Karstensen takes pleasure in acknowledging the support of Prof. J. Sumpter through a D.R.A. grant. A. Nekkal acknowledges the suppport of the Health and Safety Executive. Thanks are also due to Hibbitt, Karlsson and Sorensen Inc. for access to ABAQUS under academic license. Finally the authors wish to acknowledge Dr. C. Betegón for useful discussions.

References

Al-Ani, A.M. and Hancock,J.W, (1991), *J*-Dominance of Short Cracks in Tension and Bending, *Journal of Mechanics and Physics of Solids*, **39**, 23-43

Betegón, C. and Hancock, J. W. , (1991), Two-Parameter Characterization of Elastic-Plastic Crack-Tip Fields, *Journal of Applied Mechanics*, **58**, 104-110

B. A. Bilby and B. A. Cardew and M. R. Goldthorpe and I. C. Howard, (1986), Size Effect in Fracture, A Finite Element Investigation of the Effect of Specimen Geometry on the Field of Stress and Strain at the Tip of Stationary Cracks, Size Effects in Fracture, Inst. Mech. Eng., London UK, 37-46

Harlin, G. and Willis, J. (1988) The influence of crack size on the ductile-brittle transition. *Proc. Roy. Soc.*, **A415** 197-226.

Harlin, G. and Willis, J. (1990) The influence of crack size on the fracture behaviour of short cracks. *International Journal of Fracture*, **42** 341-355.

Hibbitt, Karlsson and Sorenson Inc., *ABAQUS Manual* Providence Rhode Island, (1992), ABAQUS V.5.3

Karstensen, A.D, Nekkal, A. and Hancock, J.W., (1995), Constraint Estimation Schemes: I Edge Crack Bars in Tension and Bending, Internal Report Department of Mechanical Engineering, Glasgow University, Glasgow G12, *Submitted for publication*

Karstensen, A.D and Hancock, J.W., (1994), A Review of Two Parameter Fracture Mechanics, Internal Report Department of Mechanical Engineering, Glasgow University, Glasgow G12

Kumar, V. and German, M. D. and Shih, C. F , (1980), Estimation Techniques for the Prediction of Elastic-Plastic Fracture of Structural Components of Neclear Systems Report to E.P.R.I, General Electic Company, Schenectady, New York

Larsson, S. G. and Carlsson, A. J. (1973), Influence of non-singular Stress Terms and Specimen Geometry on Small-Scale Yielding at Crack Tips in Elastic Plastic Material, *International Journal of Fracture*, **19**, 263-278

Leevers, P. S. and Radon, J. C. , (1983), Inherent Stress Biaxiality in various Fracture Specimen Geometries, *International Journal of Fracture*, **19** , 942-955

Miller, A.G. (1987) Review of limit loads of Structures containing defects. Nuclear Electric report and Materials. Philadelphia.

O'Dowd, N. P. and Shih, C. F. (1991a), Family of Crack-Tip Fields Characterized by a Triaxiality Parameter: Part I - Structure of Fields, *Journal of Mechanics and Physics of Solids*, **39**, 939-963

O'Dowd, N. P. and Shih, C. F. (1991b), Family of Crack-Tip Fields Characterized by a Triaxiality Parameter: Part II - Fracture Applications, *Journal of Mechanics and Physics of Solids*, **40**, 989-1015

Sherry, A.H., France, C.C and Goldthorp, M.R, (1995), Compendium of *T*-Stress Solutions for Two and Three Dimensional Cracked Geometries, *Fatigue and Fracture of Engineering Materials and Structures Ltd.* , **Vol. 18**, pp. 141-154

Shih,C.F. and O'Dowd (1992), A fracture mechanics approach based on a toughness locus. *Shallow Crack Fracture Mechanics, Toughness Tests and Applications.* The Welding Institute, Cambridge, UK.

Shih,C.F. (1983), Tables of **H**utchinson-**R**ice-**R**osengren Singular Field Quantities, Materials Research Laboratory, Brown University, RI 02912

Williams, M. L. , (1957) On the Stress Distribution at the Base of a Stationary Crack, *ASME Journal of Applied Mechanics*, **24**, 111-114

FINITE ELEMENT ANALYSIS OF A BI-MATERIAL SENT SPECIMEN UNDER ELASTIC-PLASTIC LOADING

N P O'DOWD*, P J BUDDEN†, E R J GRIFFITHS*

*Department of Mechanical Engineering,
Imperial College of Science, Technology and Medicine,
London, SW7 2BX, UK

†Nuclear Electric,
Barnett Way,
Barnwood, Gloucester, GL4 3RS, UK

1. Introduction

The fracture mechanics of homogeneous materials is now well understood at temperatures below the creep range in terms of the stress intensity factor K and the J-integral. Practical structural integrity assessments of structures containing defects may then be carried out using the R6 procedure [1] and are equivalent to a J analysis of the structure. At high temperatures, the creep equivalent of J, the so-called C* integral [2], describes the stress and strain-rate fields close to the crack tip in widespread creep conditions and hence governs crack growth. For assessments of defective structures, estimates of C* may be combined with crack growth rate data $\dot{a}(C^*)$ from simple specimen tests to estimate creep crack growth using the R5 procedure [3].

The theoretical basis of fracture mechanics at interfaces between dissimilar materials has been studied in linear elasticity and in power-law plasticity [4,5]. The crack-tip stress and strain fields are more complex than in the homogeneous case and are characterised by two parameters. J is nevertheless well-defined for cracks lying in the interface and is given by the same integral expression as in the homogeneous case. The analogy between power-law plasticity described by

$$\varepsilon_p = A\sigma^n \tag{1}$$

where ε_p is plastic strain, σ is stress and n,A are constants, and power-law creep given by

$$\dot{\varepsilon}_c = A\sigma^n \tag{2}$$

where ε_c is creep strain, then enables C* to be defined for interfacial cracks in dissimilar materials by replacing strains and displacements in the expression

J. R. Willis (ed.), IUTAM Symposium on Nonlinear Analysis of Fracture, 33–42.

for J by their respective rates.

This paper describes detailed finite element analyses on a bi-material single edge notch tension (SENT) specimen under power-law plasticity. The strain hardening coefficient n in eqn.(1) is assumed the same in both materials. The effects on the J-integral of differences in the plastic strain coefficients A between the two materials is studied.

2. Geometry, Loading and Materials Properties

The single edge notched tension (SENT) specimen is shown in Fig.1. Unit thickness T=1 and plane strain constraint are assumed. Loading is by a uniformly distributed end load $P=Tw\sigma_\infty$. A crack of length a lies on the interface between Materials 1 (y>0) and 2 (y<0). The ratio a/w of crack length to specimen width is taken to be 0.25. Similarly, the ratio L/w, where 2L is the specimen length, is equal to 2.5, which is sufficiently large to make end effects on crack tip field quantities negligible. The results for these values of a/w and L/w are representative of other a/w and L/w ratios.

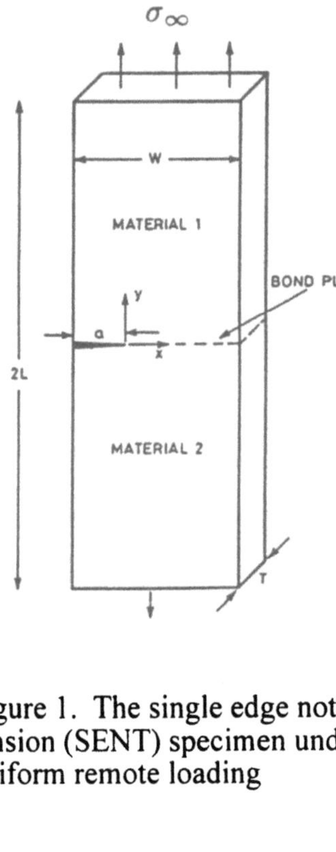

Figure 1. The single edge notch tension (SENT) specimen under uniform remote loading

Stress, σ, and total (elastic plus plastic) strain are given in each material by the linear elastic, power law hardening response:

$$\frac{\varepsilon}{\varepsilon_0} = \frac{\sigma}{\sigma_0} \qquad \sigma < \sigma_0$$

$$\frac{\varepsilon}{\varepsilon_0} = \frac{\sigma - \sigma_0}{\sigma_0} + \left(\frac{\sigma}{\sigma_0}\right)^n \qquad \sigma > \sigma_0 \tag{3}$$

where n, ε_0, and σ_0 are constants for each material with $\varepsilon_0=\sigma_0/E$, E being Young's modulus. The hardening coefficient, n, Poisson's ratio, ν, and E are the same for both materials. Then the material mismatch is quantified by the ratio $B=\sigma_{02}/\sigma_{01}$ of the normalising stress σ_0, where σ_{0i} (i=1,2) corresponds to Material i. Hence $\sigma_{02}/\sigma_{01}=1$ in the homogeneous case whereas σ_{02}/σ_{01} tends to infinity in the limit of a plastically deforming to rigid material combination.

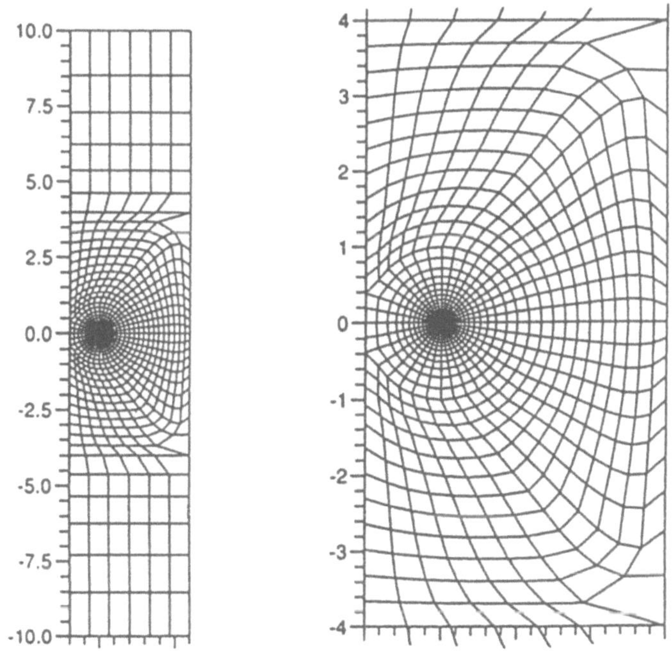

Figure 2. A representative finite element mesh (a/w = 0.25, L/w = 2.5) for the bimaterial specimen showing remote and near tip mesh.

3. Finite Element Mesh Details

The finite element analyses use FEAP, a general purpose finite element code developed at Brown University, USA. The J-integral is computed within the code using a domain integral method. Four-node quadrilateral isoparametric finite elements are used except at the crack tip where collapsed quadrilateral elements are used. A typical mesh (Fig.2) uses about 3200 nodes and 3000 elements, with 36 elements in the circumferential direction around the crack tip and a crack tip element radial dimension of 0.001a. The precise number and distribution of elements varies as a/w and L/w vary, to give accurate results.

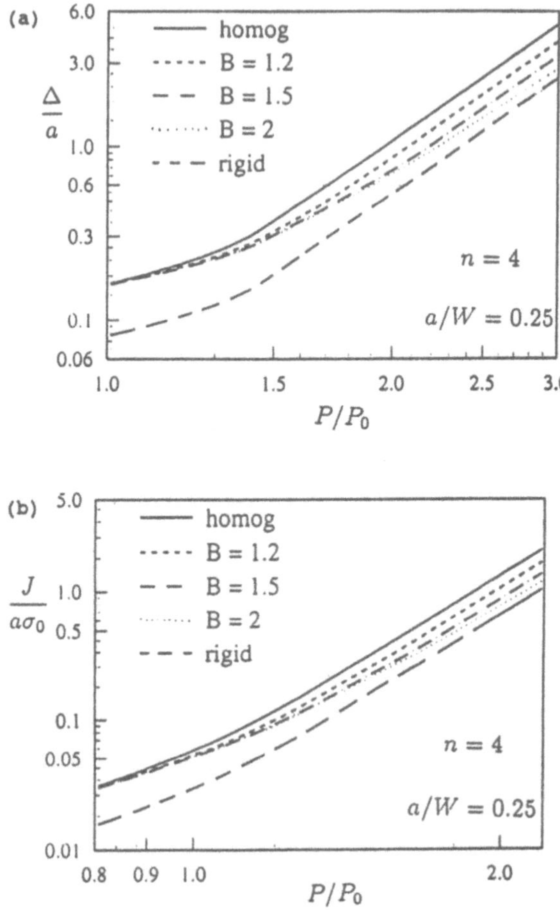

Figure 3. Variation of the normalised finite element bimaterial load-point displacement (Δ) and J values with normalised load.

4. Results

J-integral values have been computed for n=4 and 10 and a range of stress ratios B defined by

$$B = \sigma_{02}/\sigma_{01} \tag{4}$$

Figure 3 shows the normalised load-point displacement and normalised J plotted against normalised load for different mismatch ratios B with n=4. The result for a homogeneous material is indicated by the solid line and for the same material bonded to a rigid substratum by the dashed line.

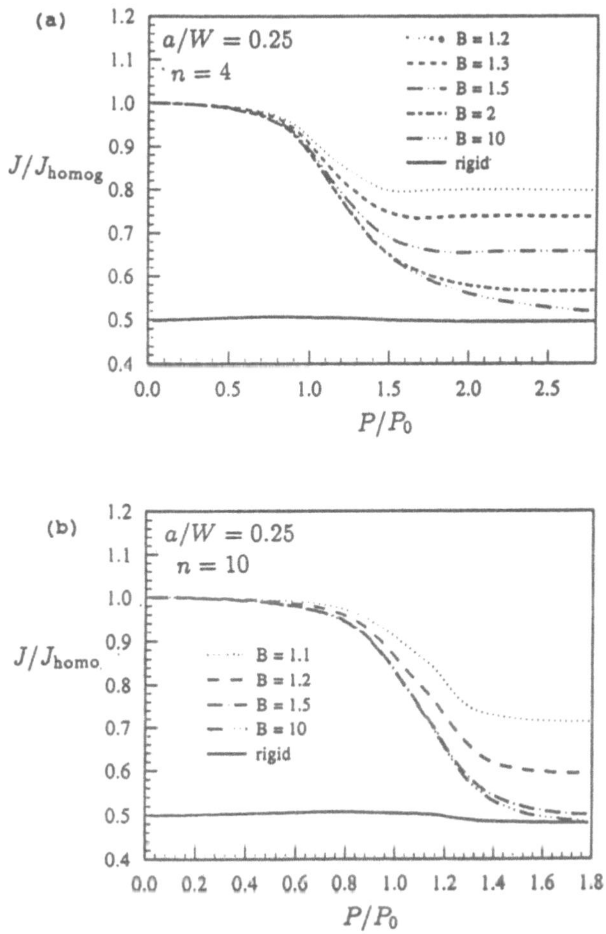

Figure 4. Variation of ratio of bimaterial to homogenous J values with load.

It can be seen that, for the rigid case, the slope of the curve is the same as for the homogeneous material in the fully-plastic region (the slope is given by n for the displacement-load curve and by n+1 for the J-load curve). At low loads, the J values for the bi-material cases correspond to the homogeneous value as the elastic properties of the two materials are identical. As the load increases and the upper material deforms more than the lower one, the response tends to a value intermediate between the homogeneous and rigid responses. Eventually all the bi-material cases appear to reach a steady state with the slope given by n for the displacement-load curve and by n+1 for the J-load curve.

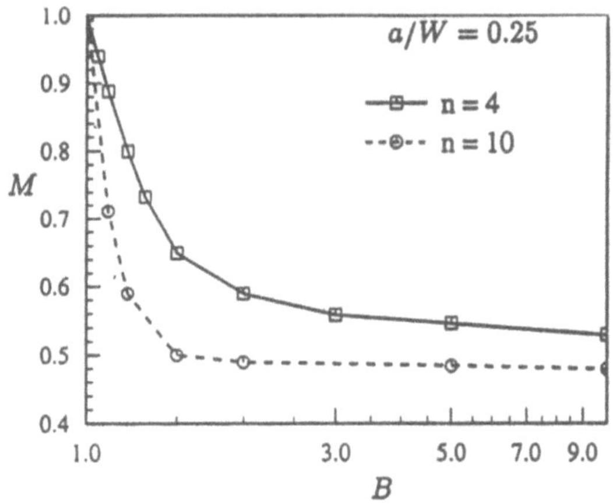

Figure 5. Finite element bimaterial results , variation of M (defined in text) against normalised load for n=4 and n=10.

In Fig.4a, the bi-material J values from Fig.3b are plotted normalised by the value of the homogeneous J at the same load level. As would be expected from the results shown in Fig.3b the ratio between J for the bi-material and the homogeneous J tends to a constant at high load. Apart from the rigid case, $B\to\infty$, the ratio is equal to unity in the elastic regime and there is a transition to fully-plastic behaviour as P/P_0 increases. For the rigid case, the ratio remains close to 0.5 in the elastic and fully-plastic regimes. In Fig.4b the ratio is plotted for an n=10 material. In this case the transition to bi-material behaviour occurs sooner for the same B value. This is because, for the same B value, the bi-material effect is stronger for the larger n. Again for the rigid case the ratio remains close to 0.5. In the fully-plastic limit, the ratio is approximately 0.48 for n=10 and 0.50 for n=4.

In Fig.5, the limiting ratio at high loads between the bi-material and homogeneous J, designated M, is plotted as a function of B for a/w=0.25. Our calculations suggest that for a/w>0.2 this may be considered to be a geometry-independent quantity.

Finally, in Fig.6 the J values obtained from the finite-element analyses are compared with those obtained using the η_p approach and also a modified EPRI-GE J-estimation approach. These methods are first briefly described.

The EPRI-GE method [6,7] is based on solutions for pure power-law plasticity. In uniaxial tension, the homogeneous pure power-law material deforms according to

$$\frac{\varepsilon}{\varepsilon_0} = \alpha\left(\frac{\sigma}{\sigma_0}\right)^n \tag{5}$$

where α is a constant. For this material, J can be represented in terms of a dimensionless function $h_1(a/w,n)$ by

$$J = \alpha\sigma_0\varepsilon_0(w-a)\frac{a}{w}h_1\left(\frac{P}{P_0}\right)^{n+1} \tag{6}$$

where P_0 is the plane strain limit load. The limit load used in this report, which is different from that used in [6,7], is due to Miller [8] and is believed to represent limit behaviour more accurately than the expression in [6,7]. The limit load is given by

$$P_0 = \frac{2}{\sqrt{3}} wT\sigma_0 f(a/w) \tag{7}$$

with

$$f(a/w) = 1 - a/w - 1.232(a/w)^2 + (a/w)^3 \tag{8}$$

for $a/w<0.545$. The h_1 function to be used in eqn.(6) is the value in [6] adjusted to account for the different limit load value.

Equation (6) applies in the fully-plastic regime where the plastic strains dominate. However, in order to compare J estimates over the full range of loading it is necessary to include the elastic contribution to J. This is given by

$$J_e = \frac{K^2(1-\nu^2)}{E} \tag{9}$$

under plane strain conditions. The total J is then given by

$$J = J_e + J_p \tag{10}$$

with J_p given by eqn.(6). At higher loads the plastic strains dominate and $J_p >> J_e$. The expression used here for J_e is somewhat different than that proposed in [6], where a modified J_e was used. However, the effect of using this modified J_e is rather insignificant and for simplicity is not used here.

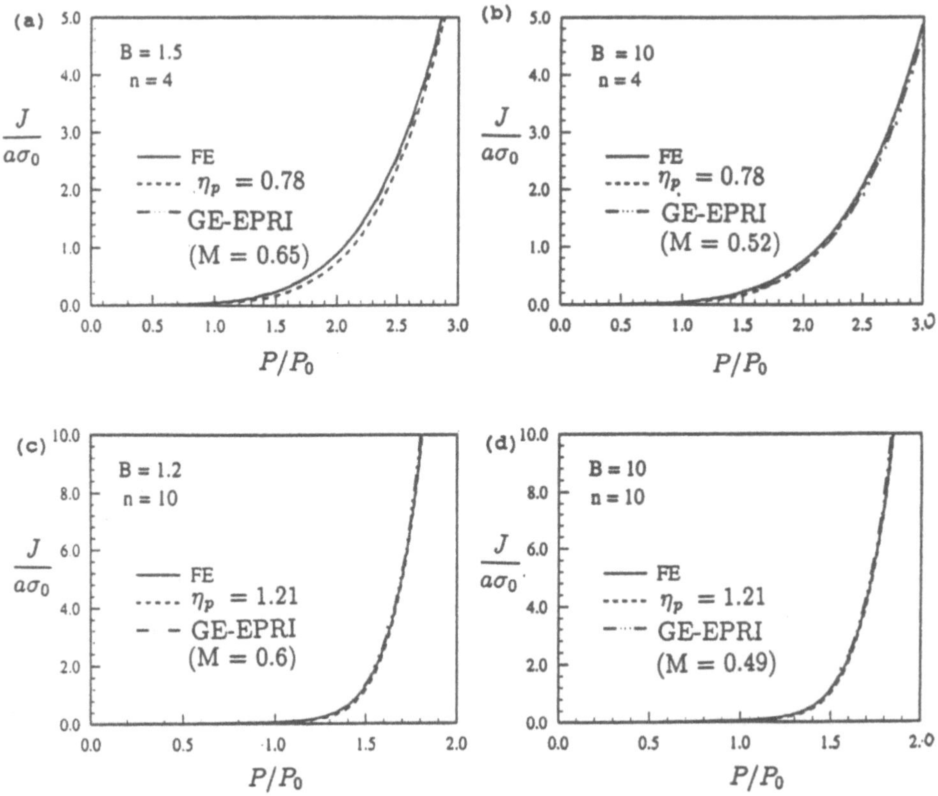

Figure 6. Finite element modified load point displacement and modified EPRI values of J are plotted for various cases.

The modified EPRI scheme to account for the dissimilar materials simply uses eqns.(6,9,10) with the h_1 function multiplied by the appropriate M value from Fig.5. It can be seen from Fig.6 that the agreement between the calculated J and the modified EPRI estimate is good over all load levels. The error is largest in the elastic-plastic regime as the ratio of bi-material to homogeneous J (Fig.4) is not constant here but depends on the magnitude of the load. The maximum error here is about 15%.

Alternatively, J may be estimated using the experimental load-point displacement method. Here, the plastic component of J, J_p, may be obtained from

$$J_p = \eta_p \frac{n}{n+1} \frac{P\Delta_p}{T(w-a)} \tag{11}$$

where Δ_p is the plastic part of the load-point displacement. From [9], it may be taken that $\eta_p = 1$ for a tension geometry (but see below). As before we may take the elastic part of J from eqn.(9) and the total J is given by eqn.(10).

Results are presented in Fig.6 for two B values and a/w=0.25 for both n=4 and n=10; the same trends have been observed for all other mismatch ratios and crack geometries examined. An important result of the numerical analysis is that the η_p values obtained for the bi-material systems are almost identical to the values obtained from a homogeneous analysis; that is η_p does not depend on the mismatch ratio but only on the n value of the material and the crack geometry. As seen in Fig.6, the agreement between the computed J and J calculated from η_p is good in all cases, though for n=4, B=1.5 the agreement is not as good as the other cases in the elastic-plastic regime.

5. Conclusions

Elastic and elastic-plastic finite-element calculations have been performed for a bi-material SENT specimen with a/w=0.25. A range of material mismatch was considered. The ratio of the bi-material J to the homogeneous J decreases to a constant value as the load increases, the constant being a decreasing function of the mismatch σ_{02}/σ_{01}. In the limit of the plastic-rigid combination, the ratio is approximately 0.5 when n=4 or 10.

The η_p values for the bi-material system are very close to those obtained for a homogenous system with the same hardening exponent. Good agreement is obtained between the numerical and estimated J using the homogeneous η_p value and the finite element load point displacement.

6. Acknowledgement

This paper is published by permission of Nuclear Electric.

7. References

1. Milne I, Ainsworth R A, Dowling A R and Stewart A T (1988), Assessment of the Integrity of Structures Containing Defects, Nuclear Electric Procedure R6 Revision 3, *Int. J. Pressure Vessel Piping*, 32, 3-104.

2. Landes J D and Begley J A (1976) A Fracture Mechanics Approach to Creep Crack Growth, in *'Mechanics of Crack Growth'*, ASTM STP 590, 128-148.

3. Ainsworth R A, Editor (1995) Assessment Procedure for the High Temperature Response of Structures, Nuclear Electric Procedure R5 Issue 2, Nuclear Electric plc.

4. O'Dowd N P, Shih C F and Stout M G (1992) Test Geometries for Measuring Interfacial

4. O'Dowd N P, Shih C F and Stout M G (1992) Test Geometries for Measuring Interfacial Fracture Toughness, *Int. J. Solids Structures*, 29, 571-589.

5. Shih C F (1991) Cracks on Bimaterial Interfaces: Elasticity and Plasticity Aspects, *Matls. Sci. Engng.*, A143, 77-90.

6. Kumar V, German M D and Shih C F (1981) An Engineering Approach for Elastic-Plastic Fracture Analysis, GE-EPRI Report NP-1931.

7. Shih C F and Needleman A (1984) Fully Plastic Crack Problems Part 1: Solutions by a Penalty Method, *J. Appl. Mech.*, 51, 48-56.

8. Miller A G (1988) Review of Limit Loads of Structures Containing Defects, *Int. J. Pressure Vessel Piping*, 32, 197-327.

9. ASTM E813-81, Annual Book of ASTM Standards, Section 3, Volume 03.01, 762-780, 1983.

GEOMETRY AND SIZE EFFECTS IN DUCTILE FRACTURE – FEM STUDIES AND THEORETICAL CONSIDERATIONS

O. KOLEDNIK
Erich-Schmid-Institut für Festkörperphysik der Österreichischen Akademie der Wissenschaften, A-8700 Leoben, Austria

AND

G. X. SHAN AND F. D. FISCHER
Institut für Mechanik der Montanuniversität Leoben and Christian-Doppler-Laboratorium für Mikromechnik der Werkstoffe, A-8700 Leoben, Austria

Abstract. In this presentation the capabilities and the limitations of the near-tip constraint parameters like the Q-stress or the stress triaxiality are discussed. It is shown that these parameters are useful but they are not sufficient to explain the constraint effects observed experimentally, not even for plane strain conditions, i.e. for side grooved specimens.

By a simple two-layer model the effects of geometry and size on the crack growth resistance are investigated. In addition to the in-plane constraint a reasonable indicator for the global out-of-plane constraint is proposed. The model brings new insight into the transferability problem.

1. Introduction

How does the crack growth resistance curve of a given material change if the geometry and/or the size of the specimen is altered? Although a huge amount of experimental data exists on this subject, e.g. see [1], and although some partial aspects are understood, [2], this question is, generally, unsolved up to now. This question is very important since it touches the fundamental problem of transferring fracture properties from specimens to structural members of arbitrary shapes and sizes.

J. R. Willis (ed.), IUTAM Symposium on Nonlinear Analysis of Fracture, 43–52.

One of the most promising ideas to cope with this problem is the transition from a one-parameter fracture mechanics to a two-parameter concept, e.g. the J-Q-theory [3,4]. It is the purpose of this presentation to highlight the capabilities but also the limitations of this concept. In the following section the J-Q-theory shall be outlined briefly.

2. The Q-Stress as a Measure of the Constraint

The basic requirement for the application of a one-parameter fracture mechanics is that one parameter, e.g. the J-integral J, characterizes the stress and strain field around the crack tip of a loaded body. This condition is fulfilled only for so-called "high constraint" crack geometries, e.g. for deeply notched bend specimens under well contained yielding conditions. In general two parameters, J and Q, are needed to describe the near-tip fields. The loading parameter J determines the deformation level and the Q-stress quantifies the constraint, i.e. the level of the hydrostatic stress, for a stationary plane strain crack [5].

Q is defined as the hydrostatic difference stress between the actual stress $\sigma_{\theta\theta}$ (resulting from a finite strain Finite Element analysis) and some reference stress $\sigma_{\theta\theta,\mathrm{ref}}$ [3–5],

$$Q \equiv \frac{\sigma_{\theta\theta} - \sigma_{\theta\theta,\mathrm{ref}}}{\sigma_0} \qquad \text{for } \theta = 0,\ r = 2\frac{J}{\sigma_0} \quad . \tag{1}$$

In Fig. 1 (from [6]) the hoop stress is plotted versus the distance r from the tip, both in non-dimensional form. The reference stress can be taken either from the HRR-field (for a power-law hardening material) or from the stress field from a small-strain FE-analysis. σ_0 denotes the yield stress and r and θ are the polar co-ordinates originating at the crack tip.

The comparison between actual and reference stresses is made at a certain distance ahead of the crack tip just beyond the process zone as given in Eq. (1), and Q is claimed to remain constant throughout a region $\frac{J}{\sigma_0} < r < 5\frac{J}{\sigma_0}$ [5,6]. Therefore, Q is a measure of the local near-tip constraint. Negative Q-stresses mean loss of constraint. For different specimen types and geometries Q-stresses have been computed [3–6]. Large negative Q-stresses appear, e.g., for bend type specimens with very short crack lengths or for tension type specimens like centre cracked panels. Further it was found that Q decreases with increasing plastic zone size. But as all these computations were made for plane geometries under plane strain conditions the Q-stresses quantify the local in-plane constraint.

In the literature several examples are presented where Q is used successfully to scale for different specimen types and geometries, e.g., the slopes of J-Δa-curves (Δa = crack extension) for a ductile A710-steel [7] or the

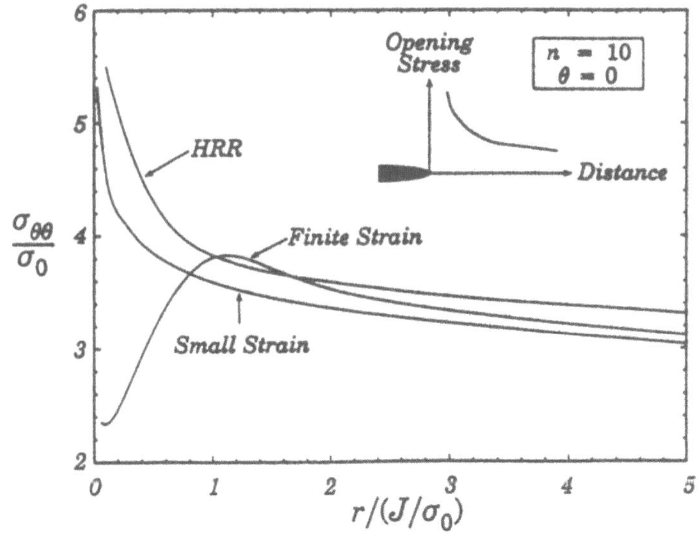

Figure 1. Computed hoop stress ahead of a crack tip is used for calculating the Q-stress, from [6].

J-values at the onset of brittle fracture of a mild steel at -50°C [8,5]. This is reasonable since with the decrease of Q the tensile mean stress within the process zone decreases. For ductile fracture this means that processes like void initiation and void growth are decelerated (leading to an enhanced crack growth resistance, i.e. a larger slope of the J-Δa-curve) and for cleavage fracture that the attainment of the critical cleavage stress is impeded (leading to a higher J at the onset of catastrophic fracture).

The T-stresses, T, [9] or the stress triaxiality, h, [10,11] are used as alternative measures of the local in-plane constraint sometimes. T is the second (r^0-)term of the Williams series expansion of the stress field for a linear elastic body and, therefore, T is strictly applicable for small-scale yielding conditions, only. For these conditions T and Q are equivalent. The stress triaxiality is defined as the ratio of the mean stress, σ_m, to the equivalent stress, σ_{eq},

$$h = \frac{\sigma_m}{\sigma_{eq}} \quad . \tag{2}$$

h is measured at a certain position ahead of the tip similar to the Q-stress. Its main advantage is that it can be used for a growing crack, too, contrary to Q and T which both originate from the fields of stationary cracks.

Are these constraint parameters sufficient to explain the geometry and size effects observed in experiments? This question will be answered in the following section.

3. The Limitations of the Local Constraint Parameters in Explaining Geometry Effects

Fig. 2a collects three J-Δa-curves of a SAE 4340 steel ($\sigma_0 = 298\,\text{MPa}$, ultimate tensile strength $\sigma_u = 1040\,\text{MPa}$, Young's modulus $E = 195\,\text{GPa}$). The curves are for CT-specimens with width $W = 50\,\text{mm}$ and thickness $B = 25\,\text{mm}$ but different initial crack length to width ratios, $a_0/W = 0.4$, 0.57 and 0.7. The dependency of the tearing modulus [12],

$$T_M = \frac{E}{\sigma_0^2}\frac{dJ}{da} \quad , \tag{3}$$

i.e. the non-dimensional slope of the J-Δa-curve, is also plotted. It should be emphasized that all curves are computed curves (for plane strain conditions) assuming a constant critical crack tip opening displacement, $\text{COD}_i = 102\,\mu\text{m}$, for initiation of crack extension and a constant local crack tip opening angle during growth of $\text{CTOA}_c = 11.5°$ [13]. The sizes of COD_i and CTOA_c were deduced from experiments, see the following section. The measured true stress vs. true strain curves were implemented pointwise in the analysis. The values of J were computed from the load vs. load-line displacement curves following the common standard procedures. Fig. 2b presents the computed normal stresses in front of the crack tip at the moment just before initiation, $J = J_i = 150\,\text{kJ/m}^2$, and Fig. 2c exhibits the local stress triaxiality at a distance of $r = 0.2\,\text{mm} \approx 2\,\text{COD}_\text{i}$ in front of the growing tip [14]. The stress triaxiality is equal for all crack lengths and it remains constant during a crack extension of 6 mm. As the normal stresses in Fig. 2b are (nearly) independent of a_0/W the Q-stresses are (nearly) constant for the three a_0/W-ratios. Nevertheless, the J-Δa-curves are different and the tearing modulus, T_M, decreases during crack extension in all cases. The conclusion is as follows: If the local in-plane constraint parameters are different this will have consequences on the slopes of the J-Δa-curves because the microscopic processes within the process zone are influenced by the different hydrostatic tension. However, the local in-plane constraint parameters are <u>not sufficient</u> for characterizing the shape of the J-Δa-curves, not even for plane strain conditions. The reason is that the global load vs. displacement behaviour of a specimen, and in turn the shape of the J-Δa-curve, are affected by the plasticity outside the near-tip region, and the amount of this plasticity is not fully characterized by a near tip constraint parameter [19].

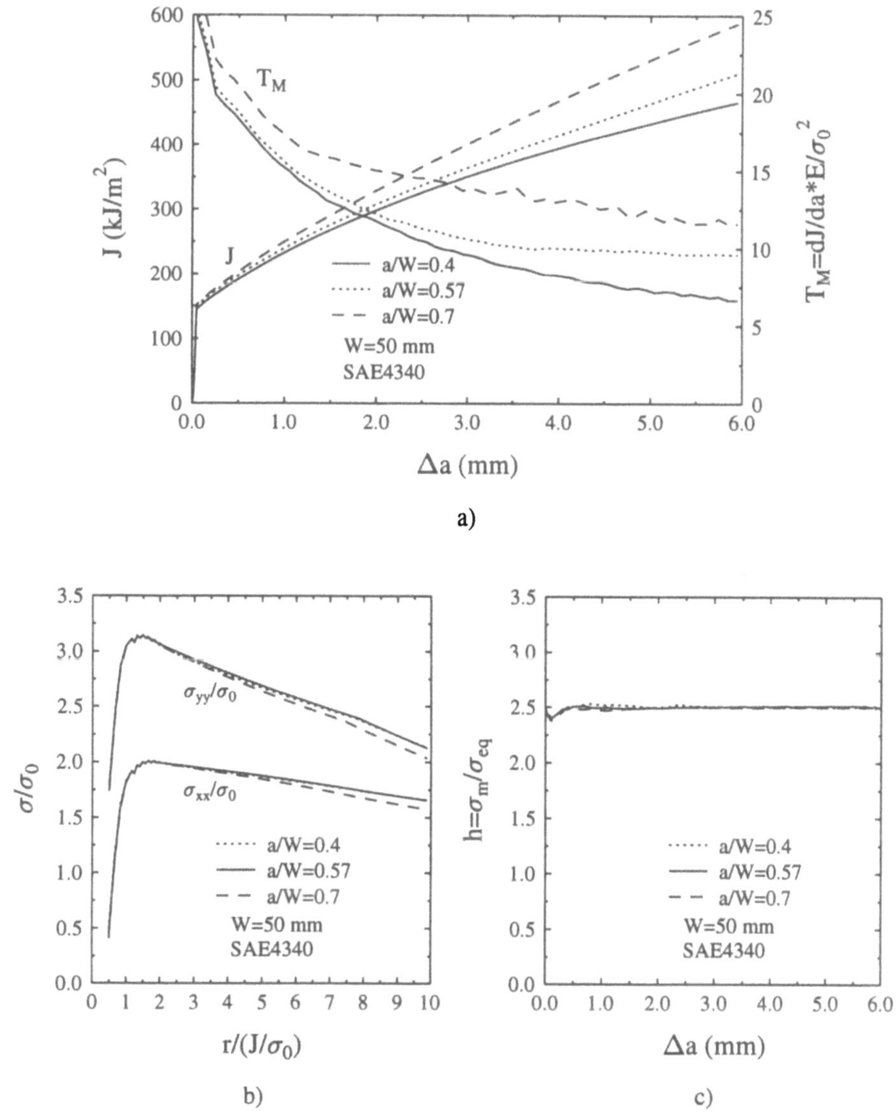

Figure 2. a) Computed J-integrals vs. crack extension curves for CT-specimens with different a/W-ratios, b) normal stresses ahead of the tip just before initiation, c) stress triaxiality measured 0.2 mm ahead of the growing crack tip. Although the Q-stresses are equal and the stress triaxiality remains constant the J-Δa-curves are different and changing their slopes.

From Fig. 2b it is seen that for $a_0/W = 0.7$ the normal stresses are a little bit smaller than for $a_0/W = 0.56$ and 0.4. So for the smallest ligament length, $b_0 = W - a_0 = 15\,\text{mm}$, the Q-stress is a little bit smaller, obviously affected by the traction free back face of the specimen. This is contrary to the figures reported in the literature where Q should slightly increase with increasing a_0/W. It is remarked that for this specimen the Irwin radius of the plastic zone at initiation is $r_y^i \approx 1.5\,\text{mm}$ and that the conditions for a valid J_{IC}-test are easily fulfilled:

$$b_0 = 15\,\text{mm} > 25\frac{\text{J}_\text{i}}{\sigma_0} = 3.8\,\text{mm} \tag{4}$$

Up to now only the in-plane constraint (for plane strain conditions) has been considered. But it is clear that the out-of-plane constraint has also a strong influence on the crack growth resistance: The J-Δa-curve of a thin specimen is steeper than that of a thick specimen. The out-of-plane constraint has a two-fold effect: On the one hand it changes the local fracture properties COD_i and CTOA_c along the crack front, on the other hand it affects the global load vs. displacement behaviour, e.g., it alters the size of the general yield load.

In the next section a simple model is presented to separate influence factors for the geometry and size effects on crack growth resistance curves. This model leads to a proposal for a measure of the global out-of-plane constraint.

4. A Two-layer Model to Separate Influence Factors for Geometry Effects

Plain sided fracture mechanics specimens are divided into two side-surface layers of total thickness B_s in plane stress and a center layer of thickness $B_e = B - B_s$ in plane strain (Fig. 3). The resistance curves of the specimens are modeled by treating fracture initiation and crack growth in the two layers separately [15]. For each layer initiation and growth are controlled by a local COD_i and a constant local CTOA_c at the moving tip. Commonly the plane stress layer behaves tougher than the plane strain layer, i.e. the COD_i and CTOA_c-values are larger.

The idea of the model is that the relative thickness of the plane stress layer, i.e. the ratio B_s/B which can be easily determined (see below), provides an indicator of the global out-of-plane constraint of the specimen. The influence of the fracture behaviour of the material can be modeled by changing the sizes of the controlling toughness parameters, COD_i and CTOA_c. The effect of the in-plane constraint can be tested by comparing the results of the plane strain analyses for different geometries and by checking the local crack tip fields to determine the above mentioned in-plane constraint parameters. The controlling toughness parameters, i.e. the $\text{COD}_i/\text{CTOA}_c$-pairs can be determined experimentally. E.g., for a low strength annealed structural steel this was done by stereophotogrammetric studies of the fracture surfaces near the specimen midsection (for the plane strain layer) and near the side surfaces (for the plane stress layer) [15]. For a SAE 4340 the results of a multi-specimen J_{IC}-test were used to plot the local crack extension (near the midsection and near the side surfaces) versus the load line displacement. From these curves the COD_i- and $CTOA_c$-values can be deduced [13].

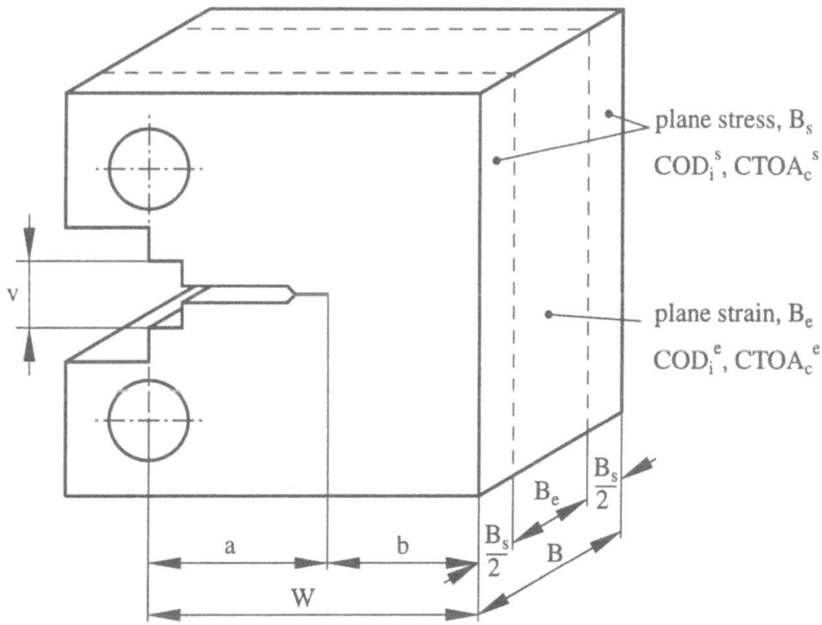

Figure 3. The two-layer model for computing resistance curves for different geometries and sizes.

The ratio B_s/B is estimated by two 2D-Finite Element analyses for plane strain and plane stress conditions and one 3D-analysis with a rather coarse mesh [16]. All three analyses are performed for a stationary crack. By comparing at a given load line displacement v the "true" load, i.e. the load of the 3D-analysis, to the loads from the plane stress and plane strain analyses the correct thicknesses of the plane stress and the plane strain layers can be determined (Fig. 4). B_s/B decreases with increasing v but soon it reaches a saturation value, $\beta_s = (B_s/B)_{\text{sat}}$. The ratio B_s/B or its saturation value β_s seem to be natural indicators of the global out-of-plane constraint.

The benefit of the two layer model is that an expensive 3D-simulation of crack extension can be replaced by two 2D-analyses plus one (cheap) 3D-analysis for a stationary crack. For details of the computations and the validation of the model, see [16, 17]. In some previous studies this two layer model was applied to investigate geometry and size effects in CT-specimens. The main results of these studies are the following:

1. For a given material and different geometries under large-scale or general yielding conditions B_s/B or β_s is a function of the ratio of the specimen thickness, B, to the ligament length, b. In a similar way the slope of the J-Δa-curve scales with the B/b-ratio if CTOA_c is held constant [17]. If B/b is small, β_s is large, the out-of-plane constraint is small and the specimen behaves like a thin specimen, i.e. $\frac{dJ}{da}$ is large.
2. For a given geometry and different materials the ratio B_s/B depends on the relative size of the plastic zone, r_y/b, and the strain hardening exponent n. The saturation value β_s is only affected by n. β_s increases, i.e. the out-of-plane constraint decreases, with increasing n [17].
3. Different patterns of geometry behaviour may appear for a given material depending on whether side-grooved or non side-grooved specimens are used [13]. High- and low-strength materials may exhibit different patterns of geometry behaviour [13,18,19].
4. The ratio $\frac{CTOA_c}{\epsilon_0}$ is crucial for the crack growth behaviour and it determines the tendency to unstable crack extension [18,19]. ($\epsilon_0 = \sigma_0/E$ is the yield strain.)

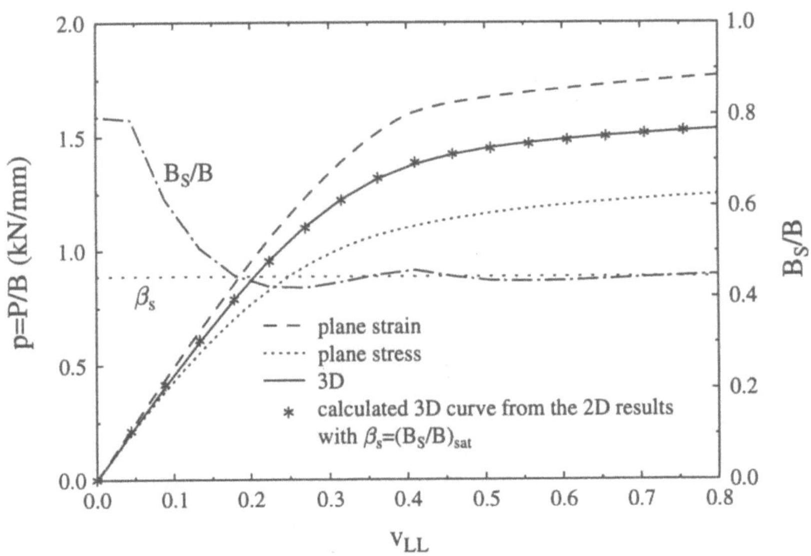

Figure 4. The relative thickness of the plane stress layer, B_s/B, is determined from computed load vs. displacement curves for a stationary crack. B_s/B decreases with increasing loading but it reaches a saturation value, β_s, if large scale yielding conditions are reached.

5. Conclusions

Local in-plane constraint parameters like the Q-stress or the stress triaxiality are useful tools but they represent only one influence factor on geometry effects among others. The out-of-plane constraint is equally important.

The near tip constraint (in-plane and out-of-plane) can probably characterize the sizes of the local fracture toughness properties COD_i and CTOA_c as the microscopic processes of fracture occur within the process zone close to the crack tip. On the contrary to fracture, plastic deformation is — apart from contained yielding conditions — not concentrated only around the crack tip. Global constraint parameters must be used to scale the global deformation behaviour. This is the reason why the local near tip constraint cannot quantify the widespread plasticity within a specimen. Even for plane strain conditions the J-Δa-curves may be different for two geometries although the near-tip constraint parameters are equal.

Our simple two layer model has brought new insight into the transferability problem and a proposal for a reasonable indicator of the global out-of-plane constraint.

References

1. Turner, C.E. (1990) A re-assessment of the ductile tearing resistance, Part I and II, in D. Firrao (ed.) *Fracture Behaviour and Design of Materials and Structures, Proc. of ECF8*, EMAS, UK, Vol.II, pp.933-968.
2. Kolednik, O.(1993) A simple model to explain the geometry dependence of the J-Δa-curves, *Int. J. Fracture* **63**, 263–274.
3. O'Dowd, N.P. and Shih, C.F. (1991) Family of crack-tip fields characterized by a triaxiality parameter - I. Structure of fields, *J. Mech. Phys. Solids* **39**, 989–1015.
4. O'Dowd, N.P. and Shih, C.F. (1991) Family of crack-tip fields characterized by a triaxiality parameter - II. Fracture applications, *J. Mech. Phys. Solids* **40**, 939–963.
5. O'Dowd, N.P., Shih, C.F. and Dodds, Jr., R.H. (1995) The role of geometry and crack growth on constraint and implications for ductile/brittle fracture, in M. Kirk and A. Bakker (eds.), *Constraint Effects in Fracture: Theory and Applications, ASTM STP* **1244**, American Society for Testing and Materials, Philadelphia.
6. Dodds, Jr., R.H., Shih, C.F. and Anderson, T.L. (1993) Continuum and micromechanics treatment of constraint in fracture, *Int. J. Fracture* **64**, 101–133.
7. Hancock, J.W., Reuter, W.G. and Parks, D.M. (1993) Constraint and toughness parameterized by T, *Constraint Effects in Fracture, ASTM STP* **1171**, American Society for Testing and Materials, Philadelphia, pp.21–40.
8. Sumpter, J.D.G. and Forbes, A.T. (1992) Constraint based analysis of shallow crack in mild steel, in: M.G. Dawes (ed.) *Shallow Crack Fracture Mechanics, Toughness Tests and Applications*, TWI, Cambridge, UK, Paper 7.
9. Betegon, C., and Hancock, J.W. (1991) Two-parameter characterization of elastic-plastic crack-tip fields, *J. Appl. Mech.* **58**, 104–110.
10. Sommer, E. (1989) Progress in the assessment of complex components. in: K. Salama, R. Ravi-Chandar, D.M.R. Taplin and P. Rama Rav (eds.) *Advances in Fracture Research, Proc. of ICF7*, Pergamon Press, Vol.3, pp.1999–2010.
11. Brocks, W. and Schmitt, W. (1995) The second parameter in J_R-curves: constraint or triaxiality, in M. Kirk and A. Bakker (eds.), *Constraint Effects in Fracture: The-*

ory and Applications, ASTM STP **1244**, American Society for Testing and Materials, Philadelphia, pp.

12. Paris, P.C., Tada, H., Zahoor, A. and Ernst, H. (1979) The theory of instability of the tearing mode of elastic-plastic crack growth, in J.D. Landes, J.A. Begley and G.A. Clarke (eds.) *ASTM STP* **668**, American Society for Testing and Materials, Philadelphia, pp.5–36.
13. Shan, G.X., Kolednik, O. and Fischer, F.D. (1995) A numerical study on the influence of geometry variations on stable crack growth in CT specimens for different materials, in M. Kirk and A. Bakker (eds.), *Constraint Effects in Fracture: Theory and Applications, ASTM STP* **1244**, American Society for Testing and Materials, Philadelphia, pp.71-87.
14. Shan, G.X., Kolednik, O. and Fischer, F.D. (1994) Numerical investigations of stable crack growth in CT-specimens of SAE 4340 steel, in M.H. Aliabadi, A. Carpinteri, S. Kalisky and D.J. Cartwright (eds.) *Localized Damage II, Computer Aided Assessment and Control*, Computational Mechanics Publications, Boston, pp.399–406.
15. Shan, G.X., Kolednik, O., Fischer, F.D. and Stüwe, H.P. (1993) A 2D-model for the numerical investigations of the stable crack growth in thick smooth fracture mechanics specimens, *Engng Fracture Mech.* **45**, 99–106.
16. Shan, G.X., Kolednik, O., Stüwe, H.P. and Fischer, F.D. (1992) A substitution method for 3D elastic-plastic FE-analyses of fracture mechanics specimens, *Engng Fracture Mech.* **41**, 625–633.
17. Shan, G.X., Kolednik, O. and Fischer, F.D. (1994) A numerical investigation of the geometry dependence of the crack growth resistance in CT-specimens, *Int. J. Fracture* **66**, 173–187.
18. Shan, G.X., Kolednik, O. and Fischer, F.D. (in press) A numerical study on the crack growth behaviours of a low and a high strength steel, *Int. J. Fracture.*
19. Kolednik, O., Shan, G.X. and Fischer, F.D. (in press) The energy dissipation rate – a new tool to interpret geometry and size effects, submitted to R.S. Piascik (ed.) *Fatigue and Fracture Mechanics: 27th Volume, ASTM STP* **1280**, American Society for Testing and Materials, Philadelphia.

MECHANISM-BASED CELL MODELS FOR FRACTURE IN DUCTILE AND DUCTILE/BRITTLE REGIMES

C. FONG SHIH, LIN XIA AND LI CHENG
Division of Engineering
Brown University, Providence, RI 02912 USA

Abstract. The fracture resistance of ferritic steels in the ductile/brittle transition regime is controlled by the competition between ductile tearing and cleavage fracture. Under typical conditions, a crack initiates and grows by ductile tearing but final failure occurs by catastrophic cleavage fracture. The zone of tearing is modeled by a layer of void-containing cells characterizing the spacing and volume fraction of voids in material elements lying in the plane of fracture. These cell elements incorporate the softening characteristics of cavity growth and its strong dependence on stress triaxiality. Under increasing strain, the voids grow and coalesce to form new crack surfaces thereby advancing the crack. Crack growth causes significant alterations in the stress field, the process zone size and directly affects the initiation of cleavage fracture. These effects are accounted for by incorporating weakest link statistics into the cell element model. The cleavage fracture model also takes into account the increase of sampling volume with crack growth and the competition between void nucleation from carbide inclusions and unstable inclusion cracking which precipitates catastrophic cleavage fracture. The model is not limited by the extent of plastic deformation nor the amount of ductile tearing preceding cleavage fracture. Load-displacement behavior, ductile tearing resistance and transition to cleavage fracture are discussed for several different test geometries and a range of microstructural parameters. The model predicts trends in the ductile/brittle transition region that agree with experimental data.

1. Introduction

This paper reviews some recent results on ductile tearing obtained by cell element models. Several new results with implications on the competition

J. R. Willis (ed.), IUTAM Symposium on Nonlinear Analysis of Fracture, 53–64.

between ductile tearing and cleavage fracture and the transition from tearing to catastrophic cleavage are discussed.

Micrographs of fractured ferritic steel test specimens provide much information on ductile tearing mechanisms and characteristic length scales. Particularly relevant are repeated observations of void growth and coalescence being confined to a macroscopically planar fracture process zone of one or two void spacings in thickness; outside this region, the voids exhibit little or no growth. Shih and Xia (1995) observed that it is advantageous to model this fracture process region by a layer of void-containing cells representative of the spacing and volume fraction of voids in material elements lying in the plane of fracture as illustrated in Fig. 1. As shown in the figure, a cell element is a three-dimensional entity with characteristic length D which contains a centered spherical void whose initial volume fraction of the cell is f_0. Salient features of a cell element are described in Section 2. Among other things, the cell element arrangement employed allows us to circumvent computational complexities arising from crack path tortuosity or non-planarity on the microscale which appear to be controlled in large part by microstructure and by micron/submicron scale deformations leading to void nucleation (Xia and Shih, 1995a, 1995b). This is not to say that crack advance away from the plane ahead of the initial crack should be disregarded. Indeed, crack growth on inclined planes and the formation of shear lips are clearly identifiable macroscopic modes of crack extension at low levels of constraint. With some additional work, the present model can be extended to handle these important situations. Such studies are in progress. For the present, crack growth is confined to the plane depicted in Fig. 1.

The aggregate of cells introduced on the fracture plane is characterized by several fracture process parameters, the two most important being f_0 and D. In the case of ferritic steel, D and f_0 are related to the spacing and the initial volume fraction of voids originating from sulphide and oxide inclusions. This cell aggregate is embedded within a conventional elastic-plastic continuum described by the usual material parameters: Young's modulus, Poisson's ratio, yield stress and strain hardening exponent. Once the fracture process parameters are calibrated using one set of experimental crack growth data, our approach permits computation of relationships among loads, displacements and crack growth, including states where stability is lost. One of the main advantages of the present computational model is the fact that the model can be used to analyze structural response under conditions of extensive yielding and large crack advance without recourse to any fracture resistance curve. Indeed, Xia, Shih and Hutchinson (1995) have successfully predicted details of the load, displacement and crack growth histories of several specimen geometries which are

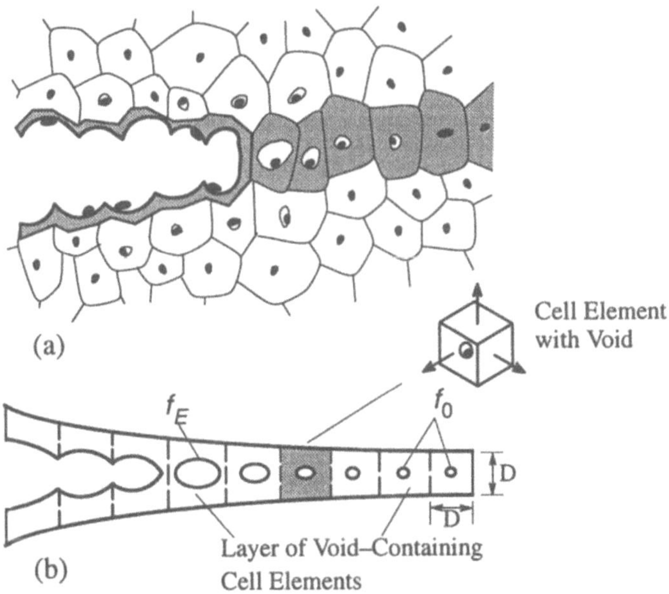

Figure 1. (a) Ductile fracture by void nucleation, growth and coalescence. (b) Fracture process region modeled by cell elements. Cell softening characteristics depend upon f_0 and stress triaxiality.

known to give rise to significantly different crack tip constraints.

Cleavage fracture data in the transition region is marked by scatter. This scatter, as well as the cleavage fracture toughness, increases with temperature. Raising the temperature also increases the amount of ductile tearing that precedes cleavage fracture. Some explanations for behaviors noted above can be found from computational studies using the above cell model in conjunction with weakest link statistics. The latter introduces an additional parameter m_s characterizing the size distribution of carbide particles. Exploratory studies to be discussed show that the augmented model predicts trends in ductile fracture resistance and ductile/brittle transition that agree with experimental data.

2. Cell Model for Ductile Tearing

A cell is a basic (smallest) material unit that contains sufficiently complete information on the essential characteristics of material separation. Cell el ements should be viewed as three-dimensional entities which are initially cubes with dimension D related to the characteristic microstructural length in the cavity growth/coalescence process. A similar point of view has been advocated by Rousselier (1987), Brocks *et al.* (1995) and others (see Xia, Shih and Hutchinson, 1995, for a more complete list of works). In our work, we employ Gurson's constitutive relation as modified by Tvergaard to describe the stress-strain behavior of a single cell element containing a centered spherical void of initial volume fraction f_0 (Gurson, 1977; Tvergaard, 1990). Because a cell element is a three-dimensional entity, its behavior depends upon tractions in all three directions (normal and parallel to the crack plane) as depicted in Fig. 1b. It is this discrete three-dimensional nature that enables cell elements to capture the critical features of ductile tearing including the strong triaxial stress effects on cavity growth and cavity-crack tip interaction on the scale of D.

Two material state parameters — the current void volume fraction f and the current flow stress of the matrix $\bar{\sigma}$ — govern a cell's stress-strain response. The void in a cell grows under increasing strain. A cell loses stress carrying capacity when the hardening of the matrix is insufficient to compensate for the reduction in the cell ligament area caused by void growth. This loss of stress carrying capacity accelerates at some level of void volume fraction corresponding to the final stage of the coalescence process. Beyond this stage, the cell no longer supports any traction and the crack advances across that cell. In our work, the final stage of material separation is modeled by a cell traction (reduction) elongation relation which takes effect when f equals f_E. The response of a cell element prior to coalescence is largely controlled by the initial void volume fraction f_0. That is, the volume fraction of large inclusions (sulphides and oxides) dictates the pre-coalescence cell behavior. However, small particles (e.g. carbides) play a critical role in the coalescence phase by nucleating microvoids which can greatly assist the process of cavity link-up. Indeed, under certain conditions the coalescence phase is primarily one of microvoid cavitation, a mode of coalescence in which relatively little energy is expended (Faleskog and Shih, 1995). For the present, the simplest way to take some account of coalescence by microvoid cavitation is by adjusting the value of f_0 and the critical value for final cell extinction, f_E.

An important concept in mechanism-based fracture mechanics is the fracture process zone (FPZ). The necessity of incorporating FPZ into ductile fracture models has also been discussed by Broberg (1996) and Cotterell and Atkins (1996). In our approach, the FPZ can be defined by the collection of cells in which the strain softening due to void growth cannot be compensated for by material strain hardening resulting in a loss of stress carrying capacity. The primary parameters controlling fracture process zone size are f_0, matrix hardening properties and crack tip constraint. The combination of high constraint and large f_0 results in an FPZ extending over many cell elements. By contrast, low constraint and a small f_0 results in an FPZ on the order of the void spacing; here the tearing process involves the interaction between the tip and one or two discrete voids ahead of it.

As already noted, the cell model can be used in the latter situation and in the former where multiple interacting voids form the FPZ.

3. Work of Fracture

Obviously, the fracture process must obey an energy balance and under small scale yielding conditions the balance relation is simply given by

$$\mathcal{G} = \Gamma \tag{1}$$

where $\mathcal{G}$ is the Griffith-Irwin energy release rate. The total work of fracture per unit area of crack advance, Γ can be partitioned into the work of the fracture process Γ_0 and the plastic dissipation in the background material Γ_P, that is,

$$\Gamma = \Gamma_0 + \Gamma_P + \Gamma_E \tag{2}$$

where Γ_E is the additional contribution taking into account the work related to changes in the process zone size and the elastic energy variation within and just outside the plastic zone due to changes in the plastic zone size.

In a ductile metal the separation process is dominated by cavity growth and coalescence with a FPZ size on the order of 10^{-4} m. The work required to rupture a voided cell of unit area in the plane of the crack defines the work of the fracture process Γ_0 which has a weak dependence on constraint within the range found to exist near the tip of a crack (Xia and Shih, 1995b). For the first increment of growth, it is argued that $\Gamma_0 \gg \Gamma_P + \Gamma_E$ so that Γ at crack initiation, Γ_I, is nearly equal to Γ_0. By contrast, Γ_P can be much greater than Γ_0 when crack extension is large compared to D, but more importantly Γ_P depends sensitively on crack tip constraint (Tvergaard and Hutchinson, 1994; Shih and Xia, 1995). In the limiting situation of steady state growth, both the process zone length and the stress field (with respect to the advancing crack tip) are unchanging so that only Γ_0 and Γ_P contribute the total work of fracture,

$$\Gamma = \Gamma_0 + \Gamma_P. \tag{3}$$

Actually a small amount of elastic energy is locked in the plastic wake but this can be added to Γ_P.

We can draw the conclusion that a resistance curve, Γ (including steady-state toughness), is not a property of the material. As a matter of fact, crack geometry can exert a strong effect on Γ even under small scale yielding as has been measured experimentally by Hancock, Reuter and Parks (1993) and computed by Tvergaard and Hutchinson (1994), Shih and Xia (1995) and Xia and Shih (1995a). Geometry effects on crack growth resistance curves are most readily apparent under large scale yielding. Such data in

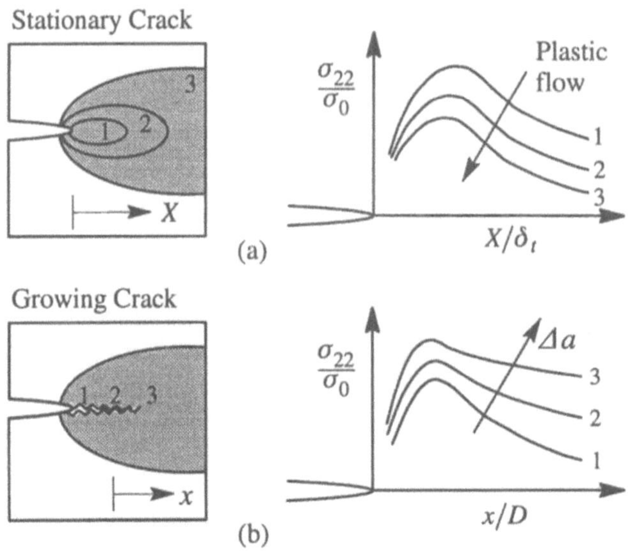

Figure 2. (a) Plastic flow (shown by shaded region) near stationary crack causes loss of constraint. (b) Crack growth increases constraint (x measured from tip of advancing crack

the form of J vs. crack advance have been reported by Joyce and Link (1995).

4. Evolution of Crack Opening Stress

The evolution (loss) of constraint as plasticity progresses from small scale yielding to full ligament yielding has been investigated by Betegón and Hancock (1991) and O'Dowd and Shih (1991). Figure 2a shows the behavior of the crack opening stress vs. distance normalized by the crack tip opening displacement for three levels of plastic flow. The results presented here and in the next section are obtained from plane strain analyses. This loss of constraint (or reduction in tensile opening stress) under increasing plastic flow has been discussed by several investigators (see O'Dowd and Shih, 1994 and references therein). The opposite trend is found for a growing crack. Figure 2b displays the crack opening stress vs. distance (normalized by D) measured from the tip of the advancing crack for three amounts of crack growth. This elevation of the stress over a physically significant distance ahead of the tip is found in typical test specimens (Xia and Shih, 1995b, 1996). A conclusive analysis showing constraint elevation under steady-state crack growth has been carried out by Varias and Shih (1993). Their steady-state result is generic in that it is not tied to any model of material

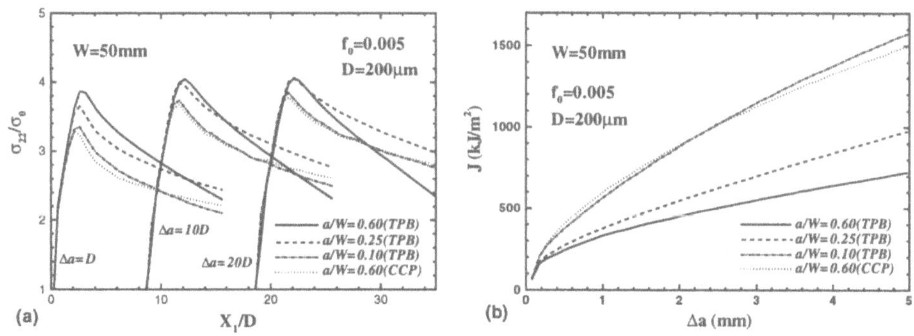

Figure 3. Comparisons of the opening stress on the plane of fracture for crack advance equal to D, 10D and 20D. (b) The relation between J and crack advance in TPB and CCP.

separation or fracture criterion. The result is valid as long as the fracture process zone is a small fraction of the plastic zone size and the latter is small compared to a characteristic crack dimension. A similar result is found to apply to rate-dependent solids (Xia and Cheng, 1996).

Figure 3a displays cell model predictions of the tensile traction across the plane of the crack for material parameters chosen to reproduce the behavior of A533B steel. Stress distributions are shown for crack growth equal to D, $10D$ and $20D$ in a center-cracked panel (CCP) and three-point-bend (TPB) specimens for three different ratios of crack length over width, a/W. The elevation of peak stresses in all the geometries considered can be clearly seen, particularly in the lowest constraint geometries (TPB, $a/W = 0.1$ and CCP $a/W = 0.6$), shown by the dash-dot curve and the dot-dot curve.

Figure 3b compares the relation between J and crack advance for the above geometries. The J values were computed using the line-integral definition on a remote contour. The strong geometry dependence of the J-resistance curves is strikingly evident. The highest resistance curves are found in the CCP and the shallow flaw TPB specimen, geometries with the lowest constraint. These two resistance curves are in close agreement and this is not unexpected in light of the similarities between the stress distributions for the two geometries (see Fig. 3a).

Contour maps of the maximum principal stress provide the most compelling description of constraint elevation with crack growth. Figure 4 displays results for the shallow flaw TPB geometry. A high stress zone, $\sigma_1/\sigma_0 > 3.5$, is non-existent at early growth (see contour map for Δa equal to D). However, high stress zones extending over distances ranging from

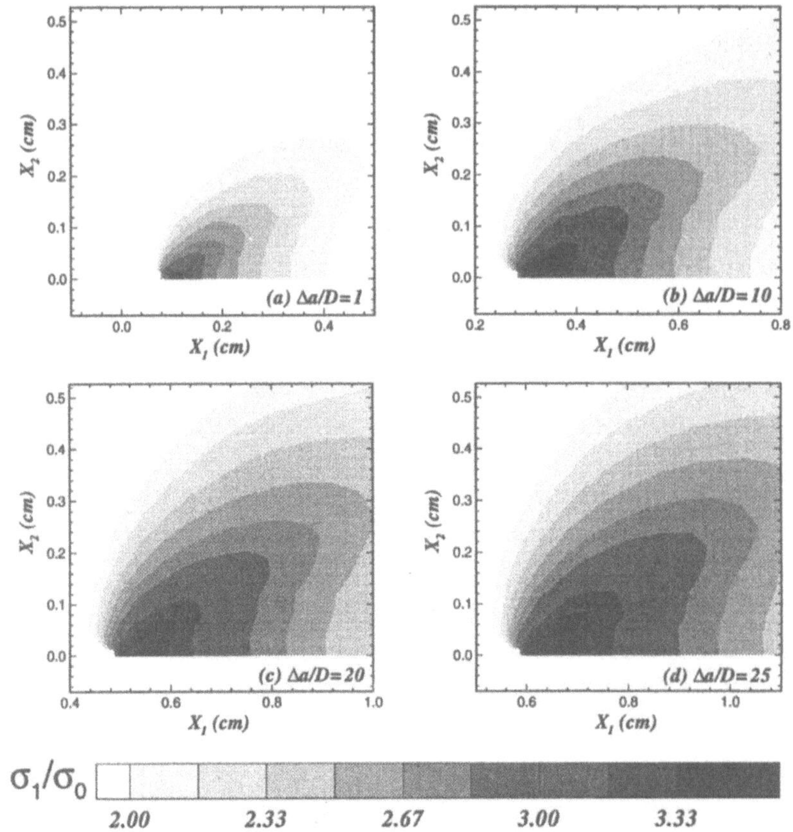

Figure 4. Contour maps of maximum principal stress for crack advance equal to D, 10D, 20D and 25D in shallow crack three-point-bend specimen, a/W=0.1, W=50mm, E/σ_o=500, N=0.1, ν=0.3; D=200μm, f_0=0.005

$10D$ to $15D$ can be found at later stages of growth, $\Delta a > 10D$ (see contour maps for $\Delta a = 20$ and $25D$). It is important to note that both the peak stress and the spatial extent of high stress increase with crack growth. Constraint evolution in the CCP and the other two TPB geometries display similar trends as can be anticipated from the stress results in Fig. 3a.

5. Statistical Model for Cleavage Fracture in the Ductile/Brittle Regime

Treatment of the initiation of unstable cleavage fracture by way of extreme value statistics has been discussed by several investigators including Beremin (1983), Mudry (1987), Wang (1991), Wallin (1993) (also see the references therein). In these studies a weakest link mechanism is assumed for cleavage fracture. That is to say, at some point during the loading a

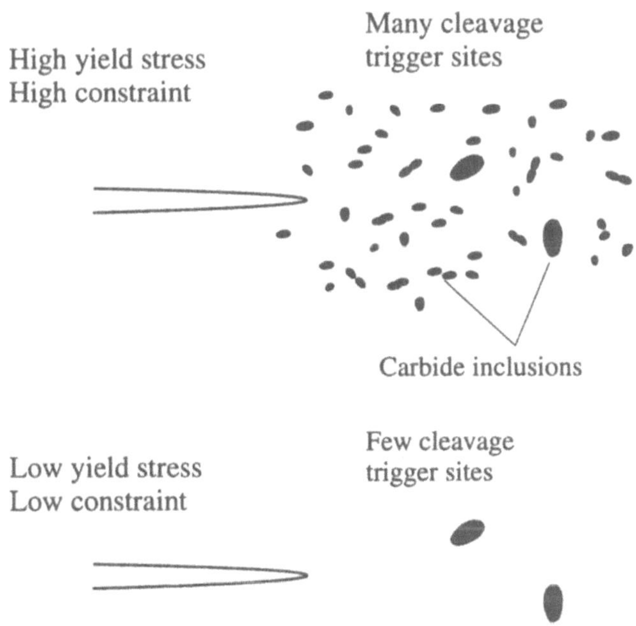

Figure 5. Distribution of carbide inclusions which are potential cleavage fracture trigger sites.

microcrack nucleates at a critical second phase inclusion and this event is sufficient to precipitate catastrophic cleavage fracture. This approach has been extended by Koers *et al.* (1995) and Xia and Shih (1996) to take full account of the ductile crack growth prior to cleavage fracture. Two effects are associated with ductile crack growth: the cumulative sampling volume is increased and the crack tip constraint is increased (see Figs. 3 and 4). This has serious implications. It suggests a significant increase in cleavage fracture probability which is taken up next.

Figure 5 illustrates the idea upon which our model is based. Under high constraint (and high yield stress), many carbide inclusions are potential trigger sites for cleavage fracture. Therefore the conditions for cleavage fracture can be met before any significant amount of ductile tearing occurs. By contrast, under low constraint the eligible cleavage fracture trigger sites are limited to large carbide inclusions which are few in number. Therefore the crack has to grow by some amount in order that the advancing stress field could sample an eligible cleavage crack nucleus. Consequently cleavage fracture in a low constraint geometry is typically preceded by some amount of ductile tearing.

The alteration of constraint and the increase of sampling volume with crack growth are fully incorporated into a weakest link cleavage model em-

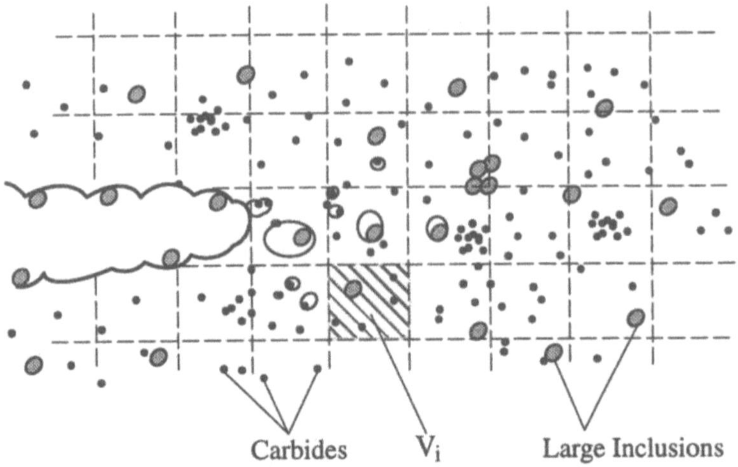

Figure 6. A statistical model for cleavage fracture in the ductile/brittle transition region. On the average each cell V_i contains one large (sulphide/oxide) inclusion and many carbides particles (Xia and Shih, 1995c).

ployed by Xia and Shih (1996). Figure 6 depicts the essential features of their model. This model also accounts for the competition between the nucleation of voids from carbide inclusions and the unstable cracking of such inclusions precipitating catastrophic cleavage fracture. A natural outcome of the model is the Weibull stress, σ_W, based on the sum of products involving cell volumes V_i and stresses averaged over these cells. The expression for the Weibull stress and its derivation is given in Xia and Shih (1996). Once σ_W is known, the cleavage fracture probability is computed using

$$P_f = 1 - \exp\left[-\left(\frac{\sigma_W}{\sigma_u}\right)^{m_s}\right] \tag{4}$$

where σ_u is a scaling stress and m_s characterizes the size distribution of cleavage fracture triggering inclusions as was already noted.

Figure 7a displays the evolution of $\bar{\sigma}_W$ ($\equiv \sigma_W/\sigma_u$) vs. J–integral for the crack geometries considered in Fig. 3. It can be seen that at the same level of applied J the cleavage fracture probability is lowest for the shallow TPB geometry and the CCP. These results can be used in the following way. Suppose cleavage fracture in a TPB specimen, $a/W = 0.6$ and $W = 50$ mm, occurs at $J_C = 200$ kJ/m^2. We wish to determine J_C for the other test specimens. Direct your attention to the solid curve in Fig. 7; for $J = 200$ kJ/m^2, $\bar{\sigma}_W \approx 0.9$. Moving horizontally across the diagram with $\bar{\sigma}_W$ fixed at 0.9 (implying equal probability of cleavage fracture) we find that $J_C = 330$ kJ/m^2 for $a/W = 0.25$, $J_C = 964$ kJ/m^2 for $a/W = 0.1$; for the CCP,

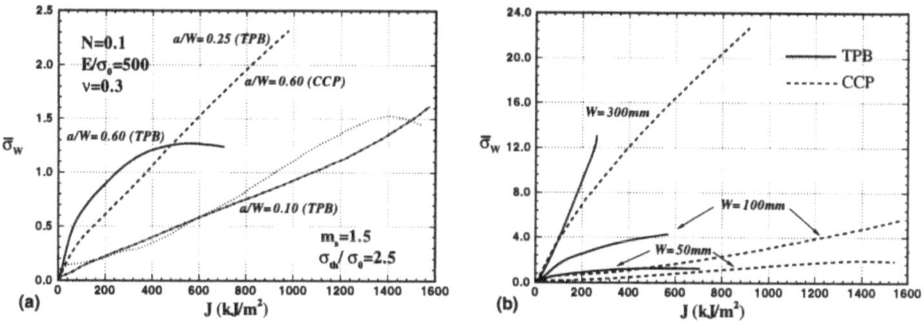

Figure 7. Geometry effect on the evolution of Weibull stress with J-integral for (a) four different a/W ratios with W=50mm and (b) three different widths, W, with a/W=0.6 and B=W/2.

$J_C = 844$ kJ/m^2. That is, the predicted cleavage toughness for the two geometries with the lowest constraint are more than four times larger!

The strong effect of specimen size on cleavage fracture can be seen in Fig. 7b which displays computed Weibull stresses for CCP and TPB specimens for $a/W = 0.6$ with W equal to 50, 100 and 300 mm. [$B = W/2$ is used here.] Contrast the strikingly different trends — $\bar{\sigma}_W$'s approach limiting values for the smallest specimens while they increase steadily for the larger geometries. The predicted trends of cleavage fracture toughness with crack size and geometry are consistent with available experimental data.

6. Acknowledgements

This investigation is supported by Grants N00167-K-0038 and N61533-93-K-0030 from David Taylor Research and Development Center funded by the Nuclear Regulatory Commission and by Grant N00014-95-1-0399 funded by the Office of Naval Research.

7. References

Beremin, F.M. (1983) A Local Criterion for Cleavage Fracture of a Nuclear Pressure Vessel Steel, *Metallurgical Transactions A*, **14A**, 2277-2287.

Betegón, C. and Hancock, J.W. (1991) Two-Parameter Characterization of Elastic-Plastic Crack-Tip Fields. *J. Appl. Mech.* **58**, 104-110.

Broberg, K.B. (1995) Critical Review of Some Methods in Nonlinear Fracture Mechanics. *Engineering Fracture Mechanics*, **50**, 157-164.

Brocks, W., Klingbeil, D., Küencke G. and Sun D.-Z. (1995) Applications of the Gurson Model to Ductile Tearing, in *Constraint Effects in Fracture, Theory and Applications, ASTM STP 1244*, American Society for Testing Materials, Philadelphia, 232-254.

Cotterell, B. and Atkins, A.G. (1996) A Review of the J and I Integrals and their Implications for Crack Growth Resistance and Toughness in Ductile Fracture. To be published.

Faleskog, J.P. and Shih, C.F. (1995) Micromechanics of Coalescence - I. Synergistic Effects of Elasticity, Plastic Yielding and Multi-Size-Scale Voids. *J. Mech. Phys. Solids.* To appear.

Gurson, A.L. (1977) Continuum Theory of Ductile Rupture by Void Nucleation and Growth: Part I - Yield Criteria and Flow Rules for Porous Ductile Media. *J. Engin. Mat. Tech.*, **99**, 2-15.

Hancock, J.W., Reuter, W.G. and Parks, D.M. (1993) Constraint and Toughness Parameterized by T. In *Constraint Effects in Fracture, ASTM STP 1171*, American Society for Testing and Materials, Philadelphia, 21-40.

Joyce, J. A. and Link, R. E. (1995) Effects of Constraint on Upper Shelf Fracture Toughness. *Fracture Mechanics: 26th Symposium, ASTM STP 1256* , American Society for Testing and Materials, Philadelphia.

Koers, R.W.J., Krom, A.H.M. and Bakker, A. (1995) Prediction of Cleavage Fracture in the Brittle to Ductile Transition Region of a Ferritic Steel, in *Constraint Effects in Fracture, Theory and Applications, ASTM STP 1244*, American Society for Testing and Materials, Philadephia, 191-208.

Mudry, F. (1987) A Local Appróach to Cleavage Fracture. *Nuclear Engineering and Design*, **105**, 65-76.

O'Dowd, N. P. and Shih, C.F. (1991) Family of Crack-Tip Fields Characterized by a Triaxiality Parameter–I. Structure of Fields. *J. Mech. Phys. Solids*, **39**, 989-1015.

O'Dowd, N. P. and Shih, C.F. (1994) The Two-Parameter Fracture Mechanics: Theory and Applications. In *Fracture Mechanics: Twenty-Fourth Symposium, ASTM STP 1207* American Society for Testing and Materials Philadephia, 21-47.

Rousselier, G. (1987) Ductile Fracture Models and Their Potential in Local Approach of Fracture. Nuclear Engineering and Design, **105**, 97-111.

Shih, C.F. and Xia, L. (1995) Modeling Crack Growth Resistance Using Computational Cells with Micro- structurally-Based Length Scales. In *Constraint Effects in Fracture, Theory and Applications, ASTM STP 1244*, American Society for Testing and Materials, Philadephia, 163-190.

Tvergaard, V. (1990) Material Failure by Void Growth to Coalescence. *Adv. Appl. Mech.*, **27**, 83-151.

Tvergaard, V. and Hutchinson, J.W. (1994) Effect of T-Stress on Mode I Crack Growth Resistance in a Ductile Solid. *Int. J. Solids and Struct.*, **31**, 823-833.

Varias, A.G. and Shih, C.F. (1993) Quasi-Static Crack Advance Under A Range of Constraints — Steady-State Fields Based on a Characteristic Length. *J. Mech. Phys. Solids*, **41**, 835-861.

Wallin, K. (1993) Statistical Aspects of Constraint with Emphasis to Testing and Analysis of Laboratory Specimens in the Transition Region, *Constraint Effects in Fracture, ASTM STP 1171*, American Society for Testing and Materials, Philadephia, 264-288.

Wang, Y.Y. (1991) A Two-Parameter Characterization of Elastic-Plastic Crack-Tip and Applications to Cleavage Fracture, Ph. D. Thesis, Department of Mechanical Engineering, MIT.

Xia, L. and Shih, C.F. (1995a) Ductile Crack Growth - I. A Numerical Study Using Computational Cells with Microstructurally-Based Length Scales. *J. Mech. Phys. Solids*, **43**, 233-259.

Xia, L. and Shih, C.F. (1995b) Ductile Crack Growth - II. Void Nucleation and Geometry Effects on Macroscopic Fracture Behavior. *J. Mech. Phys. Solids*, **43**, 1953-1981.

Xia, L. and Shih, C.F. (1996) Ductile Crack Growth - III. Transition to Cleavage Fracture Incorporating Statistics. *J. Mech. Phys. Solids*, **44**, 603-639.

Xia, L. and Cheng, L.,(1996) Quasi-Statically Steady Crack Growth in Rate-Dependent Solids — One-Parameter Fields Under Different Constraints. Submitted.

Xia, L., Shih, C.F. and Hutchinson, J.W. (1995) A Computational Approach to Ductile Crack Growth under Large Scale Yielding Conditions. *J. Mech. Phys. Solids*, **43**, 389-413.

EFFECTS OF SIZE SCALE ON FRACTURE PROCESSES IN ENGINEERING MATERIALS

JOHN F KNOTT
School of Metallurgy and Materials
The University of Birmingham

Introduction

This paper treats two aspects of size-scale. The first is concerned with the size-scale at which fracture events are analysed. Although there are substantial overlaps, four main size-scales may usefully be recognised, as follows:

i) the MACRO -scale. This is concerned with events at the "engineering" level, and with material properties treated as those of a continuum. Generally, the size-scale is upwards of a few mm but, in some situations, such as that of a single dominant crack in a high-strength steel of homogeneous microstructure, continuum concepts can be carried down to a defect size of 0.2mm.

ii) the MESO-scale. This comprises inherent "defects" or inhomogeneties, produced by processing or fabrication, which are smaller than the non destructive-testing (NDT) limit. Such defects could be grain-boundary voids in a ceramic; non-metallic inclusions in wrought metallic alloys; "brittle patches" in multi-pass welds or in dual-phase steel microstructures. A very rough estimate of the size-range is 20mm-0.2mm. In ceramics, defects of length 50μm can produce catastrophic failure at a stress of only 160 MPa. For ultra-high strength maraging steel, a defect of length 75μm could produce catastrophic failure at a stress of 2GPa, but, generally, in engineering alloys, defects of length less than 100μm are of significance only under fatigue loading.

iii) the MICRO-scale. This is associated with microstructures designed to produce a given combination of flow stress, work-hardening characteristics and fracture resistance. The average properties (such as 0.2% proof stress) may be determined by the dimensions of the metal's grains, from a few μm to more than 100μm, but the brittle particles which initially trigger off cleavage fracture are usually less than 10μm in size; the smallest size of significance is of order 10nm.

iv) the NANO-scale. This involves events at a scale less than some 2nm: a "few" atomic spacings. Typical examples concern the structure of the "core" of a dislocation; the co-ordination of nearest or next-nearest neighbours in a grain-boundary or interfacial "site" for an impurity atom; or events in the region of the tip of an atomically sharp crack as the load applied to it increases. The question here is whether a crack propagates in an "atomically sharp" manner, or whether it blunts, by the emission and gliding-away of crack-tip dislocations.

J. R. Willis (ed.), IUTAM Symposium on Nonlinear Analysis of Fracture, 65–79.

Clearly, there is some overlap between these size-scales; in particular, the boundary between "micro" and "meso" is ill-defined for composite materials.

The second aspect of scale is that of "scaling factor". Here, the problem is how a measured fracture-resistance parameter "scales" with the volume of material that is tested. It is important to recognise that a "scaling factor" may be confused by the presence of some meso-structural or micro-structural variation in the material which is not considered on the macro- or nano-scale. A simple assumption for material is that its density is uniform. Then, doubling the volume doubles the weight: quadrupling the volume quadruples the weight. If, however, small volumes were prepared by casting small testpieces and large volumes by casting large testpieces, the large pieces might contain a higher volume fraction of shrinkage porosity, confusing the simple scaling factor.

If flow strength were measured, the *force* required to produce 0.2% plastic strain would be expected to increase in a linear fashion with the cross-sectional area of the piece, but this scaling factor could be affected by *two* internal variables: any increase in porosity, as described above, *plus* the possibility of generating different microstructures (with different flow properties) as a result of differences in cooling-rate involved with the production of the larger test-volume. These examples are particularly simple and relatively easy to analyse, using non-destructive testing, quantitative metallography, micro-hardness measurements, etc. but they serve to introduce two aspects of fabricated material which have relevance to its fracture and to associated "scaling-factors": i) the presence of fabrication/processing defects (very loosely, "meso-structure", although often overlapping into the "micro-scale"), ii) spatially-localised variations in micro-structure, possessing different flow and fracture properties (these "micro-structural" variations may extend into the meso-scale: in a multi-pass weld, more-or-less into the macro-scale).

The design stress for an engineering component is conventionally taken as a material's yield strength or 0.2% proof strength divided by a "safety-factor". This is typically 1.5-2.0 for wrought steel, but 4 for cast steels under monotonic loading. The higher figure for castings reflects the fact *that some castings may contain a higher defect content than otherwise equivalent wrought material*, although this is by no means a general situation and it has been argued that design codes often discriminate unnecessarily against the use of (cheaper) castings in a number of applications [1]. The codes take account of meso-scale defect content on the basis of past experience and it is important to recognise that improvements in casting technology to reduce defect content can produce much better mechanical performance.

Under design loads, fracture is always associated with the presence of stress concentrators and measurements of fracture resistance are made, either *directly* by observing the load required to break (or energy absorbed in breaking) a testpiece containing a single, long, sharp crack or a blunt notch, of known size, or *indirectly*, by determining the fracture stress for material which is notionally homogeneous, but which actually contains a distribution of internal ("meso-scale") defects. Measurements of the energy absorbed in fracturing blunt-notched testpieces are not easy to relate to the design engineer's need for a fracture *stress* and will consequently not be discussed further in this paper. A more useful approach is based on the application of finite element stress analysis to specific blunt-notch geometries, to enable the *local fracture stress*, σ_F, to be derived directly from the ratio of applied load, P_{app}, to general yield load, P_{GY} [2,3]. For "homogeneous" material, it would be assumed that the same value of σ_F would be obtained in both small and large testpieces.

Two further macro-scale measurements are the fracture stress of a testpiece containing a single, long crack, or the fracture stress of a notionally smooth testpiece containing "inherent" (meso-scale) defects. The former is associated with fracture-toughness testing and fracture mechanics analysis [4]. Here, given "plane-strain" and "small-scale-yielding", the crack-tip stress and displacement fields at fracture are characterised by a critical value of stress-intensity-factor, K_{Ic}. For homogeneous material (K_{Ic} constant), "scaling-up" depends simply on the testpiece dimensions and compliance function Y(a/W), e.g. for a standard bend specimen, the fracture load P_F is given by:

$$P_{F(B.W.a)} = K_{IC} BW^{1/2} Y(a/W) \qquad ....1)$$

For tougher material, fracture in a small testpiece may initiate well after general yield and recourse is made to the measurement of critical, post-yield parameters: crack tip opening displacement, CTOD (δ) or J-integral, J. Here, the conditions for crack-tip dominance must be established rigorously. It is conventional to relate critical values of δ or J back to equivalent values of K and hence to contrive to "scale" from "small" to "large" dimensions, but, it is, in principle, possible to compute the failure load for "intermediate" dimensions, using numerical techniques.

The technique employed to treat the fracture stress of a body containing a distribution of inherent defects is probabilistic in nature. Conventionally, use is made of the Weibull distribution in which the cumulative fractional probability of failure, $F(\sigma)$, up to and including a given stress, σ, is given, for a test volume, V, by:

$$F(\sigma)_V = 1 - \exp\{-([\sigma - \sigma_o]/\sigma_s)^m\} \qquad ....2)$$

for $\sigma > \sigma_o$, where σ_o is the datum or "threshold" and σ_s is a "scale parameter" corresponding to the stress for which $F(\sigma)_V = 0.63$. The exponent m is termed the Weibull modulus. The effect of the test volume is incorporated by noting that, for a constant probability of failure, the failure stresses σ_a and σ_b for test volumes V_a and V_b are related through the expression.

$$V_a (\sigma_a - \sigma_o)^m = V_b (\sigma_b - \sigma_o)^m \qquad ....3)$$

In principle, the macro scale fracture stress reflects the "worst" defect configuration in the test volume. For a multiplicity of defects, the "worst" configuration incorporates a crack size distribution, a crack orientation distribution, the possibility of partial "crack-shielding", and possible toughness variations in the microstructure (e.g. variable grain-boundary cohesive strength as a function of grain-to-grain misorientation or local chemistry). By paying attention to processing, it may be possible to improve the fracture stress by reducing the overall defect-content/size-distribution, but deterministic correlations between the two are not likely to be achieved. Setting the probabilistic bounds for model systems might be possible. The Weibull distribution is an empirical curve-fit.

Cleavage Fracture in Steel

This section is devoted to a discussion of how apparent conceptual discrepancies between *Macro-scale* and *Nano-scale* treatments of rapid crack-advance by cleavage fracture can be reconciled by examination of events at the *Micro-scale*. A typical, "valid", plane-strain fracture toughness value for mild steel tested at low temperature is 40 $MPam^{1/2}$, at which temperature its yield strength is approx. 800 MPa. From the expression $EG = K^2 (1-\upsilon^2)$, where E is Young's modulus and υ is Poisson's ration, the associated critical value of energy release-rate, G, is approx. $7kJm^{-2}$. From the expression, $\delta = K^2/2\sigma_Y E$, the associated CTOD is approx. 5μm, i.e. 20,000 lattice spacings. Note, however, that if the testpiece were loaded to 39 $MPam^{1/2}$ (2.5% less than K_{1c}), the CTOD would be only some 5% smaller (as would be the plastic zone size). Unloading to zero would imply running-back of dislocations and decrease in CTOD, but it would still be 95% of approx. 15,000 lattice spacings. Reloading to 39 $MPam^{1/2}$ restores the *status quo ante*. If the unloaded specimen were supplied for testing, failure at 40 $MPam^{1/2}$ would again be anticipated, and the original values of G_{crit} and δ_{crit} would be deduced.

The Nano-scale approach to fracture follows the original model proposed by Kelly, Tyson and Cottrell [5], and subsequently developed by Rice and co-workers [6,7]. In the model, the crack is envisaged as being atomically sharp and the criterion for deciding on whether or not it extends in a brittle manner was originally couched in terms of the competition between *fracture* of the crack tip bond at the theoretical fracture strength (of approx. E/10) or *blunting* of the crack by emission of dislocations from the crack tip at the theoretical shear strength (approx. $\mu/10$). Rice's work has examined in more detail the emission and movement of dislocations away from the tip (at the Griffith stress, to avoid the thermodynamic complication of crack "healing" at lower stresses). A "saddle-point" configuration of atoms in the crack-tip region is identified (within two or three atomic spacings): beyond this, the dislocations break free and move away, to blunt the crack. The model clearly differentiates between "obviously brittle" materials, such as mica, which cleaves between silicate sheets, and "obviously ductile" materials, such as gold. Iron is found to be a "borderline" material, in which the balance between "fracture" and "blunting" is extremely close.

In some senses the Nano-scale approach satisfactorily describes the observed "ductile/brittle" behaviour of iron and steel, but it does not conform to the experimental Macro-scale observations made in low-temperature fracture toughness tests. It cannot treat effects of plastic constraint on the ductile/brittle transition temperature; it cannot explain the variation of fracture toughness with different ferrite/carbide dispersions in wrought steels or with different oxide/silicate inclusion sizes in weld metals. Recall that, for fast-running cleavage at low temperatures the Macro-scale CTOD at the fatigue-crack tip is some 20,000 atomic spacings, yet the Nano-scale approach would predict blunting and completely ductile behaviour if an atomically-sharp crack were blunted, even for a blunting of less than 10 atomic spacings.

The anomaly between the Macro-scale and Nano-scale conclusions can be largely reconciled by considering events on the Micro-scale [8]. The plastic work put in to the fracture toughness testpiece should be regarded as "precursor work" (*c.f.* the notional unloading and reloading experiment described above). The role of this "precursor work" is to generate dislocation arrays which can *initiate* micro-cracks in brittle second-phase particles, and, at the same time, create a stress/strain field ahead of the fatigue pre-crack, in which tensile stresses some 3-4 times the uniaxial yield stress can be generated. The critical event that leads to catastrophic cleavage fracture is then the *propagation* of the micro-crack nucleus under the combination of the high level of tensile stress in the plastic zone and the stress due to dislocation arrays. The latter is usually small (approx. 10%) compared with the former and the criterion for fracture is that of a critical tensile stress, σ_F, *at the site of the microcrack nucleus.*

The value of σ_F, is, in principle, identical to that measured in blunt-notched testpieces, but only if the material is *quasi-homogeneous*, i.e. if the critical distribution of micro-crack nuclei is such that equivalent sampling is experienced in the different testpieces. In the notched testpieces, values of σ_F have been found to be only weakly dependent on, if not independent of, temperature. Ritchie, Knott and Rice (RKR) [9] compared values of K_{Ic} from fracture toughness tests with values of σ_F from blunt-notched tests to deduce a "critical distance" x_C, at which a nucleus (in quasi-homogeneous material) would be located. In their mild steel, x_C was about 120μm, equal to two grain diameters, but Curry and Knott [10] later demonstrated that x_c did not relate directly to grain size but should be regarded as a statistical average of the locations of carbides above a certain size (e.g. the 90th or 95th percentile of the distribution).

Further developments have followed two routes:

i) the testing of a *set* of notched testpieces to determine a fracture stress distribution; treating this as a Weibull distribution; and combination of this with the stress distribution in the plastic zone ahead of the fatigue pre-crack to obtain the probablility of fracture at a given K-level (Beremin[11]).

ii) the testing of blunt-notched or sharply cracked testpieces of ferritic steel weld-metals, combined with detailed fractography to determine values of σ_F *at the site of the micro-crack nucleus* [12]. Application to the Macro-scale crack-tip region is then pursued *via* the *statistical distribution of the inclusion content of the material*, rather than *via* a fitted Weibull distribution to the notch fracture stress.

Both these approaches which relate Macro-scale to Micro-scale imply some variation (scatter) in the values of K_{1c} or σ_F and this topic will be treated with reference to scale effects, in the next section.

Here, attention is paid to the link between Micro-scale and Nano-scale. The anomalies inherent in the direct comparison between Macro-scale and Nano-scale are obviated once the problem is re-focused on the propagation of the micro-crack nucleus, in the idiosyncratic Macro-scale stress/strain field. The nucleus is produced in a brittle, ceramic particle (carbide, oxide, silicate) and is likely to travel at high speed, giving very little time for any thermally-activated operation of (heavily-pinned) dislocation sources ahead of the ceramic/matrix interface [8]. The microcrack as it propagates through the ceramic particle could be atomically sharp, so that the Nano-scale model is appropriate to treat the question of whether it propagates into the matrix or whether it blunts. At higher temperatures in the brittle/ductile transition region many blunted microcracks are observed in grain-boundary carbides although inclusions in weld metals tend to cavitate.

From measured values of σ_F and dimensions of microcrack nuclei, it is possible to calculate values of the local "work of fracture", γ_p. This is found to be some 9-14Jm^{-2}, approx. 4-7 times greater than the elastic work-of-fracture, but nowhere near the Macro-scale "precursor work" of 7 kJm^{-2}. An argument to support the value of 9-14 Jm^{-2} has been advanced in terms of the need to create a "mechanism" or "activated state" in the region of the atomically-sharp crack tip to allow the bond to separate fully [8]. At the force maximum in the atomic force/displacement curve, there is still an energy of interaction between the crack tip atoms and all the surrounding atoms have positive stiffness (an "iron cage" preventing separation). It is conceivable that extra (cooperative) atomic movement is needed to give sufficient compliance to allow full separation to occur and that this requires extra work. This might, for example, involve the generation and movement of crack-tip dislocations up to a "saddle- point" configuration , later able to run back out of the free surface when the crack-face is unloaded and only image-forces operate on the dislocation. The mechanism encompasses the following points:

i) the necessity for "γ_p" to increase as "a" increases to ensure propagation-controlled fracture (as evidenced by shifts in behaviour on going from torsion specimens to tension specimens to notched specimens).

ii) the independence, or weak dependence, of σ_F on test temperature. Crack-tip generation of dislocations would occur at stresses of order $\mu/10$ and so would not be affected by thermal fluctuations over the temperature range of interest. It is assumed here that matrix dislocations are fully pinned, as in annealed material. Reducing the strength of pinning can increase the value of σ_F.

Scatter In Values Of Fracture Toughness And Fracture Stress

In quasi-homogenous material, there is no reason why values of K_{1c} or σ_F should not be single-valued functions, since the appropriate maximum tensile stress available should always act on uniformly "worst" microstructure. If the probability distribution function (pdf) is a delta function, its integral, the cumulative distribution function (CDF) is a step function i.e. up to a given value K_{1c} no specimen breaks : above K_{1c}, all specimens have broken. In practice, there are random errors associated with experiments and the *central limit theorem* then predicts that the effect of the random errors is to convert the pdf to a Gaussian distribution [13] and hence the CDF to the error function (erf), normalised to unity. A good test of this postulate is to plot the CDF on normal probability paper : if a straight line is obtained, this indicates that the erf form is a good representation.

The postulate is examined by plotting on normal probability paper, CDF's relating to results of fracture toughness tests carried out on 300M steel at room temperature (for 320°C and 450°C tempers) and at -196°C (320°C temper). These are shown in Fig. 1 (courtesy Dr J E King). Similar plots are presented in Fig. 2 for G150 and G125 maraging steels at room temperature (courtesy Dr B Wiltshire); for a 0.4C 0.9Mn steel (Steel A - En8) quenched to produce a uniform martensitic microstructure and tempered for 1h at 220°C, and tested at -115°C; and for a dual-phase microstructure in a 0.15C 0.7Mn steel (Steel B - En3B), generated by holding for 1 day at 740°C and quenching, after autenitising at 910°C (courtesy Dr D J Neville [14]), also tested at -115°C. In all cases, a median ranking has been used, with the median rank of order n, F_n, derived from the close approximation [15].

$$F_n = (n - 0.3) / (N + 0.4)$$

It can be seen that for all of the fine-scale, quasi-homogenous microstructures, the erf fit is a good one, but that the behaviour is entirely different for the dual-phase microstructure, Steel B, Fig. 2. It is of interest to note that the K_{1c} distribution for G150 steel, failing by fast shear at room temperature , cannot be distinguished statistically from that for steel A, failing by transgranular cleavage at -115°C. This points up the need to combine any statistical analysis of fracture results with detailed fractography and a knowledge of operative fracture mechanisms. Another interesting point connected with the form of presentation in Figs. 1 and 2 is that it provides a simple visual means to extrapolate to obtain "lower-bound" results. Often, these are quoted as "two standard deviations below the mean", equivalent to 2.28 (%) on the CDF scale, but the suggested extrapolations to 10^{-4} (0.01%) indicated by the dashed lines seem not unreasonable, given the "tight" distribution. It is, of course, possible to predict only from the set of available data points, unless a physical model is available to justify a genuine lower bound.

Quasi-Homogeneous Microstructures

Examine first the results for the quenched-and-lightly-tempered (220°C) En8 (steel A, Fig 2) tested at -115°C and for 300M tempered at 320°C and tested at -196°C. Both steels contain approx. 0.4C and possess similar microstructures in which plate-like carbide some 0.2-0.4μm thick is located at lath boundaries, where individual laths are of the order of 1μm thick. The CDF plots suggest that random errors are of order ±2% or less and this is in agreement with Neville's assessment of ±1.6% for the Steel A tests. Values of σ_F for similar steels have been estimated as some 3150-4400 MPa, which, given the high yield stress, implies values of (σ_{max}/σ_Y) of order 2.5-3.0.

If ±2% is allowed on the value of (σ_{max}/σ_Y) at a nominal value of 3.0, i.e. 3.06 to 2.94, the corresponding spread in distance ΔX, is 0.003 $(K_{1c}/\sigma_Y)^2$. For $K_{1c} = 30$ Mpam$^{1/2}$ and $\sigma_Y = 1500$ MPa as typical figures, ΔX is only some 1.3μm. For these steels, the critical test volume is of order $B(\Delta X)^2$. The reproducibility of results is obtained because the microstructure is so fine and many units, ready to fracture at virtually identical loads, are to be found along the thickness of the testpiece. Bowen [16] has demonstrated clearly, in lower (0.15-0.2) carbon A533B steel that the microstructural factor controlling cleavage toughness in as-quenched (auto-tempered) martensites is inter-lath carbide thickness. More latitude on the value of (σ_{max}/σ_Y) implies the expectation of greater random errors in a fracture toughness test, but this is coupled with the ability of coarser microstructures to demonstrate "homogeneous" behaviour (within the less demanding constraints associated with greater random errors).

Similar principles hold for the ductile, "fast-shear" fractures associated with G150 and G125 maraging steels (Fig.2) and with 300M fractured at room temperature (Fig.1). These steels are ultra-clean from the point of view of non-metallic inclusions and "fast-shear" is associated with local decohesion of tempered carbides (or intermetallics) which are again very small particles, spaced closely. This contributes to a critical CTOD, rather than to K_{1c} as such, but the opening displacements associated with fracture are so small that the shear fracture, in mode I, produces catastrophic failure under quasi-elastic conditions. The importance of critical shear strain is, however, emphasised by the "zig-zag" fracture path and by behaviour under Mixed Mode (I/II) loading [17]. Within the limits of experiment, it is found that these fine microstructures behave in a quasi-homogeneous manner.

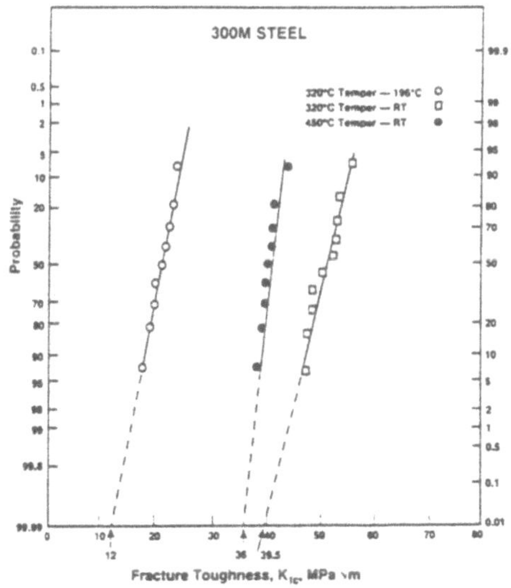

Fig.1 CDFs for 300M steel

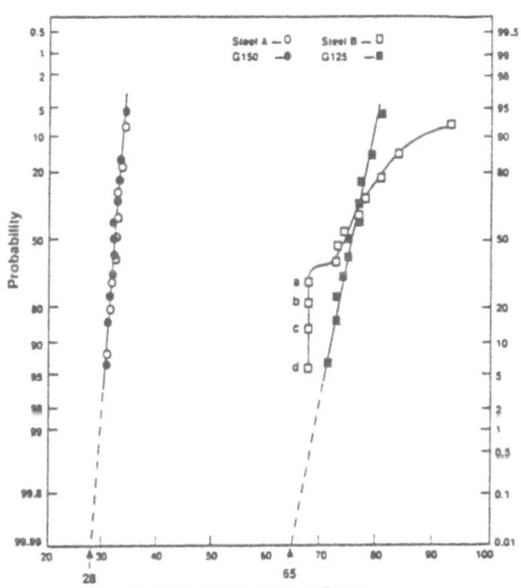

Fig. 2 CDFs for maraging steel, EN8(A) and dual-phase steel EN3B(B)

"Dual-Phase" Microstructure

The behaviour of Steel B (Fig.2) is quite different from that of the other steels having fine-scale microstructures. The fracture toughness values are widely spread, from 67.0 $MPam^{1/2}$ to 93.1 $MPam^{1/2}$ and there is a suggestion that points a, b, c and d constitute a separate distribution. Steel B has a deliberately-induced, "fine-scale" dual phase microstructure, with 68% ferrite, martensite "islands" of size 8μm and ferrite mean free path of 17μm. The "unit size" is therefore of order 8 + 17 = 25μm. For a value of K_{Ic} averaged at approx. 75 $MPam^{1/2}$ and the yield stress, taken as 600 MPa at -115°C, the "range", ΔX, taken as 0.003 $(K/\sigma_Y)^2$ is 50μm so that approx. 4 "units" are encompassed in 2-D. Inhomogeneities in distribution of units, may, however, act to increase the scatter (see later).

Neville [14] carried out similar tests using V-notched bend tests, having a notch-root radius of 250μm. The dual-phase microstructures were produced in two size-scales: "fine-scale" (73% ferrite, "island" size 6μm, ferrite mean-free-path 17μm, "unit" size 6 + 17 = 23μm) and coarse-scale (72% ferrite, "island" size 57μm, ferrite mean-free-path 128μm, "unit" size = 51 + 128 = 179μm). Specimens were fractured and values of σ_F were deduced, using the Griffiths and Owen [2] analysis. Four "fine-scale" specimens gave values of σ_F of 1534, 1540, 1544 and 1544 MPa i.e. 1541 ± 0.3% MPa. Seven "coarse-scale" specimens gave values for σ_F of 1470, 1483, 1497, 1503, 1510, 1528 and 1579 MPa, giving an average of 1510 MPa, but a range of 109 MPa, ±3.6% of the mean.

The "fine-scale" specimens failed close to general yield, where the "range", ΔX corresponding to $0.98\sigma_{max}$ is approx. one root radius (ρ) = 250μm. The "coarse-scale" specimens failed at a lower fraction of general yield, such that the "range" ΔX might be as low as $\rho/2$ = 125μm. From these results, it is clear that the fine-scale microstructure ("unit" size 23μm) exhibits quasi-homogeneous behaviour in a blunt (250μm)-notched specimen, having a (linear) "range" (at the 0.98 level) of 250μm, but demonstrates scatter in sharp-cracked tests, of range 50μm. The coarse-scale microstructure ("unit" size 179μm) exhibits scatter in a blunt-notched testpiece, having (linear) range 125-200μm. Experiments such as these help to define the conditions for quasi-homogeneous behaviour, by relating microstructural size-scales to the extent of a high-stress region ahead of a stress-concentrator.

Another important factor in a dual-phase microstructure relates to inhomogeneities in the "meso-scale" distribution of the two microstructures and their location with respect to the tip of the macro-scale crack. Hagiwara and Knott [18] heat-treated HY80 steel specimens to produce different volume fractions of ("brittle") upper-bainite in bainite/martensite microstructures. The K_{Ic} values at -142°C were observed to fall between two limits, corresponding to an upper limit for "100% martensite" (σ_F = 3125 MPa, cleavage facet size 10μm) of 57 ± 5 $MPam^{1/2}$ and a lower limit for "100% bainite" (σ_F = 2800 MPa, facet size 38μm) which was 42 ± 5 $MPam^{1/2}$. At 40% bainite, the scatter was increased, two individual values of 44.7 $MPam^{1/2}$ and 56.0 $MPam^{1/2}$ being obtained (in a small set). The (central) value of critical distance, X_C, was calculated as 62μm for martensite and 46μm for upper bainite. For a ± 10% variation in (σ_{max}/σ_Y) the "range" for this steel at $K_{Ic} \simeq 50$

$MPam^{1/2}$ is approx. 40μm, so that even the coarser facets stand a reasonable chance of being "properly" sampled and this is reflected in similar scatter for both "100%" microstructures. For the 40% bainite microstructure, it was observed that the fracture surface at X_C = 46μm (K_{1c} = 44.7 $MPam^{1/2}$) was composed primarily of 38μm "bainite-sized" facets; whereas that at X_C = 62μm (K_{1c} = 56.0 $MPam^{1/2}$) corresponded to almost completely 10μm "martensite-sized" facets. It is clear that the scatter is increased, because the lower bound corresponds to "bainite" at the critical distance: the upper to "martensite" at this position.

Studies of this sort have been recently pursued by X Zhang [19]. The steel used is A508 pressure-vessel steel and early experiments have been made on a "100% bainite" microstructure and a mixed "bainite/martensite" microstructure. Ten values of K_{1c} for each condition at a test temperature of -80°C have been determined and use has been made of plots on probability paper to examine the CDF for each microstructure. The results are plotted in Fig.3, which is highly instructive. First, the "100% bainite" results appear to confirm quite well to a Gaussian distribution and an extrapolation to 10^{-4} probability, giving a K_{1c} (min) value of 23 $MPam^{1/2}$ seems plausible. Two specimens exhibited "pop-ins", but they occurred at values of K of 27.5 $MPam^{1/2}$ and 29 $MPam^{1/2}$, both significantly higher than 23 $MPam^{1/2}$.

The results for the mixed microstructure are of particular interest. It is tempting to treat them as a single distribution and to plot the results on Weibull probability paper. This has been done in Fig.4 on the "two-parameter" basis, i.e. taking any "threshold" K_{1c} value as zero. Neville and Knott [14] have proposed a methodology to determine "threshold" values, by choosing appropriate trial values and examining "goodness of fit" to a straight line on Weibull probability paper, as decided, e.g. by "least-squares" regression analysis. The fit in Fig.4 is, however, already quite good and an examination of the points on normal probability paper (Fig.3) suggests that the data set of K_{1c} values for mixed microstructures would extrapolate back to a very low threshold; perhaps to zero (or even to negative values!); apparently to a value *lower than* 23 $MPam^{1/2}$, the 10^{-4} limit deduced *for the more brittle (bainite)* phase in the mixed microstructure. This is a result of the wider scatter for the mixed microstructure fracture toughness values , echoing the findings of Hagiawara in HY80 [18]. Zhang has not yet carried out detailed fractography, but it is expected that similar results will be obtained with respect to the "line fractions" of bainite and martensite at X_0 ahead of the crack tip. An important point is the critical "line fraction" of bainite required to produce catastrophic failure at a K_{1c} value corresponding to that for "uniform" bainite.

It is extremely important that a physical understanding of the micro-mechanisms of fracture be combined with statistical analysis, particularly for materials which have a meso (perhaps micro, or near-macro) distribution of two or more components with different fracture toughness properties. From Fig.3, it can be seen that a naive extrapolation of the CDF for the mixed microstructure could lead to a lower-bound ("threshold") value which was *lower* than the *lower-bound toughness of the more brittle constituent.* This is physically not

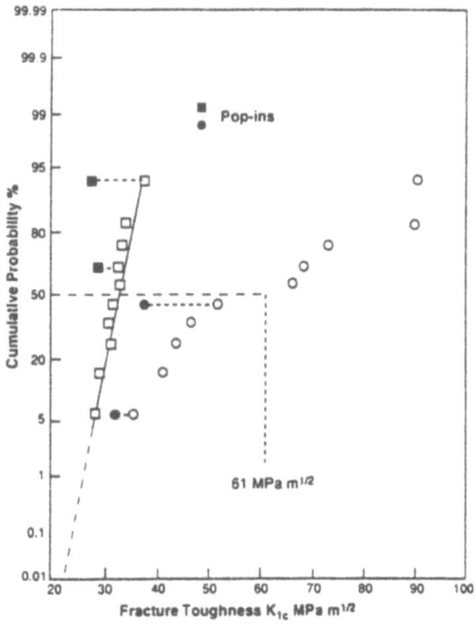

Fig. 3 CDFs for A508 in uniform and "dual-phase" microstructures

Fig.4 Weibull distribution for "dual phase"A508

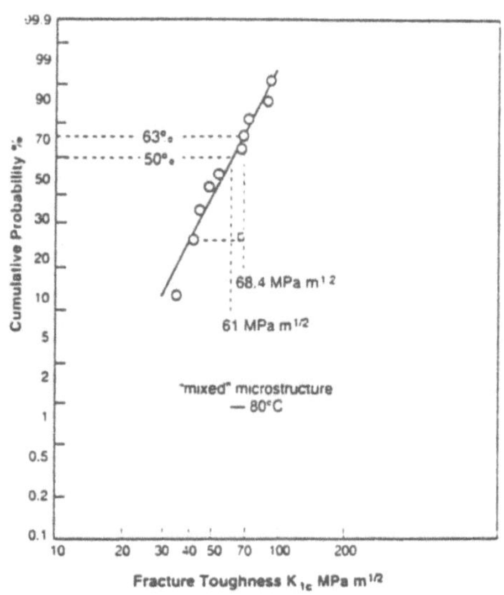

sensible. Although the Weibull analysis can produce a reasonable straight line, (Fig.4) any such analysis should be combined with *forensic fractography* to ensure that physical reality is maintained. It is of interest, in Fig.3, to note that two "pop-in" values for mixed microstructures fall within the data-set limits for the "uniform" bainite component.

Similarly, the points a, b, c and d in Fig.2 may well represent an individual set for the more brittle ferrite phase, although detailed fractography was not carried out to confirm this point.

Weld Deposits

Principles similar to those described above for wrought materials may be applied to weld metals, although the fracture initiation sites are now usually non-metallic inclusions: oxides or silicates, usually formed as a result of deoxidation processes in the weld pool [12]. These are usually small particles, at a few percent volume fraction, with distributions such that the 95th percentile is some 2μm. In his definitive work, Tweed [12] observed that some fractures were initiated from significantly larger particles (up to 13μm in size). These had unusual chemistry (containing Ca, K) and appeared to result from some of the binder, used to bind the coating to the electrode, breaking off and falling into the weld pool. These large particles were associated with lower values of fracture toughness: "outliers" from the (deoxidation product) "normal" distribution.

The fracture toughness distributions for weld metals can be examined using the CDF on normal probability paper. As for wrought material, a straight line may be expected if the material is quasi-homogeneous, i.e. if the sample volume is sufficient to sample something of the order of the 96th percentile, equivalent to 2% variation on σ_F (since σ_F depends on a $^{-1/2}$. Initial results are promising, but more data is needed to establish baselines. Variations in inclusion size (e.g. "outliers") should show as a break in the distribution (perhaps "pop-ins" will be observed) and analysis can the be combined with forensic fractography to seek the causes for these discontinuities.

Results in multipass weldments may be further confused by meso/near-macro variations in microstructure: "coarse", "as-deposited" grains and "fine", "reheated" grains. Tweed [20] has demonstrated that these changes in microstructure have a profound effect on fracture toughness, over and above the role of inclusions. The basic analysis with respect to inclusions is most easily carried out in a uniform microstructure and this may be important with respect to the quality control of the welding process. Assessment of the toughness of the multipass weld requires an approach similar to that described for mixed bainite/martensite microstructures, particularly if extrapolation to low probability is involved. Fractography must be employed to examine the micro-mechanisms of fracture, to ensure that physical sense is applied to extrapolation: a narrow scatter-band, relating to the more brittle component may give a higher "lower-bound" than the wider scatter associated with the "composite" weldment.

The scale of the "meso" distribution (a few mm per "band") is such that behaviour in notched-bar tests (a Charpy V-notch of 0.25mm root radius) is unlikely to differ from that in sharp-crack tests. It is salutary to recall the results of Newman, Benois and Hibbert [21], who demonstrated a strongly *bimodal distribution* for Charpy energy, over a range of some 60°C.

Concluding Remarks

The paper has referred to four (overlapping) *scales* of structure: *macro, meso, micro* and *nano*. It has also paid some attention to *scaling factors*. An argument has been put forward that anomalies between experimental *macro-scale* values of *fracture toughness* in steels and the predictions of *nano-scale fracture theory* can be resolved by considering events at the *micro-scale*. This focuses attention on the features of heat-treated microstructure.

Scaling factors are associated either with pieces containing a single, long dominant crack, in which case the concepts of fracture mechanics are deemed to hold, even going from post-yield to small-scale yielding, or with an array of "meso"-scale defects, in which case fracture stress distributions are analysed and size effects are related to sample volume (often not strongly backed by fractography). With macro-scale fracture toughness values being controlled by microstructural distributions in a critical "process zone" ahead of a macro crack-tip, it is of interest to explore the methodologies that should be employed to explain the distributions of fracture toughness.

Starting from first principles, a postulate is made that K_{Ic} values should initially be considered as single-valued "delta-functions" if the material is "quasi-homogeneous", by which it is meant that the fracture "unit-size" is so small in comparison with the process zone that the "worst" unit is sampled in every testpiece. The "next-worst" units will fracture at minimally higher applied fracture load than does the "worst" unit, so that an "avalanche" leads to catastrophe at notionally the same K_{Ic} value. The central limit theorem shows that random errors superimposed on the delta function produce a Gaussian distribution. The CDF (an erf) then plots as a straight line on normal probability paper and such plots have been made, using experimental results, to test the validity of the original postulate. This seems to be well validated for fine-scale microstructures and the plausibility of extrapolation to low probability (10^{-4}) can be assessed in a simple visual manner. This "lower bound" or "threshold" can be used in any scaling based on fracture mechanics.

In two-phase microstructures, the CDF is much more widely distributed (Figs.2, 3). It is possible to "straighten up" such a distribution of "macro" fracture toughness using Weibull analysis (Fig.4) but this ignores some of the essential *micro-scale* input, which requires *fractography*. From detailed examination of mixed bainite/martensite results, there is a danger that extrapolation of "mixed" results could give a *"lower-bound" lower than the "lower-bound" of the more brittle phase*. Similar concerns exist for the meso/near-macro distributions of "coarse" and "fine" microstructures in a multipass weld. It is strongly advised that in these cases the "lower-bound" be based on that of the more brittle phase.

Notched-bar testpieces will usually give more tightly-distributed fracture loads than those associated with pre-cracked testpieces, but the crucial factor is the relative sizes of the "sample volume" subjected to high stress (say, 98% or 95% of σ_{max}) and the microstructural "fracture unit".

References

1 Jackson W.J. and Wright J.C. Metals Technology **11** (1977) 425-457

2 Knott J.F. "Fracture 1977" ed. D.M.R. Taplin et al (Proc ICF4) Pergamon Vol **1,** 61-92

3 Griffiths J.R. and Owen D.R.J. J.Mech Phys Solids **19** (1971) 419-430

4 Knott J.F. "Fundamentals of Fracture Mechanics", Butterworths, 1973

5 Kelly A. Tyson W.R. and Cottrell A.H. Phil Mag **15** (1967) 567-586

6 Rice J.R. and Thomson R.M. Phil Mag **29** (1974) 73-97

7 Rice J.R. J.Mech Phys Solids **40** (1992) 239-271

8 Knott. J.F. "Advances in Fracture Resistance and Structural Integrity" ed. V.V. Panasyuk et al Pergamon 1994 pp 13-49

9 Ritchie. R.O. Knott J.F. and Rice J.R. J.Mech Phys Solids **21** (1973) 395-410

10 Curry D.A. and Knott J.F. Metal Science **13** (1979) 341-349

11 Beremin Met Trans **14A** (1983) 2277-2287

12 Tweed J.H. and Knott J.F. Acta Metall. **35** (1987) 1401-1414

13 Sokolinikoff I.S and Redheffer R.M. "Mathematics of Physics and Modern Engineering", McGraw Hill (2nd ed) 1966, pp 642-645

14 Neville D.J. and Knott J.F. J.Mech Phys Solids **34** (1986) 243-291

15 Bompas-Smith J.H. "Mechanical Survival" McGraw Hill New York 1973

16 Bowen P. Druce S.G. and Knott J.F. Acta Metall. **35** (1987) 1735-1746

17 Bhattacharjee D. and Knott J.F. Acta Metall Mater **42** (1994) 1747-1754

18 Hagiwara Y. and Knott J.F "Advances in Fracture Research" ed. D.Francois et al (Proc ICF5) Pergamon (1981) pp 707-715

19 Zhang X. current research at the University of Birmingham

20 Tweed J.H. and Knott J.F. Metal Science **17** (1983) 45-54

21 Newmann A. Benois F.F. and Hibbert, K. Schweisstechnik (Berlin)**18** (1968) 385-396

CLOSING IN ON THE CRACK TIP

JOHN W. HUTCHINSON
Division of Applied Sciences
Harvard University, Cambridge, MA 02138

1. Introduction

This paper addresses some of the issues involved in predicting fracture toughness of structural metals by linking the fracture process occurring at the crack tip through the plastic zone to the outer elastic field of a macroscopic crack. The link is extremely nonlinear giving rise to a macroscopic toughness which can be magnified many times the intrinsic work of the fracture process. The two principal fracture mechanisms considered are cleavage and void nucleation, growth and coalescence.

Almost everyone starting out to learn fracture mechanics asks the question, "How is it possible to predict fracture using elasticity theory?". The answer, of course, is that it is not possible. The critical value of crack tip intensity for a given material, called the fracture toughness, is determined by experiment. Nevertheless, going back to the first beginnings of fracture mechanics, attempts have been made to predict fracture toughness by connecting the response of a structural component or specimen containing a macroscopic crack all the way down to the crack tip where the microscopic fracture processes take place. The Dugdale-Barenblatt model was interpreted by Barenblatt (1962) to represent a crack in ideally elastic and brittle solid connecting the remote field with the atomic separation process along the crack line at the tip. When dislocations are not generated at the crack tip nor induced to move in the region surrounding the tip, the macroscopic toughness, Γ_{Ic}, for crack growth initiation approaches the atomistic work of separation. Sophisticated techniques for computing the atomistic work of separation based on quantum mechanics have recently been developed. A crude estimate for the atomistic work of separation is the theoretical lattice strength times a displacement proportional to the lattice spacing, i.e. roughly $(E/30)b$, where E is Young's modulus and b is the lattice spacing. For most metals and ceramics this is on the order of 1 to 5 J m^{-2} (see Table 1). Thus, a rough estimate for the fracture toughness of an ideally brittle material is

$$\Gamma_{Ic} \sim Eb/30 \tag{1}$$

Extrinsic factors, such as crack deflection and crack bridging by interlocking grains in polycrystals, can magnify Γ_{Ic} above the atomistic work of separation, but such effects are relatively small compared to magnification resulting from

J. R. Willis (ed.), IUTAM Symposium on Nonlinear Analysis of Fracture, 81–91.

plastic deformation at the crack tip.

There is considerable interest and effort underway to compute fundamental quantities such as the work of separation for atomic lattices and interfaces from first principle atomistic physics. This effort brings together segments of the mechanics and physics communities with the common goal of bridging the gap between the fracture events at the atomic scale and macroscopic fracture properties. Except perhaps for ideally brittle materials, the gap is not likely to be easily bridged. For most structural metals and metal/ceramic interfaces of interest, a plastic zone separates the crack tip and the outer elastic regions of the solid. As suggested in Table 1, the plastic zone can act as a huge multiplier leading to a macroscopic work of fracture which is larger than the atomic work of separation by factors of tens or even hundreds. The importance of plastic deformation in enhancing toughness was recognized years ago by the pioneers in fracture mechanics, such as Irwin and Orowan, shortly after the first fracture toughness measurements were made. Only recently, however, have efforts been made to quantify the connection between macroscopic toughness, plasticity and the 'local' work of the fracture process Γ_0. Some of the recent work on this problem will be mentioned here. An important distinction arises between a material whose fracture process is controlled at scales which are on the order of microns (see Table 1) and failure of a material or interface controlled at atomic scales. Examples of the former will be discussed first.

TABLE 1. Fracture Process, Macroscopic Initiation Toughness Γ_{Ic}, and Controlling Microstructural Scale

Fracture Process	Γ_{Ic} (J m^{-2})	10^{-3}m	10^{-6}m	10^{-9}m
Cleavage of highly brittle metals, ceramics and glasses by atomic separation (no plasticity)	1~5			b
Cleavage crystals and interfaces between metals and ceramics by atomic separation (some plasticity)	10~100			b
Cleavage of structural steels	10^2~10^4	D	S	
Void nucleation, growth and coalescence in structural metals	10^3~10^5	D	S	

(b=atomic lattice spacing, D=particle spacing and S=particle size)

2. Models of Fracture Initiation Toughness in Structural Metals

More than twenty years ago, models were proposed to predict initiation toughness for ductile structural metals for the two most important fracture process mechanisms, cleavage and void nucleation, growth and coalescence. A critical aspect of both of these processes in structural metals is the essential role of small second phase particulate components, such as brittle carbides in steel or oxide particles in aluminum, either in triggering cleavage or in nucleating voids. The particles are typically on the order of a micron in size and spaced apart by distances on the order of tens to hundreds of microns (cf. Table 1). These are the controlling scales for the fracture process in these materials. The problem of predicting the toughness of such materials does not require the mechanics analysis to bridge all the way to the atomistic scale. Of course, fundamental properties at the atomic scale influence the macroscopic properties, but their influence on the nucleation of micro-cracks or voids can be determined from analyses which decouple from the fracture toughness models, as will be made clearer below. We begin with a brief discussion of the outcome of two early models of this class.

The critical event in the cleavage of structural steels is the initiation of a dynamic micro-crack in a brittle carbide near the macroscopic crack tip. Under favorable conditions, this crack spreads from the particle, serving as the nucleus of the running cleavage crack. From then on, the crack is able to "outrun" most of the plasticity thereby avoiding crack blunting which would lead to arrest. For structural metals capable of cleaving, Ritchie, Knott and Rice (1973) assumed that initiation and spread of cracking beyond the brittle particle requires attainment of a critical stress σ_f. This is taken to be the same critical stress identified in notched bar tests in the earlier work. Ritchie et al. postulated that the stress ahead of the macroscopic crack must attain the critical stress level over distances which are comparable to the spacing D of the brittle particles. Using a plane strain, elastic-plastic analysis for the stress distribution ahead of a crack tip, these authors argued that as long as σ_f is less than between 3 to 5 times the yield stress, σ_Y, depending on the strain hardening level, the stress ahead of the crack will exceed the critical stress. The postulated condition will be met when the mode I stress intensity factor reaches

$$K_{Ic} \sim c\,\sigma_Y \sqrt{D} \tag{2}$$

Here, c is a numerical constant approximately equal to 3 but depending somewhat on strain hardening. In terms of the energetic measure of toughness, $\Gamma=(1-\nu^2)K^2/E$, where ν is Poisson's ratio, this estimate becomes

$$\Gamma_{Ic} \sim c^2(\sigma_Y{}^2/E)\,D \tag{3}$$

The range of values of listed in Table 1 are representative of values predicted by (3) for steels in the temperature range in which they are cleavable. If σ_f exceeds 3 to 5 times σ_Y the critical events cannot occur, according to this criterion, and cleavage cannot be initiated. The model regards σ_f as a parameter to be determined experimentally with either a cracked specimen or a notched specimen. Attempts to relate this critical stress to more fundamental material parameters have not been successful. The value of the model, to the extent it is valid, is that it decouples the determination of σ_f from the fracture toughness model.

When the controlling mechanism is void nucleation, growth and coalescence, Rice and Johnson (1970) considered the interaction of a nucleated void a distance D ahead of the unloaded crack tip. A simplified version of their analysis is as follows. The criterion for initiation of macroscopic crack growth states that initiation occurs when the void links up with the tip. In turn, this requires that the crack tip opening displacement $\delta \sim 0.5\, K^2/(E\sigma_Y)$, be equal to about one average void spacing D. The resulting initiation toughness is

$$\Gamma_{Ic} \sim 2\sigma_Y D \tag{4}$$

There is a dependence on the size S of the void-nucleating particles, or equivalently on the volume fraction of the particles, but this dependence is weak as long as S/D is small. Detailed finite strain computations for the interaction of a void with a crack tip have corroborated this simple relation (McMeeking 1977). Representative values from this equation are also included in Table 1 and are seen to be extremely large relative to typical values of the atomistic work of separation. The significance of a brittle to ductile transition in mechanism, from cleavage in (3) to void coalescence in (4), is evident because of the relative factor σ_Y/E. The initiation toughness can be even larger than (4) if nucleation of the voids from the second phase particles becomes the controlling step in the fracture process. As in the case of the other mechanism, atomistic considerations will come into the picture through considerations such as the strength of the interface bonding the void-nucleating particles to the metal matrix but not at the scale of the particle spacing which controls this fracture process.

3. Toughness Enhancement due to Plasticity in Small Scale Yielding

Small scale yielding pertains when the crack length and other in-plane length quantities are long compared to the size of the plastic zone at fracture. Under these circumstances the applied load experienced by the crack tip is specified by K or, equivalently, by Γ. Metals with some ductility nearly always display an increasing resistance to crack growth under plane strain conditions in the form of a crack growth resistance curve, Γ versus crack advance Δa, as sketched in Fig. 1. Following initiation of growth at Γ_{Ic}, the resistance curve rises, approaching an asymptote Γ_{SS}, corresponding to growth under steady-state conditions. Typically, the amount of growth needed to attain steady-state is several times the size of the plastic zone at that level of loading.

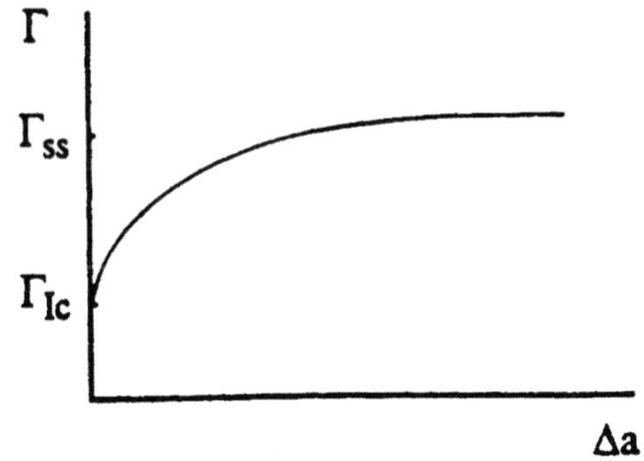

Figure 1. Typical crack growth resistance behavior when plastic flow occurs.

A study of the role of plasticity in enhancing crack growth resistance was carried out by Tvergaard and Hutchinson (1992). These authors analyzed a model which embeds a traction-separation law within an elastic-plastic continuum. The traction-separation law is considered to characterize the fracture process. The two primary parameters specifying this separation law are its peak stress, $\hat{\sigma}$, and the area under the curve, which is the work of separation per unit area, Γ_0. The material parameters characterizing the elastic-plastic continuum are its Young's modulus, Poisson's ratio, yield stress σ_Y, and strain hardening exponent N. The conventional flow (incremental) theory of plasticity is assumed to describe the metal with the Mises invariant employed to model isotropic hardening. Prior to the initiation of crack growth, the J-integral can be used to connect the remote field, whose amplitude is K or $\Gamma=(1-\nu^2)K^2/E$, to the separation zone at the tip. This procedure shows that the crack faces at the tip reduce to zero traction, corresponding to the initiation of crack advance, when Γ attains Γ_0. Thus the model says that the initiation toughness is the work of the fracture process:

$$\Gamma_{Io} = \Gamma_0 \tag{5}$$

Resistance behavior following initiation requires extensive numerical calculation, which was performed using finite element methods. From dimensional analysis, one can see that the steady-state toughness must depend on the parameters of the separation law and the elastic-plastic material according to

$$\Gamma_{ss} = \Gamma_0 \, F\left(\frac{\hat{\sigma}}{\sigma_Y}, N\right) \tag{6}$$

The numerical calculations reveal a weak dependence on other details of the shape of the separation law and on σ_Y/E and ν. The multiplicative influence of plasticity on the crack growth resistance is reflected in the appearance of the factor F in (6). Computed dependence of F for plane strain crack growth is shown in Fig. 2 for three values of the strain hardening exponent. For $\hat{\sigma}$ less than about 2.5 σ_Y, there is little crack growth resistance, i.e. $\Gamma_{ss} \cong \Gamma_{Ic} \cong \Gamma_0$, and almost no enhancement of toughness due to plastic deformation taking place in the plastic zone. (Of course, for a fracture process such as void growth and coalescence, plastic dissipation is an integral part of the work, Γ_0, of the process itself.) For peak stresses satisfying $\hat{\sigma}>2.5\ \sigma_Y$, depending somewhat on strain hardening, the enhancement of steady-state toughness due to plasticity can be appreciable. Multiplicative factors above 20 to 25 are difficult to compute with the model but, nevertheless, are clearly possible.

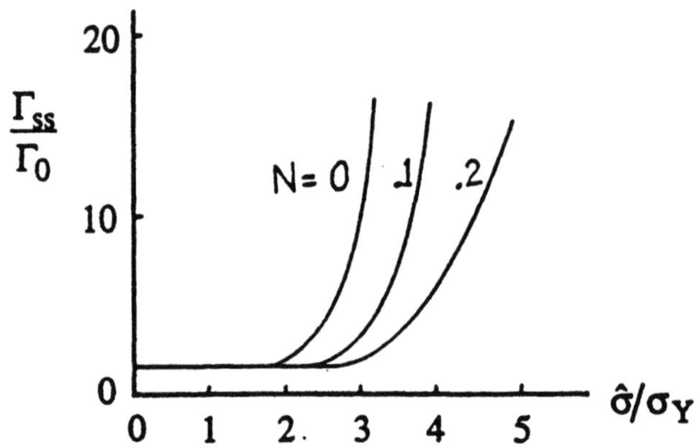

Figure 2. Ratio of steady-state toughness to the work of the fracture process for plane strain crack growth with σ_Y/E=0.003 and ν=0.3. Accurate plots are given in Tvergaard and Hutchinson (1992)

Implications of these results as applied to structural metals failing by the void coalescence mechanism are discussed in the next section. Their implication for metals and metal/ceramic interfaces whose fracture process is controlled at the atomic level by cleavage will be taken up in Section 5.

4. Crack Growth Resistance for Void Growth and Coalescence

To make contact with the results for the model discussed in the previous section, we note that the work of separation for a planar band of localized deformation containing voids is approximately

$$\Gamma_0 = \frac{1}{2}\sigma_Y D \tag{7}$$

with some dependence on the strain hardening index N but little dependence on the initial void volume fraction $f_0 \sim (S/D)^3$, as seen in Fig. 3b. The peak stress, $\hat{\sigma}$, attained in the separation of the localized band is sensitive to both N and f_0 (Fig. 3a). Use of the results in Fig. 3 for the fracture process together with the model predictions (5) and (6), enables one to predict both the initiation and steady-state toughnesses as a function of the four parameters of the material which primarily influence toughness, σ_Y, N, D and f_0. A fifth parameter, a void nucleation stress or strain, must be considered if the particles are especially well bonded to the metal matrix.

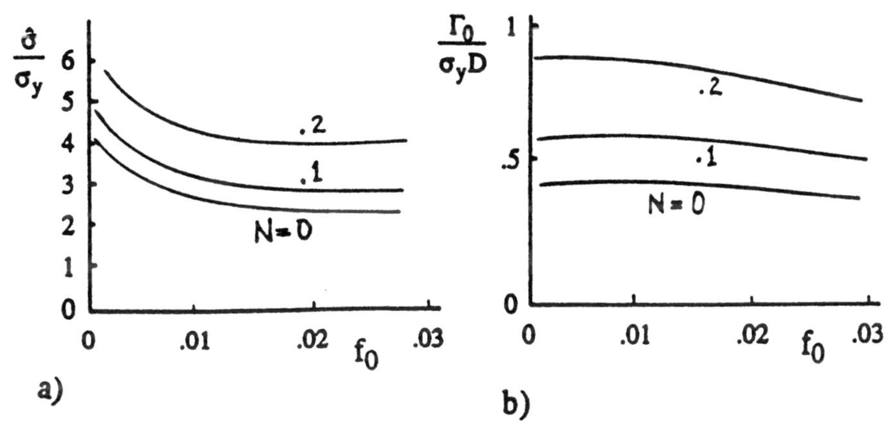

Figure 3. Trends in peak stress and work of fracture for separation of a localized band undergoing void growth and coalescence with $\sigma_Y/E=0.003$ and $\nu=0.3$. Accurate plots are given in Tvergaard and Hutchinson (1992)

By (5) and (6), the initiation toughness predicted by this model is

$$\Gamma_{Ic} = \Gamma_0 = \frac{1}{2}\sigma_Y D \tag{8}$$

which can be seen to be only one fourth of the value predicted by the Rice-Johnson model in (4). The assumption leading to (4) is that the fracture process is controlled by the interaction of a single void with the crack tip, while the implicit assumption leading to (7) and thus to (8) is that the fracture process zone extends ahead of the tip by at least several void spacings. The computations of Tvergaard and Hutchinson reveal that there are indeed two regimes: one where the length of the process zone is large compared to D when (8) is valid, and a second in accord with the Rice-Johnson model (4) wherein the fracture process zone is largely confined to the nearest void to the tip. The transition to this second regime occurs when $\hat{\sigma}/\sigma_Y$ becomes sufficiently large, or, equivalently, for example, when f_0 becomes sufficiently small (cf. Fig. 3a).

Recently a number of groups have developed computational models for crack growth in structural metals which are governed by the mechanism of void nucleation, growth and coalescence. Representative work is contained in the following publications: Needleman and Tvergaard (1987), Rousselier, Devaux, Mottet and Devesa (1989), Brocks, Klingbeil, Kunecke and Sun (1994), Bilby, Howard and Li (1993), and Xia, Shih and Hutchinson (1994). All these models use material elements ahead of the crack which contain voids and which approximately replicate the nucleation, growth and coalescence under the local conditions of stress and strain. Damage parameters such as the size and spacing of the void-nucleating particles are usually chosen such that the model reproduces a selected set of crack growth data but, in principle, they could be chosen to fit the microstructural quantities themselves. Once calibrated, the models can be used to predict crack growth and stability, together with load-deflection histories, under a wide variety of conditions. The strong size and geometry dependence of crack growth behavior under large scale yielding, which is associated with differing triaxial stress conditions near the tip, is accurately captured by these models. Thus, for the important class of structural metals failing by the void mechanism, a computational method is now more or less in place which can be used to predict the response of cracked structures and components. The models are first principles models in the sense that they bridge to the scale which controls the fracture process. More fundamental understanding at even smaller scales of the connection between the damage parameters to the features of the microstructural and the chemical make up of the material can be obtained from studies which uncouple from the crack problem.

5. Crack Growth Resistance for Atomic Cleavage

While it has been possible to bridge through the plastic zone to the fracture process zone for the void mechanism, the same cannot be said when the fracture process mechanism is atomic separation for reasons which will now be discussed. At first sight, the model of Tvergaard and Hutchinson discussed in Section 3 would seem to be applicable to this problem, if one identified the traction-separation law with that for atomic separation. One obvious possible objection is that it is most unlikely that conventional continuum plasticity theory is applicable at the scales below about 1 micron, which necessarily come into play in bridging to the atomic scale. There is an increasing body of experimental evidence that a strong size effect exists for plastic deformation at length scales below about 1 micron, with significantly increased rates of strain hardening at the smaller scales. Conventional continuum plasticity theory does not incorporate size-dependent hardening effects. Moreover, all solutions for crack tip stress distributions obtained from conventional elastic-plastic continuum theory predict that the maximum stress that can be achieved ahead of the crack tip is only about 3 to 5 times the initial yield stress of the solid, depending on the strain hardening exponent N. Such stress levels are well below the values needed to produce atomic separation in most metals. In particular, this is true for the model discussed in Section 3, as is evident from the steady-state toughness trends in Fig. 2. If one were to take the predictions of the conventional plasticity models at face value, one would conclude that plasticity eliminates the possibility of atomic cleavage as a mechanism of fracture.

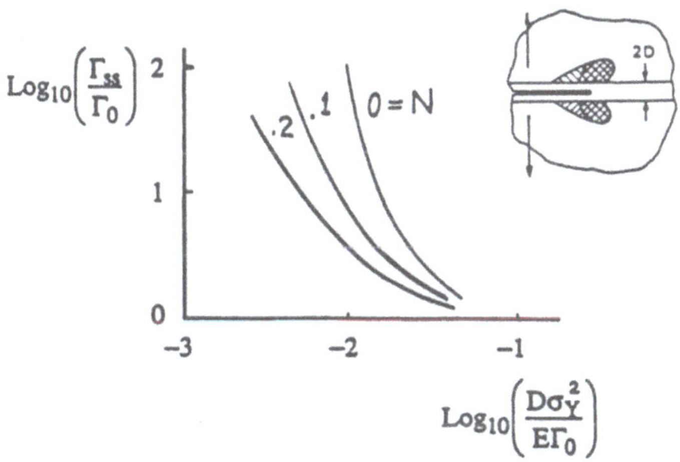

Figure 4. Trends in the ratio of steady-state toughness to the work of crack tip separation (σ_Y/E=0.003 and ν=1/3). Accurate plots are given in Suo, Shih and Varias (1993)

The thrust of the above discussion is that one must turn to theories of plasticity which incorporate more realistic hardening descriptions, or, possibly, to dislocation modeling at the smallest scales, if there is to be any hope of bridging to atomic separation. Strain gradient theories of plasticity are currently under development which incorporate scale-dependent strain hardening. Whether these will bridge to the crack tip remains to be seen. In lieu of a more realistic way of dealing with small scale plasticity, Suo, Shih and Varias (1993) proposed a model which is capable of generating the large stress values necessary at the tip to achieve atomic cleavage. They restricted consideration to metals which emit no dislocations from the crack tip, and they postulated the existence of a dislocation-free elastic strip zone along the projected crack line, as sketched in the insert of Fig. 4. The half-height of this zone is taken to be D and is imagined to have a dimension which is small compared to a micron corresponding, at least qualitatively, to the smallest average spacing between dislocations. Outside the elastic strip, the authors used conventional elastic-plastic continuum theory to represent the constitutive behavior. The existence of the elastic strip between the outer plastic zone and the tip permits the stresses in the model to rise to the high levels needed to bring about atomic cleavage. For a small scale yielding formulation, Suo et al. computed the steady-state relation between the remote loading, as measured by the steady-state toughness Γ_{SS}, and the energy release rate at the crack tip in the elastic strip, which is set to be equal to the atomic work of separation Γ_0. The resulting relation is plotted in dimensionless form in Fig. 4. Macroscopic steady-state toughnesses in excess of 10 to 100 times the atomistic work of fracture are clearly implied when the parameter $D\sigma_Y{}^2/(E\Gamma_0)$ is sufficiently small. Evidence for such a large multiplicative influence of plasticity on toughness has recently been presented for cleavage of several metal/ceramic interfaces, including a gold/sapphire interface (Reimanis, Dalgleish and Evans, 1991).

Acknowledgment This work was supported in part by the National Science Foundation (Grant No. MSS-92-02141) and in part by the Division of Applied Sciences, Harvard University.

References

Barenblatt, G. I. (1962) 'The mathematical theory of equilibrium cracks' in *Advances in Applied Mechanics*, vol. **7** (eds. H. L. Dryden and T. von Karman), 56-125.

Bilby, B. A., Howard, I. C. and Li, Z. H. (1993) 'Prediction of the first spinning cylinder test using ductile damage theory'. *Fatigue and Fracture of Engineering Materials and Structures* **16**, 1-20.

Brocks, W. Klingbeil, Kunecke, D and Sun, D.-Z. (1994) 'Application of the Gurson model to ductile tearing resistance'. To be published in *Constraint Effects in Fracture: Theory and Applications, ASTM STP 1244*, (eds. M. Kirk and A. Bakker) American Society for Testing materials, Philadelphia.

McMeeking, R. M. (1977) 'Finite deformation analysis of crack-tip opening in elastic-plastic materials and implications for fracture'. *J. Mech. Phys. Solids* **25**, 357-391.

Needleman, A. and Tvergaard, V. (1987) 'An analysis of ductile rupture modes at a crack tip'. *J. Mech. Phys. Solids* **35**, 151-183.

Reimanis, I. E., Dalgleish, B. J. and Evans, A. G. (1991) 'The fracture resistance of a model metal/ceramic interface.' *Acta Metall. Mater.* **39**, 3133-3141.

Rice, J. R. and Johnson, M. A. (1970) 'The role of large crack tip geometry changes in plane strain fracture' in *Inelastic Behavior of Solids* (eds. M. F Kanninen, et al.), McGraw-Hill, N. Y. 641-672.

Ritchie, R. O., Knott, J. F. and Rice, J. R. (1973) 'On the relationship between critical tensile stress and fracture toughness in mild steel,' *J. Mech. Phys. Solids*, **21**, 395-410.

Rousselier, G., Devaux, J.-C., Mottet, G. and Devesa, G. (1989) 'A methodology for ductile fracture analysis based on damage mechanics: an illustration of a local approach of fracture', In *Nonlinear Fracture Mechanics: Volume 11-Elastic Plastic Fracture, ASTM STP 995* (eds. J. D. Landes, A. Saxena and J. G. Merkle) American Society of Testing Materials, Philadelphia, 332-354.

Tvergaard, V. and Hutchinson, J. W. (1992) 'The relation between crack growth resistance and fracture process parameters in elastic-plastic solids'. *J. Mech. Phys. Solids* **40**, 1377-1397.

Suo, Z., Shih, C. F. and Varias, A. G. (1993) 'A theory for cleavage cracking in the presence of plastic flow', *Acta Metall. Mater.* **41**,1551-1557.

Xia, L., Shih, C. F. and Hutchinson, J. W. (1994) 'A computational approach to ductile crack growth under large scale yielding conditions', *J. Mech, Phys. Solids*,**43**, 389-413 (1995).

RELATIONS BETWEEN CRACK GROWTH RESISTANCE AND FRACTURE PROCESS PARAMETERS UNDER LARGE SCALE YIELDING

VIGGO TVERGAARD
Department of Solid Mechanics
Technical University of Denmark, Lyngby, Denmark

Abstract. Mode I crack growth under large scale yielding conditions is studied by comparing numerical plane strain analyses for four different test specimen geometries. The fracture process is represented in terms of a cohesive zone model, for which the separation work per unit area and the peak stress required for separation are basic parameters; but where also a plastic strain effect on the fracture process is incorporated. The differences between crack growth resistance curves predicted for different specimen geometries are in general agreement with the different T–stress levels obtained for different specimens. In addition, a specimen size dependence of the crack growth resistance curves is illustrated.

1. Introduction

Plastic dissipation in the material around a crack tip adds significantly to the fracture toughness, so that the value of the stress intensity factor needed to advance a crack can be much larger than that corresponding to the work of fracture per unit area of crack surface. Computations of Tvergaard and Hutchinson (1992, 1994) for mode I plane strain crack growth under small scale yielding conditions have been used to study this dependence, with the fracture process represented in terms of a traction–separation law in which the primary parameters are the work of separation per unit area, Γ_0, and the peak traction, $\hat{\sigma}$. These studies show that the plasticity induced increase of the fracture toughness depends mainly on the peak traction to initial yield stress ratio, $\hat{\sigma}/\sigma_Y$, and the strain hardening exponent, N.

With the cohesive zone model employed by Tvergaard and Hutchinson (1992, 1994) the fracture process is entirely stress dependent. But e.g. for an elastic–plastic solid the maximum stress achieved ahead of a crack–tip is

J. R. Willis (ed.), IUTAM Symposium on Nonlinear Analysis of Fracture, 93–104.

$2.97\sigma_Y$, so here the model predicts only crack blunting with no crack advance, if $\hat{\sigma}$ exceeds $2.97\sigma_Y$. Similar behaviour is found for strain hardening materials, at somewhat higher values of the limiting stress. But in the case of metals failing by the mechanism of void nucleation, growth and coalescence there is also an effect of intense plastic straining in the near vicinity of the crack tip, which accelerates the void growth process and the nucleation of more voids. To incorporate such plastic strain influence Tvergaard and Hutchinson (1995) have used a modified traction–separation law, in which the peak separation stress $\hat{\sigma}$ is reduced continuously as a function of local plastic strain for strains above a critical value ϵ_c (see also Tvergaard, 1992). The computations of Tvergaard and Hutchinson (1995) for conditions of small scale yielding have shown that the modified model allows for a better representation of very tough materials with a high value of the tearing modulus defined by Paris *et al.* (1979). It is noted that this modified model displays some mesh dependence, since the plastic strain predicted near the tip is sensitive to the element size along the line of the crack.

Predictions of crack advance by a void coalescence mechanism can be directly based on the porous ductile material model of Gurson (1977), as has been applied by e.g. Needleman and Tvergaard (1987, 1991), Rousselier (1987), Sun *et al.* (1992), Brocks *et al.* (1994) and Xia *et al.* (1995). In such finite element models the mesh must capture the interaction between the crack tip and the nearest voids, and many of the models have taken the element size to be on the order of the void spacing. As such, these models also display mesh dependence.

In the present paper, the modified traction–separation law, with a plastic strain dependence of the peak stress, is used to analyse crack growth under large scale yielding conditions. For a given set of parameters describing the fracture process results corresponding to four different specimen geometries are compared. Also the specimen size dependence of the crack growth resistance curve is studied, including specimen sizes large enough to give fully contained plastic yielding.

2. Problem Formulation and Numerical Method

Each of the four different specimen geometries considered here are analysed by numerical solutions for the region shown in Fig. 1a, with appropriate symmetry boundary conditions or load boundary conditions specified at the different edges. Here, A_0 denotes the initial crack length. For a double edge–notched tension strip (DENT) the full specimen covers the region $-H_0 \le x_2 \le H_0$, $0 \le x_1 \le 2W_0$, and uniform tensile stresses are applied at the ends. For a single edge–notched tension strip (SENT) or a single edge–notched bent strip (SENB) the full specimen covers the region $-H_0 \le x_2 \le H_0$, $0 \le x_1 \le W_0$,

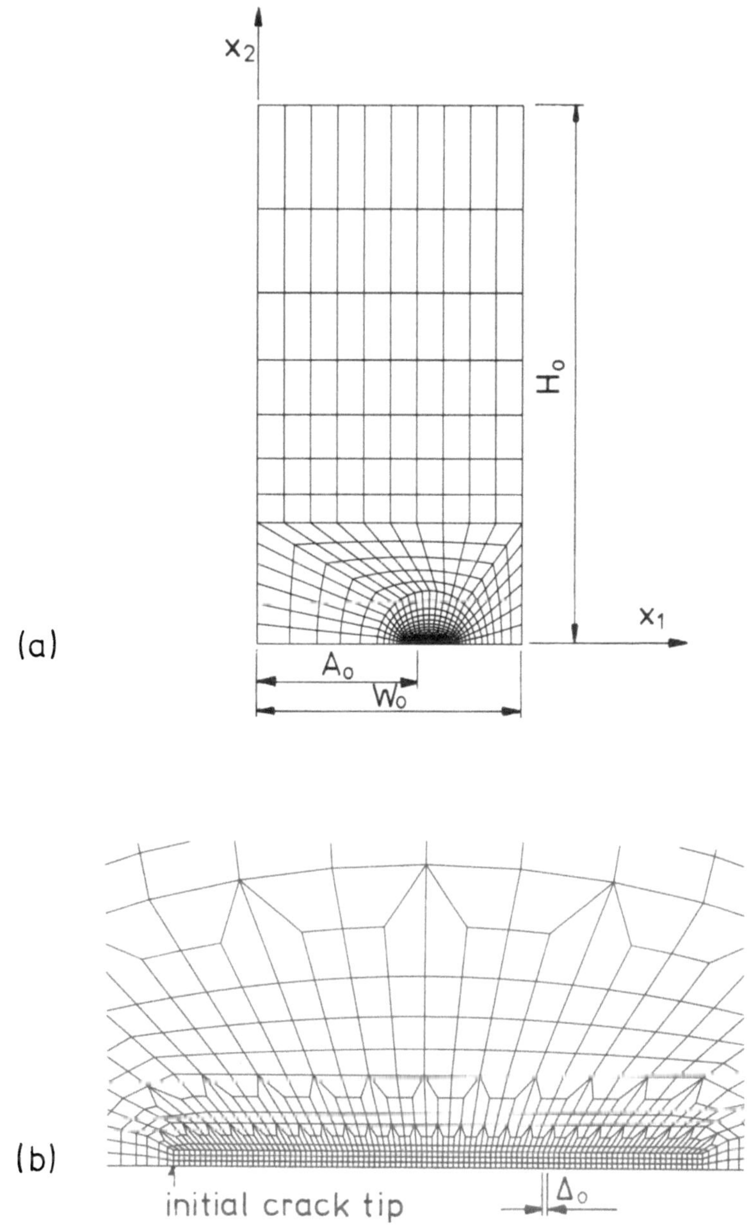

Fig. 1. Example of finite element mesh. (a) Region analysed numerically, to represent SENB, SENT, DENT or CCP specimens. (b) Refined mesh along the crack line.

with uniform tensile stresses or linearly varying pure bending stresses, respectively, applied at the ends. For a center–cracked panel (CCP) the full specimen covers the region $-H_0 \leq x_2 \leq H_0$, $-W_0 \leq x_1 \leq W_0$, and uniform tensile stresses are applied at the ends. In the the following these four specimens will be referred to by the abbreviations mentioned in parentheses.

The material is modelled as an elastic–plastic solid, using the Mises yield surface with isotropic hardening. Furthermore, the fracture process is represented in terms of a plastic strain dependent cohesive zone model, as in the investigation of Tvergaard and Hutchinson (1995) for conditions of small–scale yielding. The approximating assumption is made that plane strain conditions apply throughout the specimens analysed.

In the analyses finite strains are accounted for, using a convected coordinate, Lagrangian formulation of the field equations, in which g_{ij} and G_{ij} are metric tensors in the reference configuration and the current configuration, respectively, with determinants g and G, and $\eta_{ij} = \frac{1}{2}(G_{ij} - g_{ij})$ is the Lagrangian strain tensor. The contravariant components τ^{ij} of the Kirchhoff stress tensor on the current base vectors are related to the components of the Cauchy stress tensor σ^{ij} by $\tau^{ij} = \sqrt{G/g}\,\sigma^{ij}$. Then, in the finite–strain generalization of J_2–flow theory discussed by Hutchinson (1973), an incremental stress–strain relationship is obtained of the form $\dot{\tau}^{ij} = L^{ijkl}\dot{\eta}_{kl}$. The value of the tangent modulus at a given stress level is determined from the uniaxial true stress–logarithmic strain curve, which is taken to be specified by the power law

$$\epsilon = \begin{cases} \sigma/E & , \text{ for } \sigma \le \sigma_Y \\ (\sigma_Y/E)(\sigma/\sigma_Y)^{1/N} & , \text{ for } \sigma \ge \sigma_Y \end{cases} \tag{1}$$

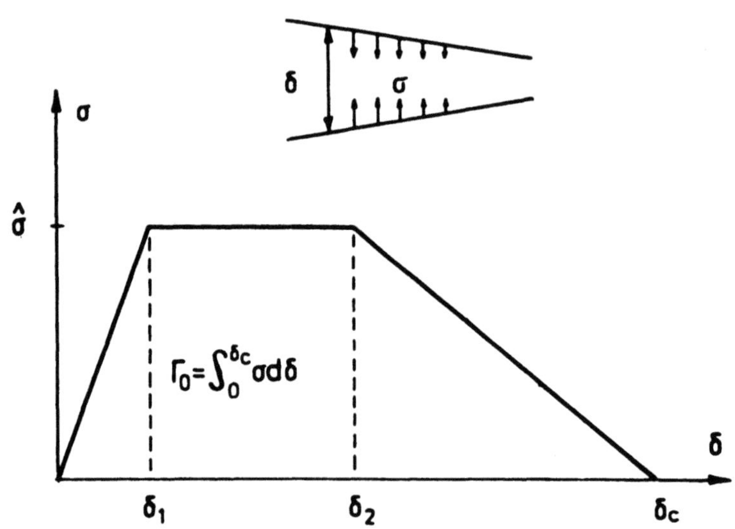

Fig. 2. Traction–separation relation for fracture process.

Here, σ_Y is the initial yield stress, E is Young's modulus, and N is the strain hardening exponent.

The traction–separation relation used by Tvergaard and Hutchinson (1992, 1994) to model the fracture process (Fig. 2) is fully specified by the work of separation Γ_0, the peak stress $\hat{\sigma}$, and the shape parameters δ_1/δ_c and δ_2/δ_c. According to this model failure initiates when the true normal stress ahead of the crack reaches the value $\hat{\sigma}$, and no crack growth is predicted at all, if the stress level $\hat{\sigma}$ is not reached. The modification of this separation law, used by Tvergaard and Hutchinson (1995) to represent a plastic strain controlled failure mechanism, is analogous to that employed by Tvergaard (1992, 1995) in studies of ductile particle debonding during crack bridging in ceramics. With this modification the peak stress $\hat{\sigma}$ in the traction–separation relation of Fig. 2 is gradually reduced when the effective plastic strain ϵ_e^P along the crack path has exceeded a critical value ϵ_c

$$\hat{\sigma} = \begin{cases} \hat{\sigma}_0 & , \text{ for } \epsilon_e^P \leq \epsilon_c \\ \hat{\sigma}_0 - \Delta\hat{\sigma}\left(\epsilon_e^P - \epsilon_c\right)/\Delta\epsilon & , \text{ for } \epsilon_c < \epsilon_e^P < \epsilon_c + \Delta\epsilon \\ \hat{\sigma}_0 - \Delta\hat{\sigma} & , \text{ for } \epsilon_e^P \geq \epsilon_c + \Delta\epsilon \end{cases} \tag{2}$$

Thus, with (2) the cohesive zone model accounts for a reduction of the material strength, which could result from plastic strain controlled nucleation of voids or from accelerated void growth near the crack tip.

The numerical solutions are obtained by a linear incremental method using a finite element approximation of the displacement fields in the incremental version of the principle of virtual work. The elements used are quadrilaterals each built–up of four triangular, linear–displacement elements. In the uniform mesh region in front of the initial crack–tip the length of one square element is denoted as Δ_0 (see Fig. 1b), and the uniform mesh region contains 100×3 quadrilaterals. Most of the computations are carried out with $\delta_c = 0.1\Delta_0$, $\delta_1/\delta_c = 0.15$ and $\delta_2/\delta_c = 0.5$, and with $W_0 = 1000\Delta_0$. The value of the effective plastic strain ϵ_e^P in (2) is calculated as the average over the quadrilateral element adjacent to the point considered in the debonding region. The edge loads at $x_2 = H_0$ are applied incrementally, and a special Rayleigh Ritz finite element method is employed to control nodal displacements within the fracture process zone (see also Tvergaard, 1990b).

Two reference quantities K_0 and R_0 are used for the presentation of results

$$K_0 = \left[E\Gamma_0/\left(1 - \nu^2\right)\right]^{1/2}, \quad R_0 = \frac{1}{3\pi}\left(\frac{K_0}{\sigma_Y}\right)^2 = \frac{1}{3\pi}\frac{E\Gamma_0}{\left(1 - \nu^2\right)\sigma_Y^2} \tag{3}$$

Here, K_0 represents the mode I stress intensity factor needed to advance the crack when plastic dissipation is negligible; i.e. the value needed to supply just the work of the fracture process Γ_0, when the modification (2) is not active. The expression for Γ_0 is

$$\Gamma_0 = \int_0^{\delta_c} \sigma \, d\delta = \frac{1}{2}\hat{\sigma}_0[\delta_c + \delta_2 - \delta_1] \tag{4}$$

The reference length R_0 scales with the size of the plastic zone when $K \cong K_0$.

The value of the J–integral is calculated on a number of contours around the crack–tip. After some crack growth the path–independence of the J–integral breaks down for contours close to the tip, but remains in a region of more remote contours. The K–values to be shown in the following are computed from J–integrals on remote contours.

3. Results

The elastic–plastic material to be considered here has the parameter values $N = 0.1$ and $\sigma_Y/E = 0.002$ in (1), and the value of Poisson's ratio is $\nu = 0.3$. For the traction–separation relation the two parameters $\hat{\sigma}_0$ and Γ_0 (or K_0) appear directly in the figures presenting the results. The first computations are carried out for the region shown in Fig. 1a, with $A_0/W_0 = 0.6$ and $H_0/W_0 = 2$, and with $\Delta_0/W_0 = 1000$, $\delta_c = 0.1\Delta_0$, $\epsilon_c = 0.05$, $\Delta\epsilon = 0.05$ and $(\hat{\sigma}_0 - \Delta\hat{\sigma})/\sigma_Y = 1.0$.

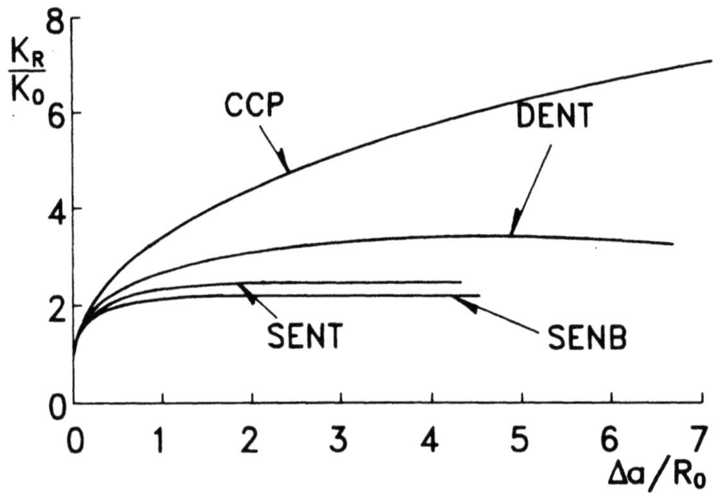

Fig. 3. Crack growth resistance curves for specimens with $\hat{\sigma}_0/\sigma_Y = 3.5$, $\epsilon_c = 0.05$ and $A_0/W_0 = 0.6$.

Fig. 3 shows crack growth resistance curves computed for $\hat{\sigma}_0/\sigma_Y = 3.5$, corresponding to the four different specimen geometries considered in this paper. It is well known that different specimen geometries give rise to different levels of T–stress at the crack–tip (Hancock, Reuter and Parks, 1991; Xia, Shih and Hutchinson, 1995). Thus, different specimen geometries will give rise to different crack growth resistance curves, even in cases where the specimen size is large enough to give small–scale–yielding (Tvergaard and Hutchinson, 1994). In the present case the length of the initial uncracked ligament satisfies the relation $W_0 - A_0 = 3.08 \ (K_0/\sigma_Y)^2$, which is within the ASTM–requirements for compact tension specimens (since $3.08 > 2.5$), so that such specimens would be large enough for testing linear elastic fracture mechanics (e.g. see Hutchinson, 1979; Carlsson, 1985). It is known that compact tension specimens show rather large constraint, $T/\sigma_Y \simeq 0.4$, and the present SENB specimens are rather similar to that; but DENT specimens have a negative T–stress and CCP specimens have a rather large negative T–stress, giving a larger plastic zone, so that larger specimens size would be required in these cases for valid K_{IC} testing. It is noted that if the specimen width W_0 was taken to be 50 mm, then for a steel the reference fracture toughness K_0 would be about 34 $MN/m^{3/2}$, which would correspond to a tough steel.

For the material parameters used in Fig. 3 previous studies (Tvergaard and Hutchinson, 1992, 1995) have shown that a steady–state fracture toughness is reached under small–scale–yielding conditions in the range of $\Delta a/R_0$ values considered, and that plastic strain has no effect on the fracture process, as ϵ_e^P remains smaller than ϵ_c. This type of behaviour is found in Fig. 3 for the SENB, SENT and DENT specimens, with different values of the steady–state fracture toughness due to the different T–stress levels resulting from the specimen geometries. For the CCP specimen the crack growth resistance is significantly higher than found for the other specimen geometries, the plastic strain controlled failure mechanism (2) does play a role, and a steady–state fracture toughness is not reached in the range studied.

In Fig. 4 the interface strength is higher, $\hat{\sigma}_0/\sigma_Y = 3.75$, while all other material parameters are identical to those in Fig. 3, and thus $W_0 - A_0 = 2.88 \ (K_0/\sigma_Y)^2$. Here, the values of ϵ_e^P near the crack–tip exceed ϵ_c slightly for the SENB, SENT and DENT specimens and much more for the CCP specimen. In this case none of the crack growth resistance curves reach their maximum in the range considered; but still the SENB and CCP specimens show the lowest and highest crack growth resistances, respectively.

In Fig. 5, for $\hat{\sigma}_0/\sigma_Y = 4.0$ and all other material parameters unchanged, the specimens satisfy the relation $W_0 - A_0 = 2.70 \ (K_0/\sigma_Y)^2$. Here, the values of ϵ_e^P near the crack–tip exceed ϵ_c significantly for all four specimen geometries, so that the plastic strain controlled failure mechanism (2) plays a strong role. It is seen that the differences between the crack growth resistance curves predicted for the four specimen geometries are much reduced compared to Fig.

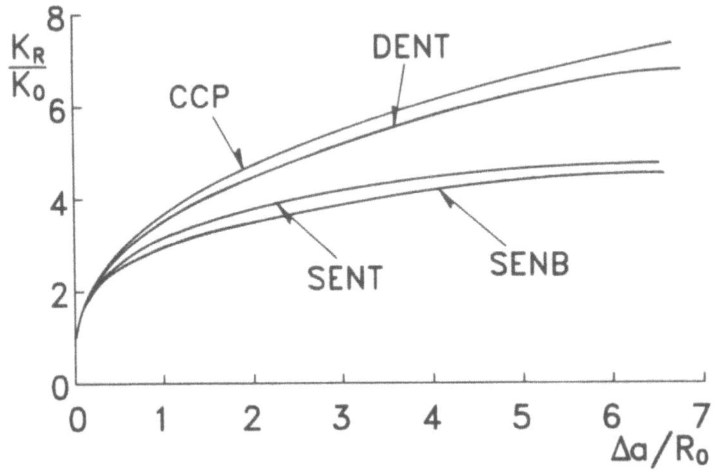

Fig. 4. Crack growth resistance curves for specimens with $\hat{\sigma}_0/\sigma_Y = 3.75$, $\epsilon_c = 0.05$ and $A_0/W_0 = 0.6$.

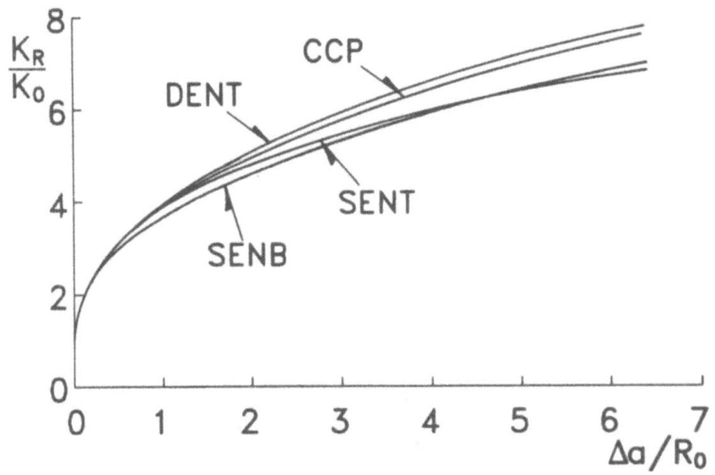

Fig. 5. Crack growth resistance curves for specimens with $\hat{\sigma}_0/\sigma_Y = 4.0$, $\epsilon_c = 0.05$ and $A_0/W_0 = 0.6$.

4, which continues the trend going from Fig. 3 to Fig. 4. This also agrees with the trend found by Tvergaard and Hutchinson (1995) for small–scale–yielding conditions that the predicted T–stress dependence of the crack growth resistance curves is reduced as the values of $\hat{\sigma}_0/\sigma_Y$ and ϵ_c are increased. In Fig. 5 the lowest curve is still that for the SENB specimen, in most of the range considered; but the curve for the CCP specimen is here slightly below that for the DENT specimen.

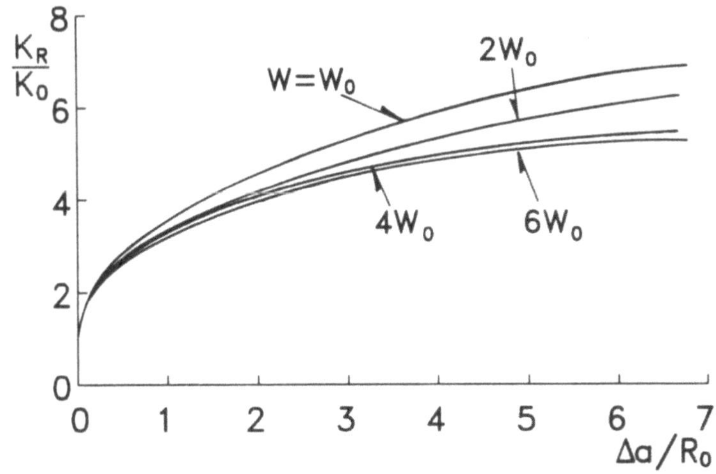

Fig. 6. Crack growth resistance curves for different size DENT specimens with identical material parameters, $\hat{\sigma}_0/\sigma_Y = 3.75$, $\epsilon_c = 0.05$, $\Delta_0/W_0 = 1000$, $\delta_c/\Delta_0 = 0.1$ and $A/W = 0.6$.

The specimen size dependence is considered in Fig. 6 by comparing crack growth resistance curves for DENT specimens of four different sizes. The material parameters are identical to those considered in Fig. 4, with $\hat{\sigma}_0/\sigma_Y = 3.75$ and with $\Delta_0/W_0 = 1000$. To introduce the different specimen sizes the dimensions A_0, W_0 and H_0 in Fig. 1 are replaced by A, W and H, respectively, with the ratios $A/W = 0.6$ and $H/W = 2$ fixed. Then, it is clear that the curve marked $W = W_0$ in Fig. 6 is identical to that for the DENT specimen in Fig. 4, and that the other three curves in Fig. 6 correspond to specimens 2, 4 and 6 times larger, respectively. It is noted that the fixed value of Δ_0/W_0 is necessary in this comparison, as it has been found (Tvergaard and Hutchinson, 1995) that the plastic strain dependent traction–separation relation (2) results in a certain mesh sensitivity of the predicted crack growth resistance curves. Thus, when this model based on the embedded cohesive zone is used for a particular material, both the parameter values specifying the traction–separation law and the mesh size along the crack growth path must be kept fixed. It is noted that the critical strain value ϵ_c in (2) is only slightly exceeded by ϵ_e^P near the tip in these four computations.

The curves in Fig. 6 show that there is a specimen size dependence, such that the level of the predicted crack growth resistance decays for increasing specimen size; but the curves for the two larger specimens differ only little. The values of the ratio $(W - A)/(K_0/\sigma_Y)^2$ for these four specimens are 2.88, 5.75, 11.50 and 17.26, respectively, which indicates, even for DENT specimens, that the sizes of the larger specimens are well into the range representing

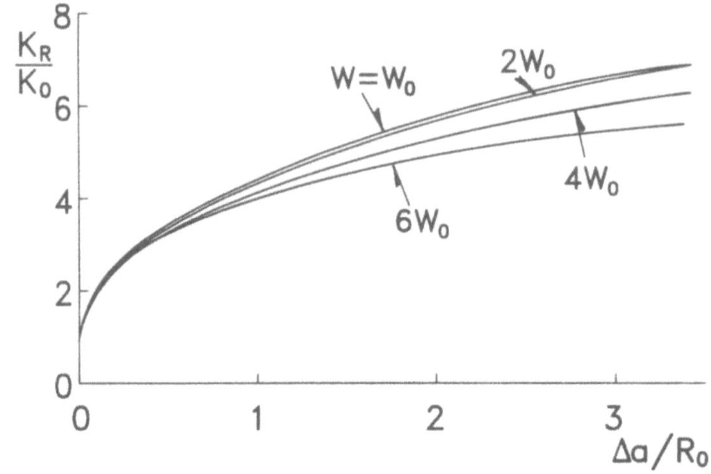

Fig. 7. Crack growth resistance curves for different size DENT specimens with identical material parameters, $\hat{\sigma}_0/\sigma_Y = 3.75$, $\epsilon_c = 0.05$, $\Delta_0/W_0 = 1000$, $\delta_c/\Delta_0 = 0.2$ and $A/W = 0.6$.

small–scale–yielding. Thus, it is assumed that the two lower curves in Fig. 6 correspond to the behaviour under conditions of small–scale–yielding. In fact, to directly compare with the small–scale–yielding results of Tvergaard and Hutchinson (1995) a computation for a large SENB specimen, $W = 6W_0$, has been carried out with somewhat different material parameters ($\sigma_Y/E = 0.003$, $\hat{\sigma}_0/\sigma_Y = 4.0$, $(\hat{\sigma}_0 - \Delta\hat{\sigma})/\sigma_Y = 0.1$ and $\epsilon_c = 0.03$), and good agreement has been found.

In Fig. 7 resistance curves are compared for the same four sizes of DENT specimens, with the same set of material parameters apart from one difference, $\delta_c = 0.2\,\Delta_0$. This double value of the critical separation δ_c also doubles the values of Γ_0 and R_0, respectively, compared to the values in Fig. 6. Thus, for the same amount of crack growth, $\Delta a/W_0$, in Figs. 6 and 7 the value of $\Delta a/R_0$ is only half as large in Fig. 7, since the plastic zone size is doubled. In Fig. 7 the values of the ratio $(W - A)/(K_0/\sigma_Y)^2$ for the four specimens are 1.44, 2.88, 5.75 and 8.63, respectively, and it is found that in all four cases the plastic strain controlled failure mechanism (2) plays a strong role. The crack growth resistance curve for the largest specimen in Fig. 7, $W = 6W_0$, may have converged towards the result for small–scale–yielding; but clearly the curve for $W = 4W_0$ has not, as this curve shows somewhat higher crack growth resistance. In the computation for the smallest specimen, $W = W_0$, the path independence of the J–integral breaks down for contours close to the growing crack, as in all computations, but here to such an extent that there is hardly any path independence for $\Delta a/R_0 > 2$. This may partly explain why the crack growth resistance curve for $W = W_0$ is only slightly above or even intersects that for $W = 2W_0$ in Fig. 7.

A comparison of crack growth computations for different specimen geometries was also carried out by Xia *et al.* (1995), using a modified Gurson model (Tvergaard, 1990a) to represent the fracture process. With a fixed set of material parameters and a fixed mesh size along the crack growth path, to represent a particular material, this model was found to reproduce important aspects of the specimen shape dependence of crack growth resistance curves. Also the alternative approach adopted in the present paper, based on a plastic strain dependent cohesive zone model, shows trends in the specimen shape dependence that agree with experimental observations (as e.g. Hancock *et al.*, 1991). Regarding a specimen size dependence of the crack growth resistance curves it is noted that also Xia *et al.* (1995) found increasing resistance for decreasing specimen size.

References

Brocks, W. Klingbeil, Kunecke, D. and Sun, D.–Z. (1994). Application of the Gurson model to ductile tearing resistance. To be published in *Constraint Effects in Fracture: Theory and Applications, ASTM STP 1244* (eds. M. Kirk and A. Bakker). American Society for Testing and Materials, Philadelphia.

Carlsson, J. (1985). Fracture mechanics (in Swedish). Royal Inst. of Technology, Stockholm.

Gurson, A.L. (1977). Continuum theory of ductile rupture by void nucleation and growth: Part I – yield criteria and flow rules for porous ductile media. *J. Engin. Materials and Tech.* **99**, 2–15.

Needleman, A. and Tvergaard, V. (1987). An analysis of ductile rupture modes at a crack tip. *J. Mech. Phys. Solids* **35**, 151–183.

Needleman, A. and Tvergaard, V. (1991). An analysis of dynamic, ductile crack growth in a double edge cracked specimen. *Int. J. Fracture* **49**, 41–67.

Hancock, J.W., Reuter, W.G. and Parks, D.M. (1991). Constraint and toughness parameterised by T . *ASTM Symposium on Constraint Effects in Fracture*, Indianapolis.

Hutchinson, J.W. (1973). Finite strain analysis of elastic–plastic solids and structures. In *Numerical Solution of Nonlinear Structural Problems* (ed. R.F. Hartung), p. 17. American Society for Mechanical Engineers, New York.

Hutchinson, J.W. (1979). Nonlinear fracture mechanics. Dept. Solid Mech., Techn. Univ. Denmark.

Paris, P.C., Tada, H., Zahoor, A. and Ernst, H. (1979). The theory of instability of the tearing mode of elastic–plastic crack growth. In *Elastic–Plastic Fracture* (eds. J.D. Landes *et al.*), 5–36, ASTM STP 668, American Society of Testing Materials.

Rousselier, G. (1987). Ductile fracture models and their potential in local approach of fracture. *Nuclear Engng. and Design* **105**, 97–111.

Sun, D.–Z., Kienzler, R., Voss, B. and Schmitt, W. (1992). Application of micromechanical models to the prediction of ductile fracture. In *Fracture Mechanics: Twenty–Second Symposium (Volume II), ASTM STP 1131* (eds. S.N. Atluri, J.C. Newman, Jr., I.S. Raju, and J.S. Epstein), pp. 368–378. American Society for Testing and Materials, Philadelphia.

Tvergaard, V. (1990a). Material failure by void growth to coalescence. *Adv. Appl. Mech.* **27**, 83–151.

Tvergaard, V. (1990b). Effect of fibre debonding in a whisker–reinforced metal. *Materials Science and Engineering* **A125**, 203–213.

Tvergaard, V. (1992). Effect of ductile particle debonding during crack bridging in ceramics. *Int. J. Mech. Sci.* **34**, 635–649.

Tvergaard, V. (1995). Cavity growth in ductile particles bridging a brittle matrix crack. *Int. J. Fracture* **72**, 277–292.

Tvergaard, V. and Hutchinson, J.W. (1992). The relation between crack growth resistance and fracture process parameters in elastic–plastic solids. *J. Mech. Phys. Solids* **40**, 1377–1397.

Tvergaard, V. and Hutchinson, J.W. (1994). Effect of T–stress on mode I crack growth resistance in a ductile solid. *Int. J. Solids Structures* **31**, 823–833.

Tvergaard, V. and Hutchinson, J.W. (1995). Effect of strain dependent cohesive zone model on predictions of crack growth resistance. *Report No. 497*, Danish Center for Applied Mathematics and Mechanics.

Xia, L. Shih, C.F. and Hutchinson, J.W. (1995). A computational approach to ductile crack growth under large scale yielding conditions. *J. Mech. Phys. Solids* **43**, 389–413.

THE EFFECT OF GEOMETRY AND SIZE ON THE GROWTH OF DUCTILE FRACTURE

B. COTTERELL
School of Mechanical and Production Engineering
Nanyang Technological University
Singapore 2263

K.Y. LAM
Department of Mechanical and Production Engineering
National University of Singapore
Singapore 0511

Z. CHEN and A.G. ATKINS
Department of Engineering
The University of Reading
Reading RG6 2A1

Abstract

Ductile crack propagation is modelled for deeply side-grooved double cantilever beam (DCB) and compact tension (CT) specimens using simple beam on elasto-plastic foundations. The side grooves dominate the constraint at the crack tip so that the deformation within the fracture process zone is one of uniaxial strain. Under these conditions there is little change in constraint with crack growth and the specific essential fracture energy is constant. In DCB specimens there is self similar crack propagation, but there is a rising J_R curve that is size dependant. In small CT specimens the crack growth is not self similar and the fracture work is not constant. Under these conditions even the initiation J is not necessarily equal to the specific essential energy.

1. Introduction

There are three problems in the development of non-linear elastic fracture for ductile metals. Firstly unloading necessarily occurs during crack propagation that

J. R. Willis (ed.), IUTAM Symposium on Nonlinear Analysis of Fracture, 105–114.

causes completely non-proportional deformation as shown experimentally (Hancock et al., 1993: Joyce and Link, 1995) and demonstrated theoretically (Xia and Shih 1995a,b). The J-integral is the rate of change in potential energy of the system that, when applied to an elasto-plastic specimen, includes the plastic work dissipated outside the FPZ as well as the fracture work dissipated within the FPZ. The J approach works for initiation because the work dissipated outside the FPZ in an elasto-plastic specimen cannot be distinguished from the energy stored in an equivalent non-linear elastic specimen and hence at initiation J is the specific fracture energy, R, at initiation. Even in the case of fracture initiation, J is only a material property if the crack grows in a self similar fashion. Barenblatt's (1962) hypotheses for the fracture work, R, being a material constant, R_0, given by

$$R_0 = \int_0^{\delta_f} \sigma \, d\delta \tag{1}$$

apply equally well to elasto-plastic fracture as they do to elastic fracture. If the size of the FPZ changes with crack growth, then $R \neq R_0$, and $J_R \neq R$ even at initiation.

The FPZ in ductile fracture can be identified with the strain-softening region where the voids are growing faster than can be compensated by strain hardening. The Gurson (1977) model, as modified by Tvergaard (1982), is accurate for the initial stages in strain softening but does not predict the final coalescence of voids by cavitation or shear localization. Without an accurate prediction for the final coalescence of the voids, which controls the specific fracture energy, R_0, it is difficult to model ductile fracture in conditions of varying constraint. Deep side-grooves have a dominant effect on the degree of constraint and it is not unreasonable to assume that under these conditions that the deformation within the FPZ is one of uniaxial strain despite the specimen geometry. With the double cantilever beam (DCB) and compact tension (CT) geometries, the deep side grooves have another benefit in that they allow an approximate analysis using the theory for beams on elasto-plastic foundation. Hence in this paper the effect of size on the fracture work R and the J_R-curve is examined. Such an analysis is not suggested for analysis of practical problems, but is used here simply to discuss the mechanics of ductile fracture.

2. Modelling ductile fracture in DCB and CT specimens with deep side grooves

Foote and Buckwald (1985) have shown that the Gross and Srawley (1966) expression for the elastic stress intensity factor, K, for a DCB specimen is accurate to within 0.3% for a specimen whose crack length to beam height ratio a/H is as small as 0.3. Rewriting the Gross and Srawley expression in terms of the energy release rate, G, gives approximately

$$G = \frac{12P^2a^2}{B^2H^3}\left[1 + 1.374\left(\frac{H}{a}\right)^2\right]. \tag{2}$$

The first term in Eq. 2 comes from the strain energy stored in bending due bending. The second comes from both the shear strain energy and rotation of the beam at the tip of the crack. Using the expression for the shear deflection of a beam (Gere and Timoshenko, 1991), the contributions from shear deformation is $0.542(H/a)^2$ (taking Poisson's ratio, ν=0.3), thus the major contribution comes from the rotation at the crack tip. If the beam has side grooves, the effect of rotation will be enhanced and therefore it is suggested that even CT specimens, provided that they have deep side grooves, can be analysed as beams on elasto-plastic foundations; the plastic deformation in the arms is modelled by the usual engineers' theory of bending. During propagation there is elastic unloading behind the crack tip which makes the load-point deflection for a crack that has propagated to a certain crack length, a, larger than the deflection for a beam with an initial crack length equal to a.

The constraint on plastic deformation assumed for the deep side notched specimens only allows plastic deformation because of the formation of voids. Hence the maximum stress in the FPZ occurs at plastic strains of the order of the yield strain. For a FPZ with an initial void volume fraction of 0.005, yield strain ϵ_0=0.002, and strain hardening exponent n=0.1, typical of a pressure vessel steel such as ASTM A533B, the maximum stress is of the order of four times the yield strength, σ_0 (Xia and Shih, 1995b). The stress falls only gently after the maximum stress is reached and it is assumed that in the FPZ that the stress is elastic up to the maximum stress, $C\sigma_0$, and is then constant until the FPZ is completely fractured (after a displacement, δ_f). Under these assumptions, the specific essential work for an infinitesimally thin FPZ is, $R_0 = C\sigma_0\delta_f$.

Two type of models have been employed to model the stress in the ligament.

2.1 SPECIMENS WITH LARGE LIGAMENTS

For specimens with large ligaments, either absolute or compared with the beam height, H, the specimen has been analyzed as a beam with a constant stress, $C\sigma_0$, in a FPZ at the tip of the crack outside of which the beam rests on an elastic foundation. The stiffness of the elastic foundation can be calculated from the B. assumption that the straining is uniaxial so that the stiffness, k, of the "foundation" is given by

$$k = \frac{(1-\nu)E}{(1+\nu)(1-2\nu)}\left(\frac{B_n}{H_n}\right), \tag{3}$$

where B_n is the width of the side grooved section, and H_n is the height of the grooves (see Fig.1a). The most important factor determining the behaviour of the specimens is their scale relative to the characteristic length, l_{ch}, of the material which is defined by

$$l_{ch} = \left(ER_0\right)/\sigma_0^2 . \tag{4}$$

A typical distribution of stress along the ligament of a DCB specimen with side grooves that reduce the specimen width to 25% is shown in Fig. 1a for a non-dimensional beam and groove height of H=0.1 and H_n=0.002 respectively (this height typically corresponds to 0.5mm for a ductile metal). The material is assumed to deform according to the true stress strain relationship

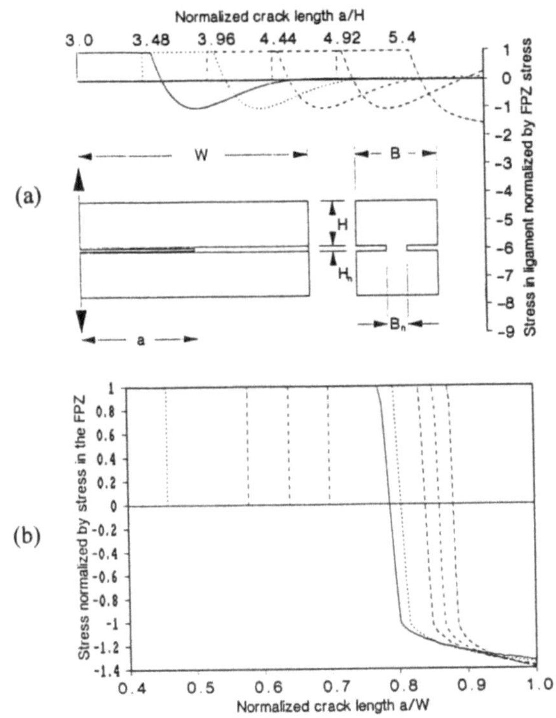

Figure 1. Stress distribution in the ligament of (a) a DCB specimen, $\bar{H}$=0.1, W/H=6, (b) a CT specimen, $\bar{H}$=0.1, W/H=1.667.

$$\sigma = \frac{\epsilon}{\epsilon_0} \qquad for\ \epsilon < \epsilon_0\,,$$

$$\sigma = \left(\frac{\epsilon}{\epsilon_0}\right)^n \qquad for\ \epsilon > \epsilon_0\,. \tag{5}$$

In Fig. 1, the strain hardening exponent, n, is assumed to be 0.1 and the strain at yield $\epsilon_0 = 0.002$. The arms of the specimen described in Fig. 1 (a), yield before the initiation of a ductile tear the moment at the crack tip being 1.72 times the yield moment at initiation rising to a maximum of 1.77 times before dropping to 0.846 when a/W = 0.9. For short crack lengths the stress distribution does not vary significantly, but as the crack approaches the back face of the specimen. The stress and displacement outside of the FPZ becomes linear and the compressive stresses become high and yielding must occur. The load-deflection curves for specimens with long ligaments are not significantly affected by the constraint factor C.

2.2 SPECIMENS WITH SMALL LIGAMENTS

For specimens that would yield in compression at the back face, it is assumed that the strain in the ligament outside of the FPZ is a linear function of its position. Yielding in compression is very sensitive to the degree of constraint, because uniaxial straining with no volume change is assumed. Therefore it has been assumed, rather arbitrarily, that in the compression region of the ligament that

$$\sigma = -C\sigma_0\left(\frac{\delta}{\delta_0}\right) \qquad for\ \delta < \delta_0,$$

$$\sigma = -C\sigma_0\left(\frac{\delta}{\delta_0}\right)^n \qquad for\ \delta > \delta_0, \tag{6}$$

where $\delta_0 = C\sigma_0 B_n/k$ is the elastic stretch in the FPZ. With this assumption there is no yielding in compression until the crack approaches the back face of the specimen when the strain distribution becomes linear. The load for a particular crack length can be found from consideration of the equilibrium of the specimen. The stress distribution along the ligament of a CT specimen (side-grooved to 25%) that yields along the back face is shown in Fig. 1b. Here strictly unloading in the zone that has yielded in compression should be considered, though unloading in the arms does significantly affect the J-integral, unloading in the compression zone is less important and has been neglected. Here the load is very obviously dominated by the constraint factor C. Since no experiments have been carried out on CT specimens with deep side grooves the validity of this modelling has not been proved.

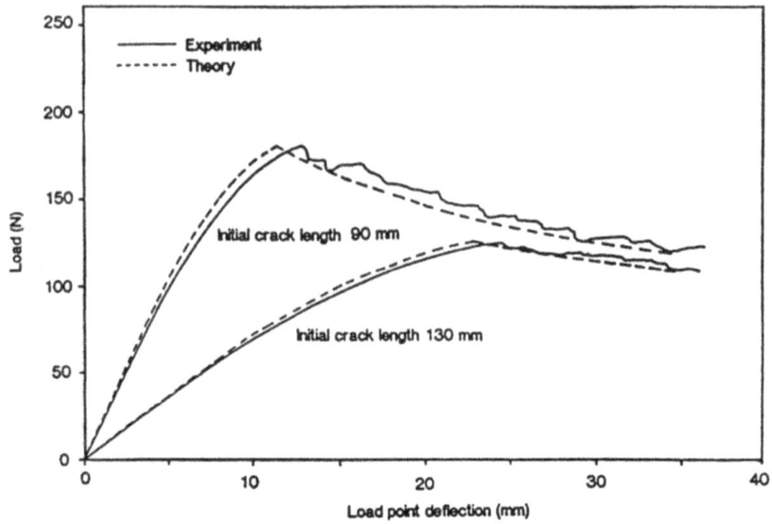

Figure 2. Load deflection curves for two DCB α-brass specimens ($\sigma_o = 230$ MPa, $\epsilon_o = 0.00217$, n = 0.33, $\nu = 0.33$, $B_n/B = 0.063$)

3. Comparison of experimental results with model

To date experiments have only been made on DCB specimens with very large ligaments. The load-deflection curves for two α-Brass specimens with initial crack lengths of 90 and 130 mm are shown in Fig. 2. The arms in both of these specimens yield before a ductile tear is initiated. The Young's modulus, and plastic behaviour of this material was found from independent tensile tests. The constraint factor C was assumed to be 4 and the critical crack opening displacement $\delta_f = 0.0815$ mm (equivalent to specific fracture energy $R_o = 75$ kJ/m^2) was found by determining the best fit to the results obtained for the shorter of the two crack lengths. The loads agree extremely well, whereas the deflections are slightly underestimated by the model. This under estimation may be because the effective height of the groove should be slightly greater than the height of the machined groove.

4. Fracture work and the J_R-curve

Crack propagation in the DCB specimen is almost perfectly self-similar with the FPZ translating ahead of the crack tip with only insignificant changes in its shape caused by a change in the ratio of the bending moment to shear force at the crack tip. However, in the CT geometry or when the crack in a DCB specimen approaches the back face, there can be significant changes in the size of the FPZ. When the crack propagation is steady-state, the fracture work, R (defined as the work performed within the FPZ to increase the crack by a unit area) is identically equal to the specific essential work, R_o, and the value of the J-integral at initiation, J_{Ic}, is identical to R_o. Under such conditions the fracture work, $R=R_o$, is constant during crack propagation. However, if the crack propagation is not steady-state, the work of fracture $R \neq R_o$. If the concept of a finite width FPZ is employed, the work already performed on the pupative FPZ as it enters the FPZ must be considered as well as the work performed within it. During a crack extension of da, the FPZ propagates by $(da+da_p)$, where a_p is the length of the FPZ. The work already performed per unit length on this new portion of the FPZ is $0.5C\sigma_o\delta_o$, where $\delta_o=C\sigma_oB_n/k$ is the elastic stretch in the FPZ. Thus the work of fracture is given by

$$\begin{aligned} R &= \int_a^{a+a_p} C\sigma_0 \frac{\partial}{\partial a}(\delta - \delta_0)\, dx + \frac{C\sigma_0\delta_0}{2}\left(1 + \frac{da_p}{da}\right) \\ &= R_0' + C\sigma_0 \frac{\partial}{\partial a}\left\{\int_a^{a+a_p} (\delta - \delta_0)\, dx + \frac{\delta_0 a_p}{2}\right\} \end{aligned} \tag{7}$$

where $R_o'=C\sigma_o(\delta_f - \delta_o/2)$. This definition of the essential work of fracture, for a finite width FPZ, is slightly different to R_o to insure that during steady-state crack growth $J_{Ic}=R_o'$. When the FPZ approaches the back face of the specimen and enters material that has already been strained in compression the work already performed in the pupative FPZ should be included, but such large crack growths are probably not important. The plastic component of the J-integral for deep side grooved CT specimens can be calculated from the η-factor given by Merkle and Corten (1974) using the correction for crack growth given by Ernst et al. (1981). The plastic component of the J-integral for the DCB specimen can also be analyzed using an η_p-factor = 1.08 based on the crack length, a, rather than the ligament so that

$$J_p = \frac{\eta_p}{B_n a}\int_0^{\Delta_p} P d\Delta_p \tag{8}$$

where Δ_p is the plastic load-line deflection (Cotterell et al 1995). The elastic component, J_e, can be determined from the elastic compliance of the specimens for both CT and DCB specimens.

The FPZ size, the work of fracture, R, and J_R are shown for two DCB specimens (with side grooves that reduce the thickness to 25%) of non-dimensional height ($\overline{H} = H/l_{ch}$) of 0.1 and 0.2 are shown in Fig. 3. It can be seen that the apparent crack growth resistance, implied by the J_R-curves, increases with crack extension and depends upon the size of the specimen whereas the essential fracture work, R, remains constant. As J_R reaches a maximum the model of section 2.1 ceases to be appropriate.

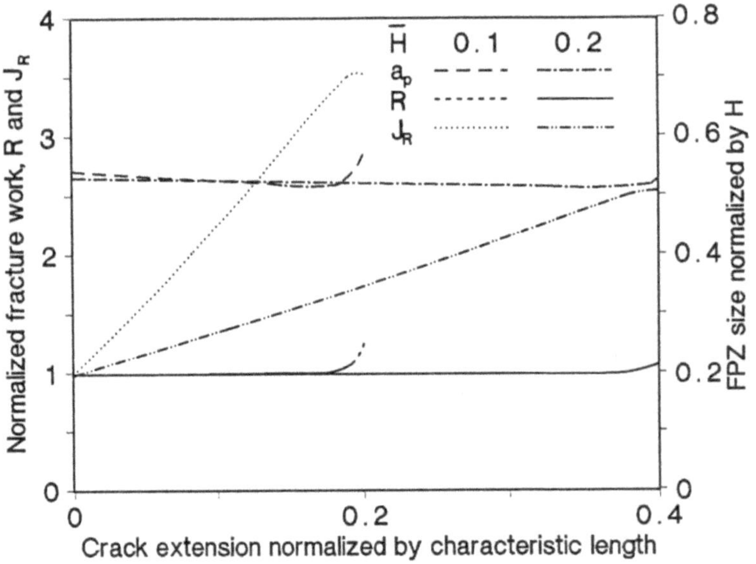

Figure 3 Fracture work in DCB specimems, W/H=6, $\bar{H}$=0.1, 0.2

Unloading in the arms has a large effect on the deflection and J_R as can be seen from Fig. 4 where the load deflection curve obtained during crack propagation is compared with the locus of the initiation load-deflection for different crack lengths

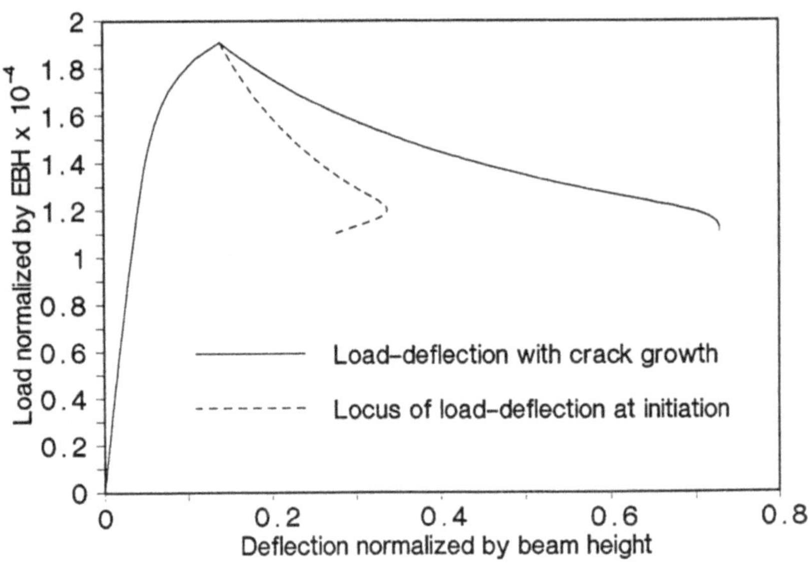

Figure 4. Load-deflection curve for DCB specimen ($\bar{H}$=0.1, W/H=6, B_n/B=0.25, a_o/W=0.5)

The results for the CT geometry are shown in Fig. 5 for specimens whose non-dimensional heights $\bar{H}$ are 0.1 and 1. Here the larger specimen has been analyzed using section 2.1, but the smaller specimen has to be analyzed according to section 2.2. The results for $\bar{H}$=1.0 are very similar to those for the DCB specimen, but for $\bar{H}$=0.1 the size of the FPZ decreases significantly with crack growth causing the essential fracture work, R, to be less than R_o and J_i to be greater than R_o. In the smaller specimen J_R only decreases slightly with crack growth because the stress distribution is very similar to that assumed by Merkle and Corten (1974) in their estimation of the η-factor..

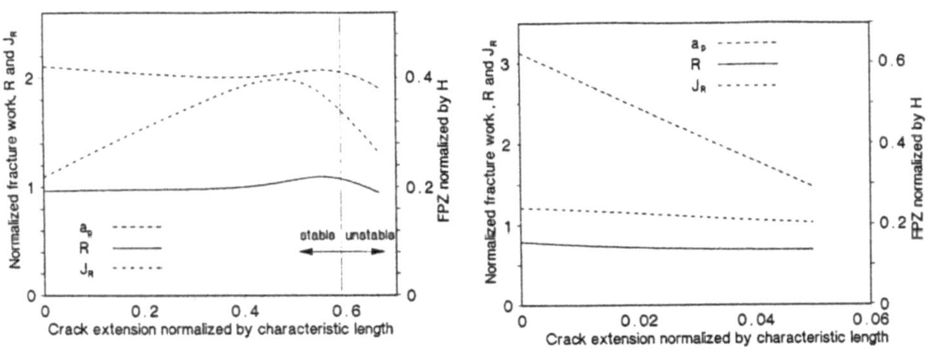

(a) (b)

Figure 5. Fracture work in CT specimens B_n/B=.25, (a) H=1.0, a_o/W=0.3; (b) H=0.1, a_o/W=0.4.

5. Conclusions

The ductile fracture of DCB and CT specimens, with deep side grooves, can be modelled approximately by the use of the engineers' theory of bending treating the side-grooved section as a foundation to the beam. This model enables the relationship between the essential work of fracture within the FPZ and the J_R-curve to be examined. It has been assumed that the deep-side-grooves dominate the constraint at the crack tip so that the FPZ deforms uniaxially. With this assumption, the stress-displacement relationship within the FPZ is unique and hence the specific essential energy R_o is a constant. Except at initiation most of the crack growth resistance implied by the J_R-curve includes large amounts of work performed outside of the FPZ which depend upon the size and geometry of the specimen. This conclusion is not new, but here the essential fracture work performed within the FPZ has been clearly identified. In fracture mechanics we should be seeking material constants, the J_R-curve is not a material constant. While the work of fracture in the FPZ depends upon constraint, it is a better candidate for a fracture parameter than the J integral. However, in very small specimens the work of fracture in the FPZ need not be identical to the specific essential work R_o. The most fundamental parameter is the stress-displacement relationship within the FPZ. However, at the moment there is no simple model that gives accurately the complete stress-displacement curve.

Acknowledgements

Support for this research from the European Economic Community under Contract CI1*-CT93-0327 is gratefully acknowledged.

References

Barenblatt, G.I. (1962) The mathematical theory of equilibrium cracks in brittle fracture, *Adv. Appl. Mech.* 7, 55-129.

Cotterell, B., Chen, Z. and Atkins, A.G. (1995) An η-factor for calculation of the J-integral in the double cantilever beam specimen, to be published.

Ernst, H.A., Paris, P.C. and Landes, J.D. (1981) Estimations on J-integral and tearing modulus T from a single specimen test record, in R. Roberts (ed.), *Fracture Mechanics - 13th Symposium*, ASTM STP 743, American Society for Testing and Materials, Philadelphia.

Foote, R.M.L. and Buckwald, V.T. (1985) An exact solution for the stress intensity factor for a double cantilever beam, *Int. J. Fract.* **29**, 125-134.

Gere, J.M. and Timoshenko, S.T. (1991) *Mechanics of Materials,* 3rd SI edition, Chapman and Hall, London.

Gross, B. and Swraley, J.E. (1966) Stress Intensity Factors by Boundary Collocation for Single-Edge Notch Specimens Subjected to Splitting Forces, NASA TN D-3295.

Gurson, A.L. (1977) Continuum theory of ductile rupture by void nucleation and growth: Part 1 - Yield criteria and flow rules for porous ductile media, *J. Engng. Mater. Tech.* **99**, 2-15.

Hancock, J.W., Reuter, W.G. and Parks, D.M. (1993) Constraint and toughness parameterized by T, in E.M. Hackett, K.H. Schwalbe and R.H. Dobbs (eds.), *Constraint Effects in Fracture*, ASTM STP 1171, American Society for Testing and Materials, Philadelphia.

Joyce, J. A. and Link, R.E. (1995) Effects of constraint on upper shelf fracture toughness, in W.G. Reuter, J.H. Underwood and J.C. Newman (eds.), *Fracture Mechanics - 263h Symposium*, ASTM STP 1256, American Society for Testing and Materials, Piladelphia (in press).

Merkle, J.G. and Corten, H.T. (1974) A J integral analysis for the compact specimen, considering axial forces as well as bending effects, *J. Pres. Ves. Tech.* **96**, 286-292.

Tvergaard, V. (1981) Influence of voids on shear band instabilities under plane strain conditions, *Int. J. Fract. Mech.* **17**, 389-406.

Xia, L. and Shih, C.F. (1995a) Ductile crack growth - I. A numerical study using computational cells with microstructurally-based length scales, *J. Mech. Phys. Solids* **43**, 233-259.

Xia, L. and Shih, C.F. (1995b) Ductile crack growth - I. Void nucleation and geometry effects on macroscopic fracture behaviour, *J. Mech. Phys. Solids* **43**, 1953-1981.

INTERFACIAL CRACKING IN THERMOMECHANICALLY LOADED ELASTOPLASTIC BIMATERIALS

K.P. HERRMANN AND T. HAUCK
Laboratorium für Technische Mechanik
University of Paderborn
Pohlweg 47-49
D-33098 Paderborn, Germany

1. Introduction

The failure analysis of thermally loaded two-phase compounds, in material science usually denoted as bimaterials, requires the determination of stress states due to applied nonstationary temperature fields as well as the consideration of the mismatch of the mechanical and thermal material constants. The structural performance of material compounds is essentially affected by existing defects of various kinds. Regarding the formation of yielding zones in ductile materials, which particularly arise in the vicinity of defects, the utilization of elastic-plastic constitutive equations is necessary. Various publications address the problem of mechanically strained interface cracks in elastic-plastic bimaterials. For the case of small scale yielding (SSY), HRR-like stress field structures have been found [1, 2]. Proceeding contributions based on the deformation theory have provided asymptotic stress fields, where the leading term of those asymptotic stress fields is parameterized by the J-integral [3, 4]. From the mechanical point of view, an energy balance in the crack tip area identifies the J-integral as a crack driving force [5, 6]. In this paper, the quantitative characterization of different self-stress states in the vicinity of an interfacial crack tip is performed by using the J-integral, where the influence of temperature gradients close to the interface crack tip is of most interest.

J. R. Willis (ed.), IUTAM Symposium on Nonlinear Analysis of Fracture, 115–124.

2. Basic Equations

The calculation of stress states in thermally loaded bimaterials is based on a so-called sequentially coupled solution of the heat transfer and thermal stress initial boundary value problem. Firstly, the heat transfer problem is mathematically described by Fourier´s law as follows

$$\nabla\left[\lambda \nabla T(x_i,t)\right] = Q(x_i) + \rho c_p \frac{\partial T(x_i,t)}{\partial t} \tag{1}$$

where $Q(x_i)$ denotes the given distribution of heat sources, λ the coefficient of thermal conductivity, ρ the material density and c_p the specific heat coefficient. The transient response analysis of the heat transfer problem yields the inhomogeneous, time-dependent temperature field $T(x_i, t)$. Secondly, the analysis of the thermally induced self-stress state in the bimaterial is performed based on the incremental theory of plasticity by using the subsequent constitutive material equation

$$d\varepsilon_{ij} = \frac{1+\nu}{E} ds_{ij} + \frac{1-2\nu}{3E}\delta_{ij} d\sigma_{kk} + \delta_{ij}\alpha dT; \quad (\sigma_e \le \sigma_0) \tag{2}$$

$$d\varepsilon_{ij} = \frac{1+\nu}{E} ds_{ij} + \frac{1-2\nu}{3E}\delta_{ij} d\sigma_{kk} + \frac{3}{2}\frac{d\sigma_e}{E_p \sigma_e} s_{ij} + \delta_{ij}\alpha dT; \quad (\sigma_e \ge \sigma_0) \tag{3}$$

where $d\varepsilon_{ij}$ defines the incremental strain tensor, s_{ij} the deviatoric part of the stress tensor, σ_{kk} the trace of the stress tensor, σ_e the effective stress, σ_0 the yield stress, E Young´s modulus, ν Poisson´s ratio, E_p the plastic tangential modulus, α the thermal expansion coefficient and dT the incremental temperature change.

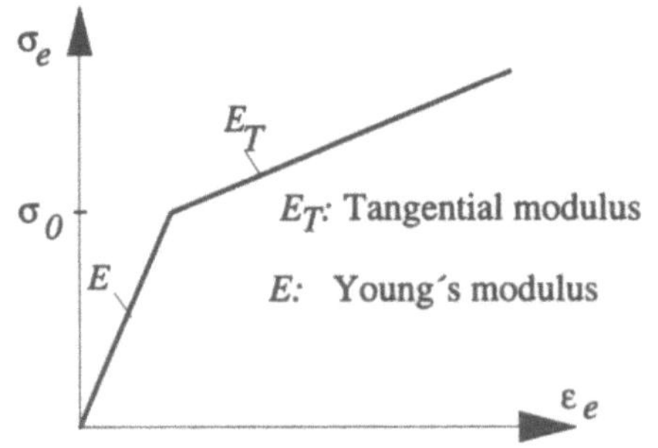

Figure 1. Bilinear stress-strain relation

For this material law, von Mises´s yield condition, an associated flow rule and isotropic hardening is assumed. Figure 1 shows the applied bilinear relation between the effective stress σ_e and the effective strain ε_e.

The solution of the initial boundary value problem is performed by utilizing the finite element software system SOLVIA, where most of the included routines needed for the present problem refer to a publication by Snyder and Bathe [7].

3. *J*-integral

3.1 DEFINITION

The quantitative characterization of different self-stress states is performed by the *J*-integral, where the *J*-integral values are obtained from corresponding finite element calculations of the associated stress and deformation fields.

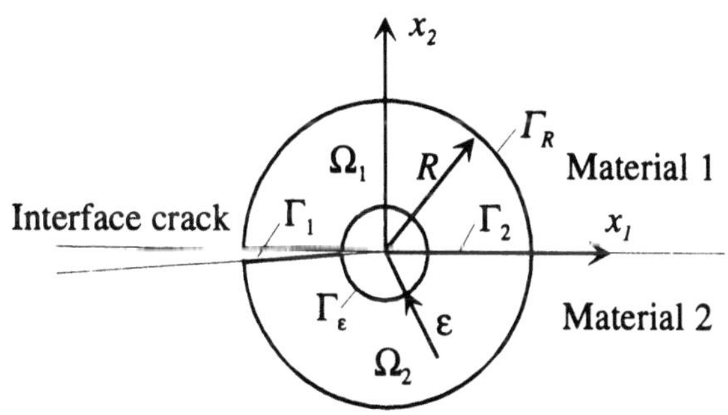

Figure 2. Contour- and domain integration

The basic approach for the vectorial J_k -integral to be introduced in order to describe a virtual crack extension in the x_k-direction reads

$$J_k = \lim_{\varepsilon \to 0} \int_{\Gamma_\varepsilon} b_{kj}\, n_j \, d\Gamma \tag{4}$$

where b_{kj} is Eshelby´s energy momentum tensor and n_j is the unit vector perpendicular to the line element $d\Gamma$ of the contour Γ_ε. According to the references [5, 6], Eshelby´s energy momentum tensor b_{kj} is given by

$$b_{kj} = W\delta_{kj} - \sigma_{ij} u_{i,k} \,; \quad W = \int_{\varepsilon_{ij}^m} \sigma_{ij} d\varepsilon_{ij}^m \,; \quad d\varepsilon_{ij}^m = d\left(\varepsilon_{ij} - \alpha T \delta_{ij}\right) \tag{5}$$

where W denotes the strain energy density and u_i the displacement vector. The projection of the J_k-vector on the direction of the virtual crack extension m_k yields the magnitude of Rice´s *J*-integral, namely

$$J = m_k J_k \,. \tag{6}$$

3.2. EQUIVALENT DOMAIN INTEGRAL REPRESENTATION

The evaluation of the J-integral is performed by applying the equivalent domain integral method (EDI) [8]. By the help of the weight function $s(x_i)$

$$s = \frac{R - |x_i|}{R} \tag{7}$$

and invoking Gauss's divergence theorem it is possible to convert the contour integral (4) into the following domain- and contour integral [8]:

$$J_k = - \int_{\Omega_1 \cup \Omega_2} \left(s b_{kj}\right)_{,j} d\Omega + \int_{\Gamma_1 \cup \Gamma_2} s \Delta b_{kj} n_j d\Gamma \tag{8}$$

By restricting to monotonic applied loads and by neglecting volume forces the final expression

$$J_k = - \int_{\Omega_1 \cup \Omega_2} \left[W s_{,k} - \sigma_{ij} u_{i,k} s_{,j} - \sigma_{mm} \alpha T_{,k} s\right] d\Omega + \int_{\Gamma_1 \cup \Gamma_2} s \Delta b_{kj} n_j d\Gamma \tag{9}$$

for the the J_k-integral is gained.

The remaining contour integral in formula (9), in which, as should be reminded, the energy momentum tensor is discontinuous along the crack Γ_1 and the interface Γ_2, vanishes, if the crack propagates along the ligament Γ_2 as well as if no loads are subjected to the crack surfaces Γ_1 [9]. The calculation of the domain integral is carried out in the curvilinear coordinates of the applied isoparametric finite elements and by using the Gauss-Legendre quadrature formula.

4. Application

4.1. THERMOMECHANICALLY LOADED INTERFACE CRACK

The bimaterial sheet damaged by an interface crack and depicted in Fig. 3 is considered. In addition, a heat source Q of constant power supply $Q_0 \left[J/s\,m^3\right]$ after initiation at time $t = 0$ and located in a small circular domain at the given position (r_Q, ϑ_Q) in one of the respective bimaterial components is assumed to act.

For completeness of the thermal boundary value problem, at the external boundary a convective heat-transfer is defined, according to

$$n_j T_{,j} = \frac{\beta}{\lambda}[T_u - T]; \quad (x_i \in \Gamma) \tag{10}$$

where T_u denotes the environmental temperature and β is the heat convection coefficient. On the crack surfaces the so-called natural boundary condition

$$n_j T,_j = 0; \quad (x_i \in \Gamma_1) \tag{11}$$

holds true; that means the crack surfaces behave like thermal insulators. At initial time $t = 0$ a constant temperature $T(x_i, t = 0) = T_0$ is prescribed.

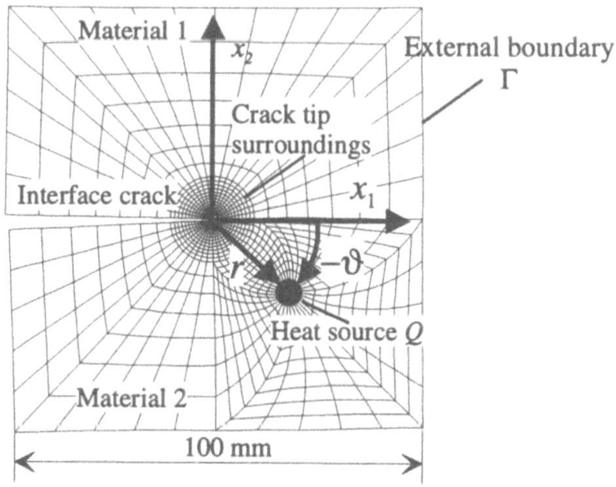

Figure 3. Discretized bimaterial structure

The external boundary of the bimaterial and the crack surfaces are assumed to remain stress-free, that means

$$\sigma_{ij} n_j = 0; \quad (x_i \in \Gamma \cup \Gamma_1). \tag{12}$$

Further at the interface the continuity conditions for the traction vector and the displacement vector, respectively

$$\sigma_{ij}^{I} n_j = \sigma_{ij}^{II} n_j; \quad u_i^{I} = u_i^{II}; \quad (x_i \in \Gamma_2), \tag{13}$$

has to be fulfilled. Moreover, plane stress state calculations have been carried out for zero initial stresses. Table 1 assembles the used material constants for the selected bimaterial.

TABLE 1. Material constants

Material	λ [W/Km]	c_P [J/kgK]	ρ [kg/m^3]	β [W/Km^2]	E [MPa]	ν [1]	α [1/K]	σ_0 [MPa]	E_T [MPa]
1	113	376	8440	200	9000	0.35	$2.1 \cdot 10^{-5}$	50	300
2	15	502	7800	200	210000	0.29	$1.19 \cdot 10^{-5}$	100	7000

4.2. RESULTS

For the thermal stress analysis under investigation, the heat source Q is assumed to be situated at various positions of a circle centered in the interface crack tip and with the coordinates ($r_Q = 25$ mm, $-90° \leq \vartheta_Q \leq 90°$). The fan $90° \leq \vartheta_Q \leq 270°$ was excluded from the analysis due to a possible interpenetration of the crack surfaces. Figure 4 shows the almost stationary temperature distribution in the bimaterial for the chosen heat source positions ($r_Q = 25$ mm, $\vartheta_Q = \pm 45°$) at the time $t = 600$ s. For the stationary case, the thermal energy supplied by the source is removed from the bimaterial by the convective heat flux along the external boundary. As thermally isolated crack surfaces are prescribed, the heat flux within the bimaterial is being redirected by the crack and accounts for the heat flux vector square root singularity at the interface crack tip [10].

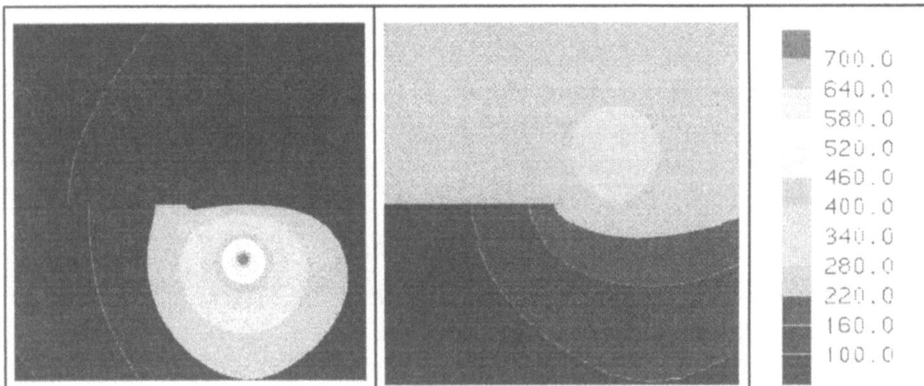

$\vartheta_Q = -45°$, $r_Q = 25$ mm | $\vartheta_Q = +45°$, $r_Q = 25$ mm | T / °C

Figure 4. Bimaterial-$T(x_i, t = 600$ s) distribution (lateral length = 100 mm)

The most important features of the analysis of the interface crack are the distributions of the temperature and of the stress tensor components, respectively, nearby the crack tip. The magnitudes and the directions of the temperature gradients strongly depend on the ϑ_Q-coordinate of the heat source location. The temperature as well as the temperature gradients take maximum values for $\vartheta_Q < 0°$ due to the smaller heat conduction coefficient λ_2 of the material 2, cf. Fig 5.

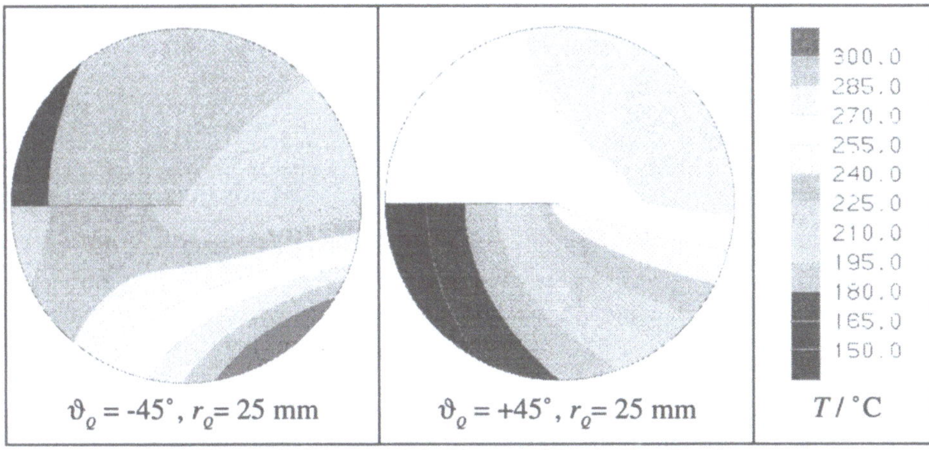

Figure 5. Bimaterial-$T(x_i, t = 600$ s) distribution at the crack tip surroundings (radius = 10 mm)

Figures 6-8 depict the distribution of the stress tensor components related to the previously presented temperature distribution. It is concluded from the structure of the stress field contour plots that the direction of the temperature gradients plays a dominant role for the stress distribution. The near-tip singular stress field is superimposed on nonsingular stresses due to the thermal mismatch along the interface, which is especially visible for the case $\vartheta_Q < 0°$. It is noteworthy that in this part of the bimaterial the stress fields are most influenced by the position of the heat source, because the source produces maximum values of the temperature gradients for $\vartheta_Q < 0°$.

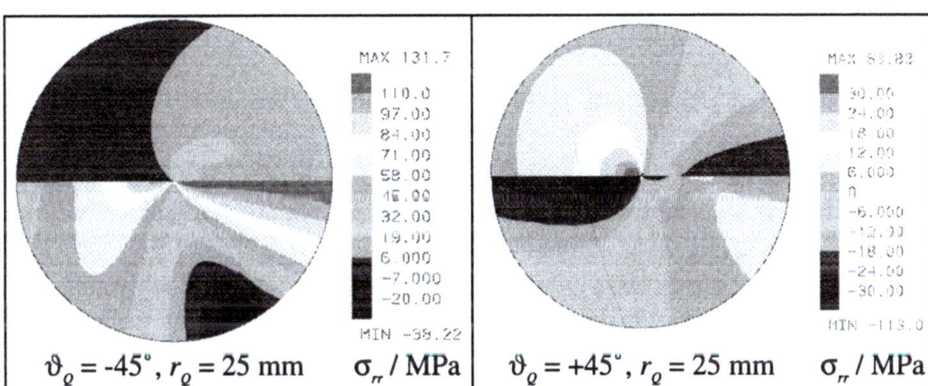

Figure 6. Bimaterial-$\sigma_{rr}(x_i, t = 600$ s) distribution at the crack tip surroundings (radius = 10 mm)

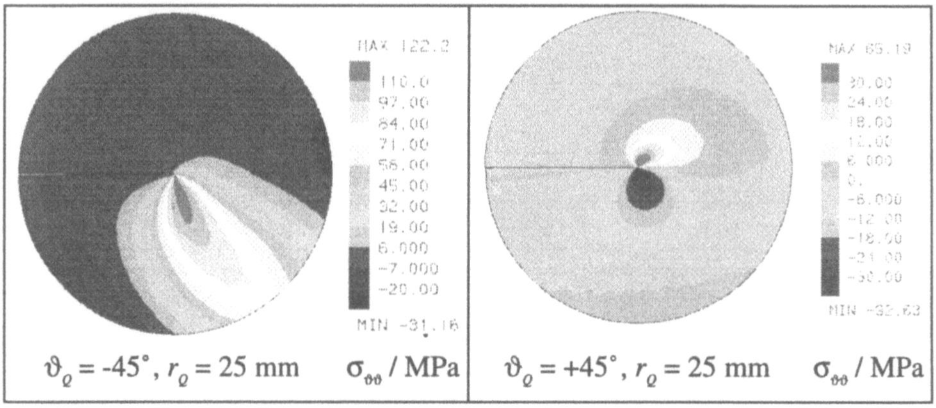

Figure 7. Bimaterial-$\sigma_{\vartheta\vartheta}(x_i, t = 600$ s) distribution at the crack tip surroundings (radius = 10 mm)

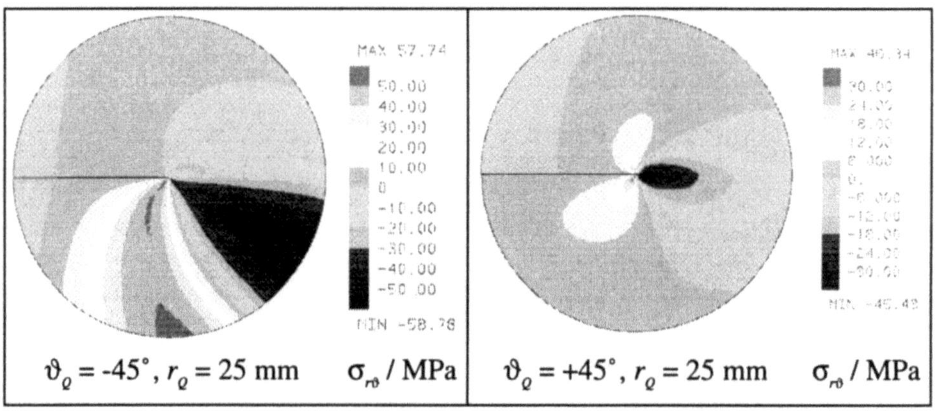

Figure 8. Bimaterial-$\sigma_{r\vartheta}(x_i, t = 600$ s) distribution at the crack tip surroundings (radius = 10 mm)

The worst-case loading of the bimaterial can only be determined by calculating fracture mechanical parameters. In this contribution, the assessment of the self-stress states in the cracked bimaterial is realized by calculating Rice's J-integral. A special postprocessor adopting the presented theory of section 3 was developed in the Laboratorium für Technische Mechanik. The numerical integration was conducted for various circular domains centered at the crack tip ($2 \text{ mm} \leq R \leq 10$ mm) by using the results of the finite element calculations. Figure 9 shows the J-Integral in dependence on the radius R of the integration domain for the heat source location ($r_Q = 25$ mm, $\vartheta_Q = -30°$).

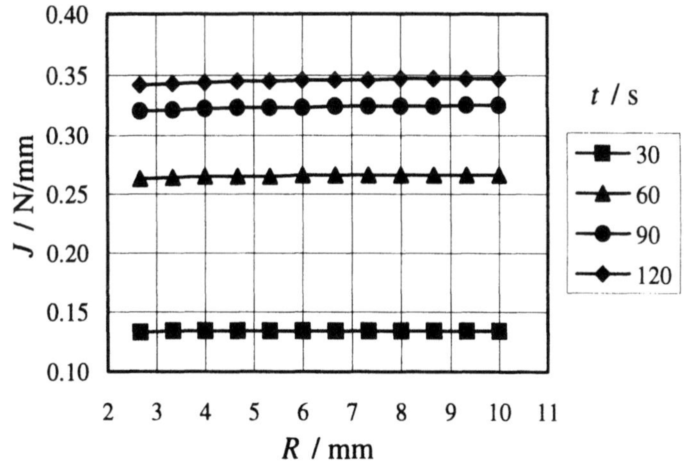

Figure 9. $J = J(t, \vartheta_Q = -30^\circ, R)$

As can be recognized from the diagram given above the values of the J-integral do not depend on the chosen domain of integration. Therefore, the J-integral is appropriate for a further analysis.

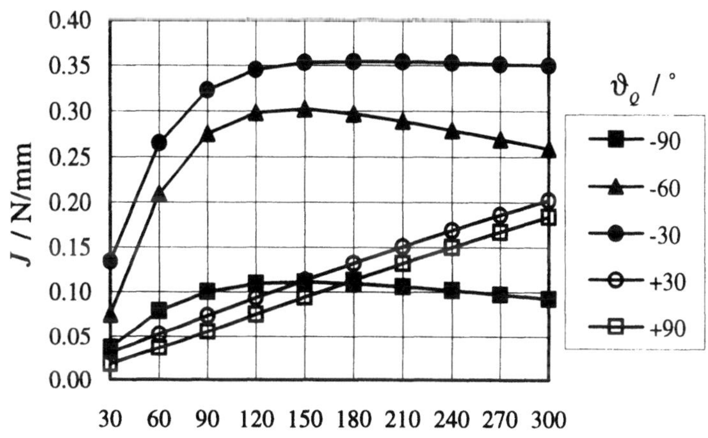

Figure 10. $J - J(t, \vartheta_Q)$

Figure 10 depicts the transient behaviour of the J-integral versus varying heat source positions ϑ_Q. For the cases of heat source positions $\vartheta_Q < 0^\circ$, the maximum J-values arise before the stationary state is reached. On the other hand, for the cases $\vartheta_Q > 0^\circ$, the J-integral depends on time in an approximately linear manner and, in addition, almost does not change for heat source coordinate variations $0^\circ \leq \vartheta_Q \leq 90^\circ$. Assessing the loading cases under investigation, the worst-case heat source position $\vartheta_Q = -30^\circ$ can be identified.

5. Conclusion

In this contribution the investigation of a thermally self-stressed elastoplastic bimaterial damaged by interface cracks has been performed. The calculation of the corresponding self-stress states is based on the so-called sequentially coupled solution of the heat transfer and thermal stress initial boundary value problem. The thermal interface crack problem was solved by applying the finite element method. For this investigation, the influence of temperature gradients or heat fluxes, respectively, on the self-stress state was of most interest. It was recognized that the self-stress state at the interface crack tip is influenced by the position of the heat source and, in addition, apart from the thermomechanical material constants, by the heat conduction properties of the bimaterial components. The usage of the J-integral renders possibly the assessment of thermally induced self-stress states at interface crack tips in elastoplastic bimaterials, from which worst case conclusions can be obtained.

Acknowledgment- The support of the German Research Foundation (DFG) is gratefully acknowledged by the authors.

6. Literature

[1] Shih, C.F., Asaro, R.J. (1988) Elastic-plastic analysis of cracks on bimaterial interfaces, Part I: Small-scale yielding, *J. App. Mech.* **55**, 299-316.

[2] Shih, C.F., Asaro, R.J. (1988) Elastic-plastic analysis of cracks on bimaterial interfaces, Part II: Structure of small-scale yielding fields, *Journal of Applied Mechanics* **56**, 763-779.

[3] Zywicz, E., Parks, D.M. (1992) Small-scale yielding interfacial crack-tip fields. *J. Mech. Phys. Solids* **40**, 511-536.

[4] Sharma, S. M., Aravas, N. (1993) On the development of variable-separable asymptotic elastoplastic solutions for interfacial cracks, *Int. J. Solids Structures* **30**, 695-723.

[5] Eshelby, J. D. (1951) The force on an elastic singularity, *Philosophical Magazin* **33**, 87-112.

[6] Buggisch, H., Gross, D.; Krüger, K., H. (1981) Einige Erhaltungssätze der Kontinuumsmechanik vom J-Integral-Typ, *Ingenieur-Archiv* **50**, 103-111.

[7] Snyder, D. M., Bathe, K. J. (1981) A solution procedure for thermo-elastic-plastic and creep problems, *Nuclear Engineering and Design* **64**, 49-80.

[8] Nikishkov, G.P., Atluri, S. N. (1987) An equivalent domain integral method for computing crack tip integral parameters in non-elastic, thermo- mechanical fracture, *Engng. Fract. Mech.* **26**, 851-867.

[9] Herrmann, K. P., Hauck, T. (1995) Thermisch beanspruchte Grenzflächenrisse in elastoplastischen Medien, *27. Vortragsveranstaltung des DVM-Arbeitskreises Bruchvorgänge,* Köln-Porz, 14.-15. Februar 1995, DVM "Bruchmechanik von Verbundwerkstoffen und Stoffverbunden ", S. 343-351

[10] Dreilich, L. (1985) *Beiträge zum thermo-elastischen und krummlinigen Rißproblem*, VDI Fortschrittsberichte, Reihe 18, **21**,VDI Verlag, Düsseldorf.

CONDITIONS OF CRACK ARREST BY INTERFACES

JEAN LEMAITRE
Université Paris 6
L.M.T. 61, avenue du Président Wilson
94235 Cachan - FRANCE

Abstract : When a crack in a layer reaches an interface with another layer, it may be arrested or at least retarded by the phenomenon of debonding of the interface and reinitiation of a new crack in the second layer. A Shear Lag analysis associated to Damage Mechanics gives closed form solutions for the crack arrest conditions or the number of cycles of fatigue to crack initiation which determines the retardation, as functions of the design parameters of multimaterials.

1. Introduction

The dream of any good designer of structures is to avoid cracks or to stop their growth if they may be present or at least to reduce the crack growth rate as much as possible.

An old drastic repair of a cracked structure consists in drilling a small hole at the crack tip to remove the process zone and to oblige the crack to undergo a period of reinitation before it can continue to grow. In aeronautics, for example, riveted stiffeners on panels give a much bigger retardation effect on fatigue crack growth than built-in stiffeners because the crack in the riveted panel must reinitiate a new crack in the stiffener to continue to grow, which is not the case in the built-in stiffener where the crack continues to grow with a small retardation only due to the bigger thickness (Lemaitre 1974).

In multimaterials, the interface may play the same role provided it is not too strong so that the crack cannot pass through instantaneously and not too weak to avoid a complete debonding of the interface. Stress solutions for cracks terminating at interfaces have been obtained in the prior work of Williams (Zak and Williams 1963), then by Erdogan (Erdogan and Biricikoglu 1973) and more recently by Ballarini and Huo (1991). The conditions of bifurcation of cracks in interfaces of multilayered materials have been much studied (He and Hutchinson 1989) (Jensen, Hutchinson and Kyung-Suk Kim, 1990) but not their reinitiation in the following layer probably because the location and the state of stress governing the crack reinitiation are difficult to analyze (Suo and Hutchinson 1990). In fact, a simple Shear Lag analysis associated to Damage Mechanics gives enough information to obtain a closed form

J. R. Willis (ed.), IUTAM Symposium on Nonlinear Analysis of Fracture, 125–133.

solution of the loading or the number of cycles of fatigue which induces a crack reinitiation in the second layer of a two layered material having a crack loaded in mode I in the first layer. From these approximate solutions it is possible to deduce qualitative results regarding the influence of design parameters such as the geometry, the loading, the material properties and the quality of the interface on the crack reinitiation. It may help to optimize multimaterials against fracture.

2. Mechanism

Consider the two layered specimen of figure 1 where the layer A has a crack perpendicular to the interface I and is loaded in mode I by a remote stress σ_∞. Several sets of experiments have shown that when σ_∞ increases monotonically or when it varies periodically as in a fatigue process, there is first a debonding of the interface over a length of the order of magnitude of the thickness of the layer A and then a reinitiation of the crack in the second layer B may occur or not depending upon the geometry and the material properties of the layers and the interface (Vidonne 1995).

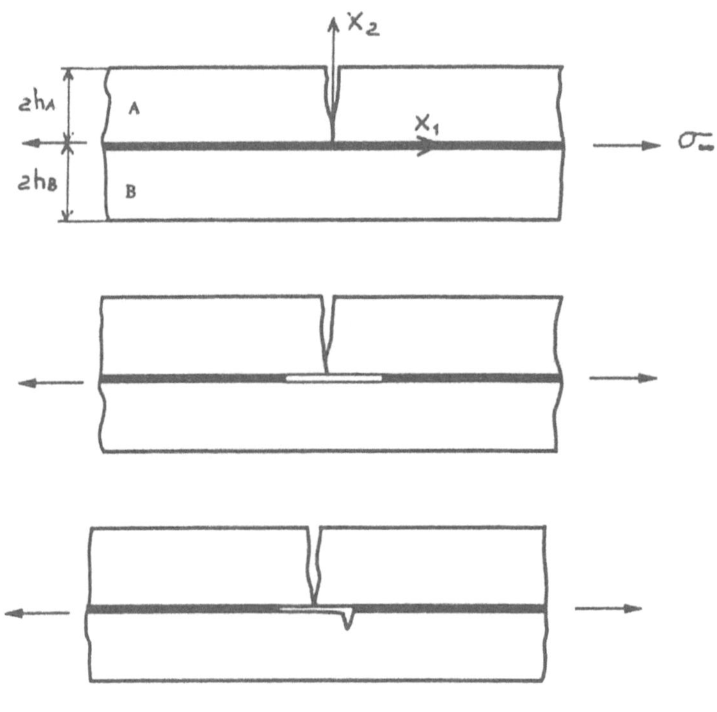

Fig. 1 Reinitiation of a crack in a two layered specimen

The overload or the number of cycles of fatigue necessary to reinitiate the crack act as retardation effects in the failure process of the multimaterial and if no reinitiation occurs the interface acts as a crack arrestor. Let us see how the design parameters govern this phenomenon.

3. Analysis of the Debonding of the Interface

Suppose that the crack in the layer A has reached the interface. We neglect any bending effect and consider that the two layers are loaded in pure tension in order to apply the pure Shear Lag analysis to find the state of stress in the interface (Volkersen 1938). Using fracture mechanics to define the debonding length as in (Hutchinson and Suo 1992) may also be used but it does not help to define the conditions of crack reinitiation in the layer B.

The two layers are elastic with Young's moduli E_A and E_B, the interface considered has zero thickness, has a shear modulus G_I and a debond shear stress τ_C. The equilibrium equations of the multimaterial and of each layer give rise to the following differential equation for the shear stress in the interface σ_{12}^I (Lemaitre 1992).

$$\frac{d^2\sigma_{12}^I}{dx_1^2} - \lambda^2\sigma_{12}^I = 0$$

with $\lambda = \left(\frac{G_I E_\infty}{2h_A h_B E_A E_B}\right)^{½}$, $E_\infty = \frac{E_A h_A + E_B h_B}{h_A + h_B}$ and $\frac{d\sigma_{11}^B}{dx_1} = -\frac{\sigma_{12}^I}{2h_B}$.

For the boundary conditions considered :

$$x_1 = 0 \quad \rightarrow \sigma_{11}^B = \sigma_\infty \frac{h_A + h_B}{h_B}$$

$$x_1 \rightarrow \quad \infty \rightarrow \sigma_{11}^D = \sigma_\infty .$$

The solution is : $\sigma_{12}^I = 2h_A \lambda \sigma_\infty \exp(-\lambda x_1)$.

An approximation for the debonding zone is the length l_D over which the calculated stress σ_{12}^I is larger than the debond critical value τ_C.

Writing $\sigma_{12(x_1=l_D)}^I = \tau_C$ gives $\frac{l_D}{2h_A} = \frac{1}{\mu_1\sqrt{2}} \text{Ln}\left(\sqrt{2}\frac{\sigma_\infty}{\tau_C}\mu_1\right)$

with $\mu_1 = \left(\frac{h_A}{h_B}\frac{G_i E_\infty}{E_A E_B}\right)^{½}$.

Several qualitative conclusions may be drawn from this equation :

- There is no debonding if :

$$\frac{\sqrt{2}\sigma_\infty}{\tau_C}\mu_1 \leq 1 \quad \text{or} \quad 2\frac{\sigma_\infty^2}{\tau_C^2}\frac{h_A}{h_B}\frac{G_I}{E_A E_B}\frac{E_A ha + E_B h_B}{h_A + h_B} \leq 1.$$

This is a condition for crack arrest but with weak accuracy due to the approximation of the Shear Lag theory which gives a finite stress at $x_1 = 0$.

- For classical values of engineering parameters such as $\sigma_\infty/\tau_C, h_A/h_B, E_A/E_B, G_I/E_A$, the debonding length is of the order of magnitude of a few thicknesses of the cracked layer.

$$0 < \frac{l_D}{2h_A} \leq 10$$

- The debonding length decreases if h_A/h_B decreases, or if σ_∞/τ_C decreases, but it is not much influenced by the elasticity modulus.

These results have been checked by a Finite Element Analysis (Vidonne 1995).

4. Analysis of the Crack Reinitiation Conditions

The same Shear Lag analysis may be applied to find the state of stress in the layer B after a debonding of length l_D. The same equations apply with a translation of l_D (or $-l_D$) of the x_1 axis.

- for $0 < x_1 < l_D$, the layer B is subjected to a pure tension :

$$\sigma_{11}^B = \sigma_\infty \frac{h_A + h_B}{h_B}$$

- for $x_1 \geq l_D$, the layer B is subjected to a shear stress equal, by continuity, at its upper surface $x_2 = 0$, to the shear stress in the interface :

$$\sigma_{12}^B = \sigma_{12}^I = 2h_A \lambda \sigma_\infty \exp(-\lambda[x_1 - l_D])$$

and a tension stress, which is the solution of

$$\frac{d\sigma_{11}^B}{dx_1} = -\frac{\sigma_{12}^I}{2h_B}$$

with the boundary condition $\sigma_{11(x_1 \to \infty)}^B = \sigma_\infty$:

$$\sigma_{11}^{B} = \sigma_{\infty}\left[1 + \frac{h_A}{h_B}\exp\left(-\lambda\left[x_1 - l_D\right]\right)\right].$$

Damage Mechanics now provides the concept of the Damage Equivalent Stress which governs the phenomenon of crack initiation (Lemaitre 1992). It is deduced from the Strain Energy Density Release Rate which is the associated variable to the Damage Variable D defined as the surface density of microcracks or microcavities in a Representative Volume Element :

$$\sigma^* = \sigma_{eq} R_v^{1/2}$$

where σ_{eq} is the von Mises Equivalent stress :

$$\sigma_{eq} = \left(\frac{3}{2}\sigma_{ij}^{D}\sigma_{ij}^{D}\right)^{1/2}, \quad \sigma_{ij}^{D} = \sigma_{ij} - \sigma_H\delta_{ij}, \quad \sigma_H = \frac{1}{3}\sigma_{hk}$$

and R_v the triaxiality function :

$$R_v = \frac{2}{3}(1+\nu) + 3(1-2\nu)\left(\frac{\sigma_H}{\sigma_{eq}}\right)^2$$

ν being the Poisson's ratio.

It is straightforward to calculate the Damage Equivalent Stress at the upper surface of the layer B as a function of x_1, h_A/h_B and the group parameter μ_2 defined by

$$\mu_2 = \frac{G_I E_\infty}{E_A E_B}.$$

Figure 2 shows, as an example, that σ^* is maximum at the debond tip and its expression for $x_1 = l_D$ is :

$$\frac{\sigma^*}{\sigma_\infty} = \left(\left[1 + \frac{h_A}{h_B}\right]^2 + 6\frac{h_A}{h_B}\mu_2^2\right)^{1/2}\left(\frac{2}{3}\left[1+\nu_B\right] + \frac{1}{3}\left[1-2\nu_B\right]\left[1 + 6\frac{h_A}{h_B}\frac{\mu_2^2}{\left(1+\frac{h_A}{h_B}\right)^2}\right]^{-1}\right)^{1/2}.$$

This proves that the crack reinitiation in the layer B will occur at the point $\left(x_2 = 0, x_1 = l_D\right)$.

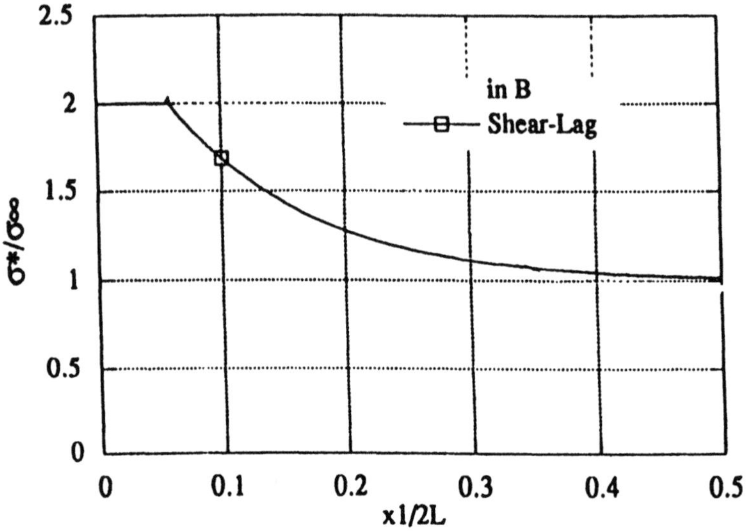

Fig. 2 Damage equivalent stress evolution (L being a reference length)

Another interesting result shown in figure 3 is that σ^*, despite its nice formula, is quasilinear with h_A/h_B and does not depend very much upon the material parameter μ_2 at least in the range of their engineering design values. The simple formula

$$\frac{\sigma^*}{\sigma_\infty} \approx 1 + \frac{h_A}{h_B} \quad \text{with} \quad R_v \approx 1$$

is a good approximation and in fact a lower bound by comparison with a Finite Element Analysis (Vidonne 1995).

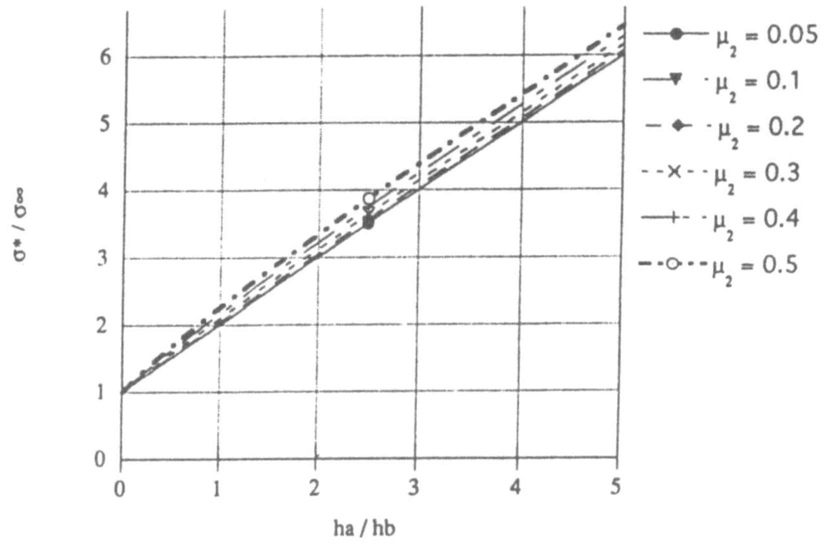

Fig. 3 Damage equivalent stress function of the design parameters

Having σ^*, damage mechanics provides conditions for crack reinitiation. Assuming the material of the layer B to be quasi-brittle, that is elastic at the macro-scale but elasto-plastic and damageable at the micro-scale, a two scales micromechanics analysis of an elasto-plastic and damageable inclusion embedded in an elastic matrix gives the following kinetic law of evolution of the isotropic damage D (Lemaitre 1992) :

$$\dot{D} = \frac{\sigma_f^2}{2ES}\left[\frac{2}{3}(1+\nu) + 3(1-2\nu)\left(\frac{\sigma_H}{\sigma_f}\right)^2\right]\dot{\varepsilon}_{eq} \quad \text{if} \quad \varepsilon_{eq} \geq \varepsilon_D$$

$$D = D_C \leq 1 \rightarrow \text{mesocrack initiation}$$

where the accumulated strain rate $\dot{\varepsilon}_{eq} = \left(\tfrac{2}{3}\dot{\varepsilon}_{ij}^D \dot{\varepsilon}_{ij}^D\right)^{1/2}$ may be replaced by its value as a function of the von Mises stress rate through the law of elasticity:

$$\dot{\varepsilon}_{eq} = \frac{2}{3}\frac{1+\nu}{E}\left|\dot{\sigma}_{eq}\right|.$$

σ_f is the fatigue limit of the material, ε_D a damage threshold taken here as equal to zero, and S a damage strength of the material. Then,

$$\dot{D} = \frac{\sigma_f^2(1+\nu)}{3E^2S}\left[\frac{2}{3}(1+\nu) + 3(1-2\nu)\left(\frac{\sigma_H}{\sigma_f}\right)^2\right]\left|\dot{\sigma}_{eq}\right|.$$

Using the rough approximation $\sigma^* = \sigma_\infty(1 + h_A/h_B)$ which corresponds to a one dimensional state of meso stress at the tip of the debonding $x_1 = l_D$ (but a three dimensional state at the microscale) :

$$\sigma^* = \sigma_{eq}$$

$$\dot{D} = \frac{\sigma_{fB}^2(1+\nu_B)}{3E_B^2 S_B}\left[\frac{2}{3}(1+\nu_B) + \frac{1}{3}(1-2\nu_B)\frac{\sigma_\infty^2\left(1+\dfrac{h_A}{h_B}\right)^2}{\sigma_{fB}^2}\right]\left(1+\frac{h_A}{h_B}\right)\left|\dot{\sigma}_\infty\right|.$$

This formula shows that a low damage rate $dD/d\sigma_\infty$ is obtained if :

- the thickness of the uncracked layers h_B is large in comparison to the thickness h_A of the cracked layer (this is the case of a coated system),
- the elasticity modulus E_B and the damage strength S_B of the uncracked layer material are large but if its fatigue limit σ_{fB} is low.

The conditions of a crack reinitiation are derived from the integration of this damage rate equation for any history of loading $\sigma_\infty(t)$.

* Case of static loading

An obvious integration gives the critical remote stress σ_∞^C corresponding to the mesocrack initiation $D = D_C$ as the solution of :

$$\frac{\sigma_{fB}^2(1+\nu_B)}{3E_B^2S_B}\left[\frac{2}{3}(1+\nu_B)\sigma_\infty^C+\frac{1}{9}(1-2\nu_B)\frac{\sigma_\infty^{C3}\left(1+\frac{h_A}{h_B}\right)^2}{\sigma_{fB}^2}\right]\left(1+\frac{h_A}{h_B}\right)=D_C$$

with the same qualitative conclusions to maximize σ_∞^C as to minimize the damage rate $dD/d\sigma_\infty^*$.

* Case of fatigue loading

Consider a periodic loading defined by $0 \le \sigma_\infty \le \sigma_{\infty M}$ which induces fatigue of the layer B after debonding of the interface. From the damage rate equation, the damage per cycle is :

$$\frac{\delta D}{\delta N}=\frac{\sigma_{fB}^2(1+\nu_B)}{3E_B^2S_B}\left[\frac{2}{3}(1+\nu_B)\sigma_{\infty M}+\frac{1}{9}(1-2\nu_B)\frac{\sigma_{\infty M}^3\left(1+\frac{h_A}{h_B}\right)^2}{\sigma_{fB}^2}\right]\left(1+\frac{h_A}{h_B}\right)$$

and the number of cycles N_R to reach a crack reinitiation corresponding to $D = D_C$ is

$$N_R=\frac{3E_B^2S_BD_C}{\sigma_{fB}^2(1+\nu_B)\left(1+\frac{h_A}{h_B}\right)}\left[\frac{2}{3}(1+\nu_B)\sigma_{\infty M}+\frac{1}{9}(1-2\nu_B)\frac{\sigma_{\infty M}^3\left(1+\frac{h_A}{h_B}\right)^2}{\sigma_{fB}^2}\right]^{-1}$$

This result where the number of cycles to crack reinitiation is a cubic inverse function of the stress amplitude may be not quantitatively realistic but it shows the main trends of the influence of the design parameters to obtain a crack retardation defined by N_R as large as possible.

5. Conclusion

The dream of the good designer of multimaterials relating to crack arrest or crack growth retardation is alive if :

- the possible cracked layer has a small thickness in comparison to its substrate,

- the elasticity modulus and the damage strength of the substrate are large but its fatigue limit is low,

- the debonding length of the interface which immediately follows the crack reaching the interface is of the order of magnitude of several thicknesses of the crack layer, it decreases if the debonding shear stress increases but is not much influenced by the modulus of elasticity of the layers and by the shear modulus groups as $G_I E_\infty / E_A E_B$.

6. References

Ballarini R. and Luo H.A. (1991). Green's Functions for Dislocations in Bonded Strips and Related Crack Problems, *Int. J. Fract.*, **50**, 239-262.

Erdogan F. and Biricikoglu V. (1973). Two bonded half planes with a crack going through the interface, *Int. J. Eng. Sci.*, **11**, 745-766.

He M.Y. and Hutchinson J.W. (1989). Crack deflection at an interface between dissimilar elastic materials, *Int. J. of Solids and Structures* **25**, 1053-1067.

Hutchinson J.W. and Suo Z.(1992). Mixed mode cracking in layered materials, *Advances in Applied Mechanics* **29**, 63-191.

Jensen S., Hutchinson J.W. and Kyung-Suk Kim (1990). Decohesion of a cut prestressed film on a substrate, *Int. J. of Solids and Structures* **26**, 1099-1114.

Lemaitre J. (1974). Prévision de la progression des fissures de fatigue dans les structures minces renforcées, *Revue de physique Appliquée* **9**, 667-672.

Lemaitre J. (1992). *A course on Damage Mechanics*, Springer Verlag

Lemaitre J., Leckie F. and Shermann D. (1992). Crazing of Laminates, *European J. of Mechanics A Solids* **11**, 289-304.

Suo Z. and Hutchinson J.W. (1990). Interface crack between two elastic layers, *Int. J. Fract.*, **43**, 1-18.

Vidonne M.P. (1995). Endommagement et rupture des interfaces dans les multimatériaux, Thèse de Doctorat Université Paris 6, L.M.T. Cachan.

Volkersen M.P. (1938) Die Nietkraftverteilung, *Zugbeanspruchten mit Konstanten Laschenquerschritten, Luftfahrtforschung* **15**, 41-47.

Zak A.R. and Williams M.L. (1963). Crack point stress singularities at bi-material Interface, *J. Appl. Mech.*, **30**, 142-143.

DYNAMIC CRACK GROWTH IN INELASTIC SOLIDS WITH APPLICATION TO ADIABATIC SHEARBANDS

SIA NEMAT-NASSER
Center of Excellence for Advanced Materials
Department of Applied Mechanics and Engineering Sciences
University of California, San Diego
La Jolla, CA 92093-0416

1. Introduction

Asymptotic solutions of the stress and strain fields near the tip of a steadily advancing crack in an elastic-plastic solid, have been worked out by Slepyan (1976) for Mode II, and by Gao and Nemat-Nasser (1983a,b, 1984) for all three fracture modes, as well as for elastic, power-law hardening (elastic-plastic) material models; see also Nemat-Nasser and Obata (1990). Adiabatic shearbands in similar model materials have been examined analytically, mostly as one-dimensional problems; see, for example, Clifton *et al.* (1984), Wright and Batra (1985), Wright and Walter (1987), Burns (1990), Walter (1992), and Olmstead *et al.* (1994). There are many features in common between a dynamically growing crack and an advancing adiabatic shearband in an elastic-plastic solid. Here, some of these are briefly examined, focusing on near-field asymptotic solutions of these problems. Of particular interest is the effect of the assumed constitutive model on the structure of the asymptotic solutions, especially the nature of the temperature field.

It turns out that the near-tip asymptotic solution of a steadily advancing crack or adiabatic shearband, critically depends on the manner by which the material's flow stress may depend on temperature and strain rate. For a perfectly plastic model, the strain and temperature will be logarithmically singular, as the tip is approached. Similar results hold when the flow stress is power-law hardening, but assumed to be independent of the strain rate and temperature. On the other hand, if the flow stress is assumed to depend on temperature, all field quantities will be regular near the tip of the crack or shearband. The situation again changes if the flow stress is assumed to depend on strain rate by a power law, but remains independent of the temperature and strain. The field quantities now become power-law singular. If, in addition to the strain-rate

J. R. Willis (ed.), IUTAM Symposium on Nonlinear Analysis of Fracture, 135–140.

dependency, the flow stress also depends on temperature, the asymptotic near-tip solution again becomes regular. The paper therefore demonstrates how important the assumed constitutive model is in defining the most essential structure of the corresponding solutions.

2. Basic Equations

2.1. FIELD EQUATIONS

As a unified approach, consider the field equations common to both the dynamic crack-growth and adiabatic shearband problems. These are the equations of linear momentum and energy, namely,

$$\nabla \cdot \boldsymbol{\sigma} = \rho \frac{d\mathbf{v}}{dt}, \quad \rho\, c \frac{dT}{dt} = k\, \nabla^2 T + \dot{W}^p, \tag{1,2}$$

where $\boldsymbol{\sigma}$ is the Cauchy stress, ρ is the mass-density, $\mathbf{v}$ is the velocity, T is the temperature, c is the heat capacity, k is the conductivity constant, and $\dot{W}^p$ stands for the rate of plastic work per unit current volume. In terms of the plastic strain rate (deviatoric), $\dot{\boldsymbol{\varepsilon}}^p$, this work is given by $\dot{W}^p = \boldsymbol{\sigma} : \dot{\boldsymbol{\varepsilon}}^p$.

2.2. CONSTITUTIVE RELATIONS

For the constitutive relations, we consider an elastic-plastic model, and examine various flow-stress relations. The plastic strain rate, $\dot{\boldsymbol{\varepsilon}}^p$, is defined by

$$\dot{\boldsymbol{\varepsilon}}^p = \dot{\gamma}\, \boldsymbol{\sigma}' / 2\tau, \tag{3}$$

where $\boldsymbol{\sigma}'$ is the deviatoric stress, and

$$\dot{\gamma} = (2\, \dot{\boldsymbol{\varepsilon}}^p : \dot{\boldsymbol{\varepsilon}}^p)^{1/2}, \quad \tau = (\frac{1}{2} \boldsymbol{\sigma}' : \boldsymbol{\sigma}')^{1/2}, \tag{4,5}$$

are the effective plastic strain rate, and the effective stress, respectively. The flow stress is a relation between τ, $\dot{\gamma}$, and temperature T. We consider five different cases, as follows:

$$\tau = \begin{cases} k_0 & \text{perfect plasticity} \\ k_0\, \gamma^{1/n} & \text{power-law hardening} \\ k_0\, g(T) & \text{thermal softening} \\ k_0\, \dot{\gamma}^m & \text{strain-rate hardening} \\ k_0\, \dot{\gamma}^m\, g(T) & \text{more general}. \end{cases} \tag{6}$$

In this work, we confine attention to a simple case where g is a linear function of temperature,

$$g = (1 - a\, T). \tag{7}$$

This may be viewed as the two leading terms of the Taylor expansion of an exponential function of temperature, $g = \exp\{-aT\}$, where a is generally very small, *i.e.*, $a = 0(10^{-4})$.

In view of (3-5), note that the rate of plastic work becomes

$$\dot{W}^p \equiv \boldsymbol{\sigma} : \dot{\boldsymbol{\varepsilon}}^p = \tau\dot{\gamma} \geq 0. \tag{8}$$

3. Steady-State Solution

When the shearband or crack is advancing in the direction of, say, the x_1-axis, see Figure 1, at a constant velocity V, the material time derivative becomes

$$\frac{d}{dt} \equiv -V\frac{\partial}{\partial x}, \tag{9}$$

and the basic equations in the moving coordinates (x, y), become

$$\nabla \boldsymbol{.} \boldsymbol{\sigma} = -\rho V \frac{\partial \mathbf{v}}{\partial x}, \tag{10}$$

$$-\rho c V \frac{\partial T}{\partial x} = k\nabla^2 T + \tau\dot{\gamma}. \tag{11}$$

We now examine the structure of the solutions of these equations with $k = 0$ (no heat conduction), in a region $r_0 < r < R_0$ close to the tip of the running crack or the shearband. We compare these solutions for the five flow stress models of (6).

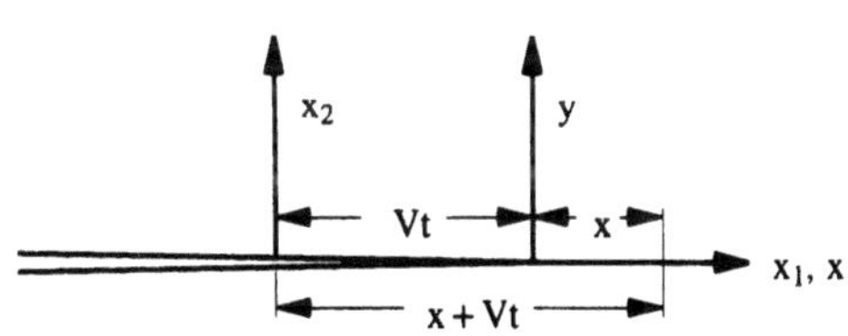

Figure 1. The coordinate systems for steadily growing crack or shearband: x_1, x_2 are the stationary, and x, y are the moving coordinates

3.1. THE $\tau \leq k_0$ MODEL

Normalize the flow stress using k_0 as the unit of stress. Then, $\tau = 0(1)$. From (5,10) it follows that $\partial\mathbf{v}/\partial x = 0(1/r)$, so that $\dot{\gamma} = 0(1/r)$, where r measures length from the tip of the shearband or the crack. From this and the incompressibility condition, it now follows (Gao and Nemat-Nasser, 1983a) that the strain and strain rate have the form

$$\varepsilon = 0(\ln(R_0/r)), \quad \dot{\varepsilon} = 0(1/r). \tag{12,13}$$

From (11) with k = 0, it now is deduced that

$$T = 0(\ln(R_0/r)). \tag{14}$$

Hence, according to this model, the temperature field must have a logarithmic singularity at the tip of the running crack or shearband. Zehnder and Rosakis (1991) report sharp variations in the temperature field ahead of a running crack, which seems to support this result. For this case, Gao and Nemat-Nasser (1983a) provide detailed solutions for the stress and deformation fields based on this model, and give illustrative examples for all three fracture modes; see also Hori and Nemat-Nasser (1989). For Modes I and II, they show that the conditions of the non-negative plastic work lead to the existence of a stress discontinuity at certain angles, $\pm\,\theta^*$, while the strain field remains continuous; for a systematic development of the field variables across a moving discontinuity in an elastic-plastic material; see Nemat-Nasser and Gao (1988). Recent finite-element calculations by Varias and Shih (1995) seem to support the existence of the stress discontinuity.

3.3. Power-law Hardening

Similar results are obtained when a power law work-hardening is assumed. Consider the stress-strain relation defined by

$$\gamma = \tau/E + c_0(\tau - \tau_0)^n, \tag{15}$$

where E is the Young modulus, c_0 and τ_0 are material constants, and n is the work-hardening exponent. This is equivalent to the power-law hardening given in (6). Then the strain, stress, and temperature fields are singular as follows:

$$\gamma = 0(\ln(R_0/r))^{n/(n-1)}, \quad \tau = 0(\ln(R_0/r))^{1/(n-1)},$$
$$T = 0(\ln(R_0/r))^{(n+1)/(n-1)}. \tag{16-18}$$

These results can be verified starting with the basic field equations and following the procedure outlined in Gao and Nemat-Nasser (1983b).

3.4 THE $\tau \le k_0\, g(T)$ MODEL

When linear temperature softening is assumed, the structure of the solution changes completely. In this case, the yield condition

$$\tau = k_0(1 - aT) \tag{19}$$

shows that $\tau = 0(1)$ and $T = 0(\ln(R_0/r))$ are not consistent. Assuming $\mathbf{v} = 0(r^N)$, from (10) and (11), with k = 0, it follows that N = 0. Hence the leading terms in the field quantities are all functions of the angle θ only. There is no singular dependence on the radial distance, r, from the crack or shearband tip in this case.

3.2 THE $\tau \le k_0 \dot{\gamma}^m$ MODEL

Set $\mathbf{v} = 0(r^N)$ and note that the field equations then take on the following form:

$$\mathbf{v} = r^N \mathbf{V}(\theta), \quad \boldsymbol{\sigma} = r^N \mathbf{S}(\theta),$$
$$\tau = r^N S(\theta), \quad \dot{\gamma} = r^{N/m}\Gamma(\theta),$$
$$T = r^{N((m+1)/m)+1} F(\theta). \tag{20-25}$$

From the field equations, it now follows that

$$N = -m/(1-m), \tag{26}$$

for this case. Note that, this leads to a weak temperature singularity, when m is small, *i.e.*,

$$T = 0(1/r^{2m}). \tag{27}$$

3.5 THE $\tau \le k_0 \dot{\gamma}^m g(T)$ MODEL

This is the model used by Wright and Walter (1996). Using expressions (20-25) in the field equations, together with the flow stress, $\tau = k_0 \dot{\gamma}^m g(T)$, it follows that

$$N = m/(1+m). \tag{28}$$

From (28), it is seen that there is no singularity in this case. Thus, this model does not seem to accord with the temperature field measured by Zehnder and Rosakis (1991) for the running crack.

Acknowledgements

This work has been supported by the US Army Research Office (ARO) under contract No. DAAL-03-92-G-0108 with the University of California, San Diego.

4. References

Burns, T.J. (1990) A Mechanism for Shear Band Formation in the High Strain-Rate Torsion Test, *J. Applied Mechanics* **112**, 836-844.

Clifton, R.J., Duffy, J., Hartley, K.A. and Shawki, T.G, (1984) On Critical Conditions for Shear Band Formation at High Strain Rates, *Scripta Met.* **18**, 443-448.

Gao, Y.C. and Nemat-Nasser S. (1983a) Dynamic Fields Near a Crack Tip Growing in an Elastic-Perfectly-Plastic Solid, *Mechanics of Materials* **2**, No. 4, 47-60.

Gao, Y.C. and Nemat-Nasser,S. (1983b) Near-Tip Dynamic Fields for a Crack Advancing in a Power-Law Elastic-Plastic Material: Modes I, II and III, *Mechanics of Materials* **2**, No. 4, 305-317.

Gao, Y.C. and Nemat-Nasser, S. (1984) Mode II Dynamic Fields Near a Crack Tip Growing in an Elastic-Perfectly-Plastic Solid, *J. Mechanics and Physics of Solids* **32**, 1-19.

Hori, M. and Nemat-Nasser, S. (1989) Steady Crack Growth in Elastic-Plastic Work-hardening Solids: Near-tip Fields and Stress Discontinuity," Proceedings of the ICF-7, Texas March 20-24, 1989 *Advances in Fracture Research*, 333-339.

Nemat-Nasser, S. and Gao, Y.C. (1988) Discontinuities in Elastic-Plastic Solids, *Mechanics of Materials* **7**, 215-229.

Nemat-Nasser, S. and Obata, M. (1990) Some Basic Issues in Dynamic Crack Growth in Elastic-Plastic Solids, *J. Fracture* **42**, 287-300.

Olmstead, W.E., Nemat-Nasser, S. and Ni, L. (1994) Shear Bands as Surfaces of Discontinuity, *J. Mechanics and Physics of Solids* **42**, 697-709.

Slepyan, L.I. (1976) Crack Dynamics in an Elastic-Plastic Body *Izv. AN SSSR, Mekh. Tver. Tela* **11**, 126.

Varias, A.G. and Shih, C.F. (1995) Dynamic Steady Crack Growth in Elastic-Plastic Solids - Propagation of Strong Discontinuities, *J. Mechanics and Physics of Solids* **42**, No. 11, 1817-1848.

Walter, J.W. (1992) Numerical Experiments on Adiabatic Shear Band Formation in One Dimension, *J. Plasticity* **8**, 657-693.

Wright, T.W. and Batra, R.C. (1985) The Initiation and Growth of Adiabatic Shear Bands, *J. Plasticity* **1**, 205-212.

Wright, T.W. and Walter, J.W. (1987) On Stress Collapse in Adiabatic Shear Bands, *J. Mechanics and Physics of Solids* **35**, 701-720.

Wright, T.W. and Walter, J.W. (1996) The Asymptotic Structure of an Adiabatic Shear Band in Antiplane Motion, *J. Mech. Phys. Solids* **44**, No. 1, 77-97.

Zehnder, A.T. and Rosakis,A.J. (1991) On the Temperature Distribution at the Vicinity of Dynamically Propagating Cracks in 4340 Steel, *J. Mech. Phys. Solids* **39**, 385-415.

DYNAMICALLY GROWING SHEAR BANDS IN METALS:

A STUDY OF TRANSIENT TEMPERATURE AND DEFORMATION FIELDS

A. J. ROSAKIS, G. RAVICHANDRAN AND M. ZHOU
Graduate Aeronautical Laboratories
California Institute of Technology
Pasadena, CA 91125, U.S.A.

1. Introduction

The objective of our work is (1) to establish the critical conditions for initiation and growth of shear bands in pre-notched plates subjected to asymmetric impact loading (dynamic mode II, see Fig. 1); and (2) to investigate and characterize the transition in the modes of failure when such plates are subjected to a variety of loading rates (impact velocities). These failure modes which may feature either dynamic shear band growth, dynamic crack propagation or both, (see Fig. 1), have been observed in structural materials such as the high strength steels and heat-resistant titanium alloy considered in the present work. The approach is to study both the dynamic mechanical deformations and the processes of heat generation and thermal softening. This necessitates the simultaneous use of high speed optical and infrared diagnostics in the experiments. In parallel to the experimental study, full-scale thermomechanical finite element simulations are conducted to assist the development of criteria for shear band initiation and propagation. The calculations make use of constitutive parameters measured in house through material testing in a variety of strain rates. Such a combined experimental and numerical approach enables us to make direct comparisons between measurements and predictions obtained using various material constitutive and failure models.

2. Results

Materials studied include C-300 (a high strength maraging steel), Ti-6Al-4V and HY-100. However, C-300 has so far been the primary material of focus. Experiments and numerical simulations have shown that there is a strong dependence of temperature increase and speed of shear band

J. R. Willis (ed.), IUTAM Symposium on Nonlinear Analysis of Fracture, 141–150.

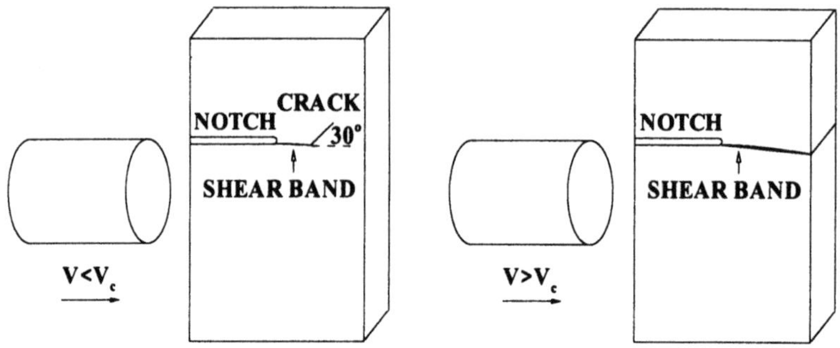

FIG. 1 ASYMMETRIC IMPACT CONFIGURATION AND FAILURE MODES

propagation on loading rate or impact velocity, as demonstrated in the results summarized below.

2.1 Temperature Rises

Figure 2 shows the measured temperature field generated by a propagating shear band in a C-300 specimen. The 3-D contour plots reveal the spatial structure of the temperature field generated by the propagating shear band. The spike in the middle indicates the location of the band. The width of the zone of intense heating is approximately 200-300 μm.

Figure 3 is a summary of the maximum temperatures observed inside propagating shear bands corresponding to different impact velocities for C-300 and Ti-6Al-4V. Clearly, the maximum temperatures increase monotonically with the impact velocity for both materials. For C-300, the highest temperature observed approaches 1700 K (1,427 °C) or 90% of its melting point. This is the first time that such high shear band temperatures have ever been measured in a laboratory setting. The temperatures observed here for the C-300 steel are substantially higher than previously reported by Duffy and Chi (1992) for shear bands in other types of steels. Also, these temperatures are substantially higher than those predicted by previous finite element analyses of the shear banding process. The data in the plot demonstrate a strong dependence of the maximum shear band temperature on the loading rate or the impact velocity.

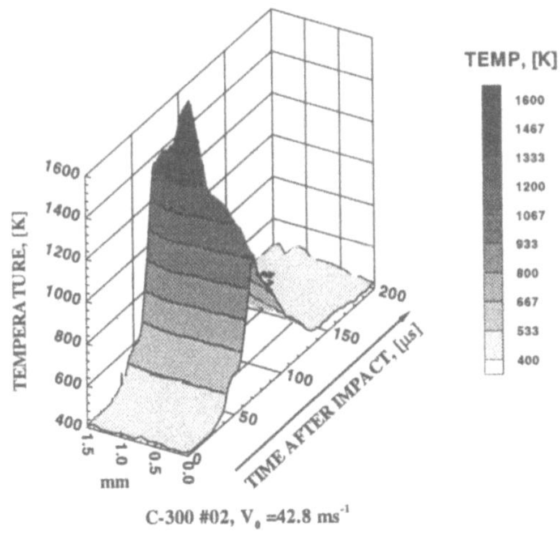

FIG. 2 TEMPERATURE PROFILE AROUND A PROPAGATING SHEAR BAND

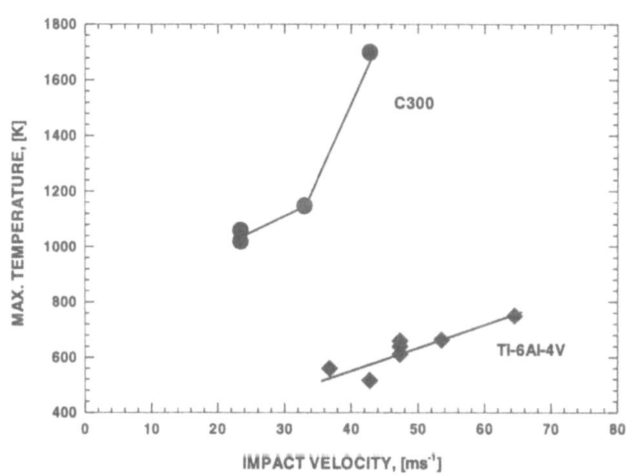

FIG. 3 MAXIMUM TEMPERATURE AS A FUNCTION OF IMPACT VELOCITY

2.2 Shear Band Propagation Speed

The speed of shear band propagation is studied using high speed photography at framing rates up to 2×10^6 frames/second. Shear band speed histories during tests at three different impact velocities are shown in Fig 4. The shapes of curves indicate that the shear bands do not propagate at constant speeds. The speed is lower initially and reaches a maximum

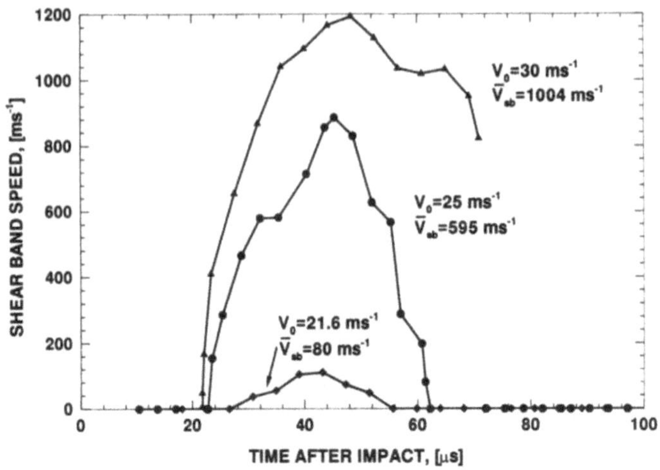

FIG. 4 SHEAR BAND SPEED HISTORIES FOR C-300 AT DIFFERENT IMPACT VELOCITIES

for each of the impact velocities. The highest speed observed is close to 1,200 ms^{-1}. This value is the highest shear band speed ever recorded in a laboratory environment and is approximately 38% of the shear wave speed of C-300 steel. The dependence of the shear band speed on impact velocity is shown in Fig. 5. Both the maximum speed and the average speed measured for each impact velocity are shown. This figure clearly shows a dramatic increase of shear band speed over a short region of impact velocity. The minimum impact velocity required to initiate a shear band is approximately 20 ms^{-1}. The shape of the curves indicates a tendency to saturate by shear band speed with further increase in impact speed. The strong dependence of shear band speed on impact velocity at lower impact velocities and the tendency to saturate at higher impact velocities observed in the experiments have also been predicted by the boundary layer solution of Gioia and Ortiz (1995).

The results shown in Figs. 3-5 also demonstrate that higher speeds correspond to higher rates of deformation and therefore translate into higher temperatures within the shear bands.

3. Numerical Simulations

Finite element simulations of the experiments are carried out using a coupled thermomechanical finite deformation formulation of LeMonds and Needleman (1986) and Needleman (1989). The effects of strain and strain rate hardening as well as thermal softening are accounted for. The

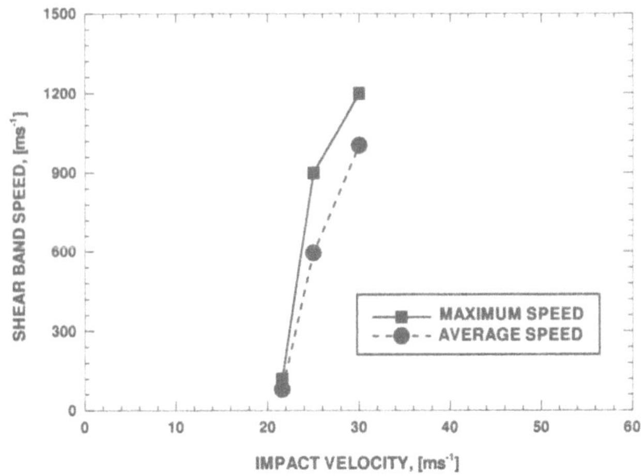

FIG. 5 **SHEAR BAND SPEED AS A FUNCTION OF IMPACT VELOCITY, (C-300)**

adjustable parameters used in the constitutive law have been obtained in house by means of extensive constitutive testing using Hopkinson bar experiments. Full details can be found in Zhou *et al.* (1996a,b).

3.1 Shear Stress

Figure 6 shows the distributions of shear stress σ_{12} at two different times after impact. The impact velocity is 25 ms^{-1}. This is an intermediate velocity at which both shear banding and fracture are observed in the experiments, as illustrated in Fig. 1. The existence of a severe shear stress state (mode-II conditions) is seen. This loading is responsible for the initiation and propagation of shear bands. The simulations show that the initiation of the shear band occurs at approximately 22 μs. This initiation time corresponds to the activation of a critical equivalent strain criterion in Zhou *et al* (1996b). After the critical strain is reached in a particular element, the stress-carrying capability of the material is assumed to follow that of a Newtonian fluid, carrying both a hydrostatic pressure component and a viscous stress component. The use of such a constitutive relation for materials inside the shear band is motivated by their ability to sustain pressure, by the continued dissipation through deformation and frictional forces and by the high temperature values (90% of the melting point of C-300 steel) observed in the experiments as discussed above.

The calculated time of initiation is consistent with the initiation time observed in experiments at this impact velocity by Zhou, Rosakis and

Ravichandran (1996a). The detachment of the area of intense shear stresses indicates the propagation of the shear band. As the shear band propagates the size of the area in front of the band tip, where intensive shear exists, increases and reaches a maximum at approximately 44-45 μs. The size of this area and the intensity of the shear stress inside it decrease gradually following this point. By 65 μs, the shear mode of loading in the shear band tip region has ceased to exist, (not shown). As a result, the shear band decelerates and eventually arrests. The calculated shear stress level in the area in front of the tip is approximately 1.3 GPa. This is in accord with the estimate of Mason, Rosakis and Ravichandran (1994) based on CGS interferometry and a line plastic zone model. The calculation also captures the initiation of a crack from the tip of the arrested shear band later in the deformation. For details, see Zhou, Ravichandran and Rosakis (1996b).

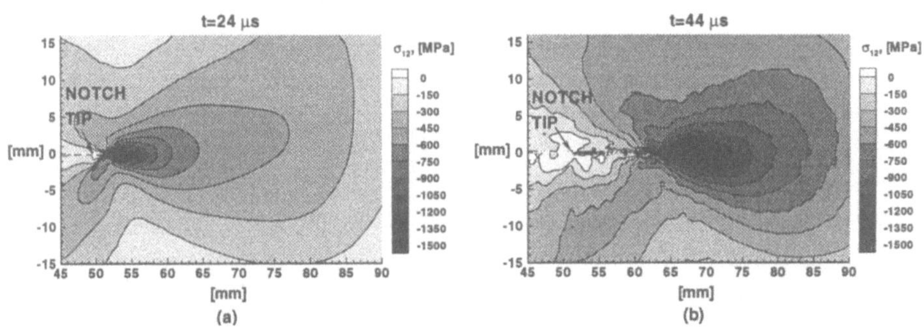

FIG. 6 SHEAR STRESS DISTRIBUTIONS DURING SHEAR BAND PROPAGATION

3.2 Shear Band Length Histories

Figure 7 is a summary of the calculated length histories of propagating shear bands at three impact velocities. The corresponding experimental results are also shown. Clearly, these two sets of results are in good agreement. The curves show a dramatic increase in the length and speed of shear band propagation with increasing impact velocity. In addition, the profiles indicate that the shear band speed is lower initially. It increases and reaches a maximum at approximately 45 μs. This corresponds to the time when the size of the area of intense shear stress in front of the shear band tip is maximum. The speed decreases after 45 μs. While the calculated curves show slightly more variations in band speed over the course of propagation under each impact velocity, the calculated average speeds match the average speeds measured in experiments. Also, both the calculated curves and the measured curves show the same dependence on impact velocity. Furthermore, the calculated and the measured curves clearly demonstrate that the duration of shear band propagation is between approximately 20-60 μs after impact, consistent with the experimental observations.

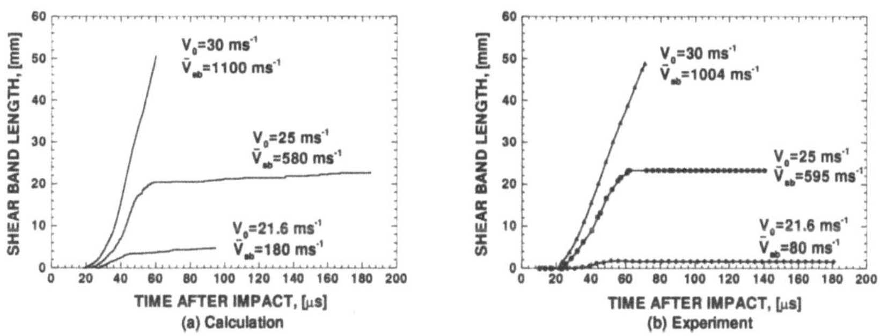

FIG. 7 COMPARISON OF CALCULATED AND MEASURED SHEAR BAND LENGTHS

3.3 *Temperature Rises*

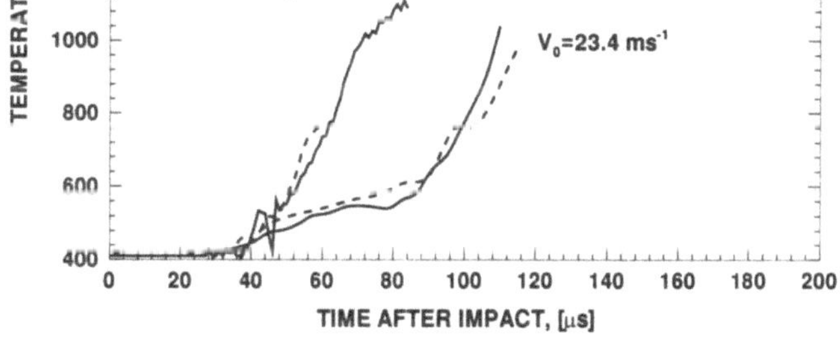

FIG. 8 TEMPERATURE INCREASES AT DIFFERENT IMPACT VELOCITIES

Figure 8 compares the calculated temperature profiles and the corresponding profiles measured in experiments. The profiles correspond to temperatures in the middle of the shear bands. They are obtained by focusing a detector at a point 3 mm in front of the notch tip before impact. For details, see Zhou *et al.* (1996a). During the deformation the shear band tip propagates toward and passes through the point of observation. The temperature profiles discussed here are the temperature histories recorded by that detector. Each profile shown corresponds to a particular impact velocity. Temperature profiles at two impact velocities (25 ms^{-1} and 30 ms^{-1}) are shown. Both the experimental profiles and the calculated profiles show relatively slow initial increases followed by quicker increases, consistent with the distribution shown in Fig. 2. There is a good agreement between the measured and the calculated curves. The curves show higher temperatures and higher rates of increase for higher impact velocities.

3.4 Shear Band Toughness

There is a need to characterize material resistance to the initiation and propagation of shear bands. The concept of a shear band toughness was recently proposed by Grady (1992). A realistic understanding of the issue calls for the study of the energy required for band initiation and propagation. The generalized J-integral for dynamic conditions (Moran and Shih, 1987a,b) is used in the analysis here.

Since the size of the area with large plastic deformation is relatively small (small scale yielding), J can be regarded as a measure of the driving force for shear band propagation. In Fig. 9, the instantaneous J value is plotted as a function of instantaneous shear band speed for the four impact velocities. Although the data show a certain amount of scatter, a universal trend of increasing J with increasing shear band speed is observed. It should be noted that the data forming this universal curve correspond to four calculations which involve drastically different shear band speed histories. This may point to the existence of a material-dependent relation between driving force (J) and instantaneous shear band speed. Beyond a shear band speed of approximately 500 ms^{-1}, the driving force (J) appears to reach a saturation value equal to approximately 250 KJm^{-2}. The value of J at initiation (approximately 100 KJm^{-2}) is consistent with that measured by Mason *et al* (1994) using CGS interferometry. The results presented here are only an initial attempt to approach the issue of shear band toughness by suggesting a possible correlation between a measure of the driving force and shear band propagation speed. Further analyses are

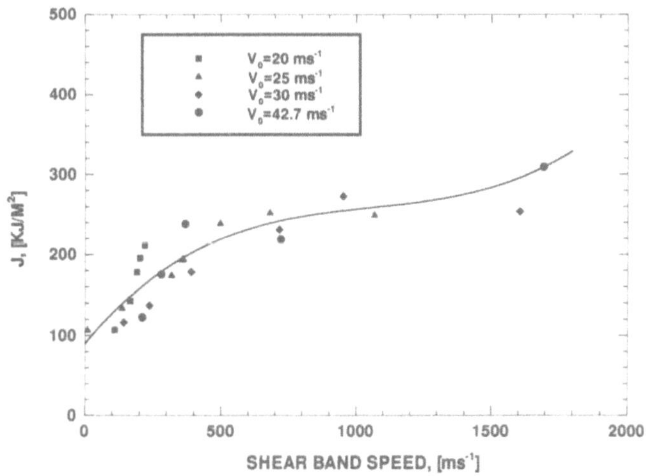

FIG. 9 J-INTEGRAL AS A FUNCTION OF SHEAR BAND SPEED AT DIFFERENT IMPACT VELOCITIES

certainly needed in order to achieve a better understanding of material resistance to the initiation and propagation of shear bands.

4. Conclusions

Controlled shear band initiation and propagation are obtained by using the stress and deformation fields at the tip of an asymmetrically impact-loaded notch as a trigger. Temperature fields around propagating shear bands and the speed of shear band propagation are studied experimentally and numerically. Both the experiments and the numerical simulations demonstrate strong dependences of shear band temperature and speed of propagation on loading rate or impact velocity. There is a good agreement between the measured and the computed temperature profiles and shear band speeds. Full details of this work can be found in Zhou *et al.* (1996a,b). Related work on double notch C-300 specimens is described in Zhou *et al* (1996).

Acknowledgement

The authors gratefully acknowledge support from the Army Research Office through grant No. DAAH 04-93-G0037 under Dr. K. Iyer. We are also grateful to Dr. Y. Rajapakse for support from the Office of Naval Research through grant No. N00014-90-J-1340. The computations were carried out on a Cray C98 computer at the Goddard Space Center and a Cray Y-MP2E/232 computer at the Jet Propulsion Laboratory. Many helpful discussions with Prof. M. Ortiz of Caltech are also acknowledged.

References

Duffy, J. and Chi, Y. C., (1992), On the Measurement of Local Strain and Temperature during the Formation of Adiabatic Shear Bands, *Mat. Sci. Engng*, **A157**, pp. 195-210.

Gioia, G. and Ortiz, M., (1995), The Two-dimensional Structure of Dynamic Shear Bands in Thermoviscoplastic Solids, *J. Mech. Phys. Solids*, to be published.

Grady, D. E., (1992), Properties of an Adiabatic Shear-band Process Zone, *J. Mech. Phys. Solids*, **40**, No. 6, pp. 1197-1215.

LeMonds, J. and Needleman, A., (1986), Finite Element Analyses of Shear Localization in Rate and Temperature Dependent Solids, *Mech. Mat.*, **5**, pp. 339-361.

Needleman, A., (1989), Dynamic Shear Band Development in Plane Strain, *J. Appl. Mech.*, **56**, pp. 1-9.

Mason, J. J., Rosakis, A. J. and Ravichandran, G., (1994), Full Field Measurements of the Dynamic Deformation Field Around a Growing Adiabatic Shear Band at the Tip of a Dynamically Loaded Crack or Notch, *J. Mech. Phys. Solids*, **42**, No. 11, pp. 1679-1697.

Moran, B. and Shih, C. F., (1987a), Crack Tip and Associated Domain Integrals from Momentum and Energy Balance, *Engng. Fract. Mech.*, **27**, pp. 615-642.

Moran, B. and Shih, C. F., (1987b), A General Treatment of Crack Tip Contour Integrals, *Int. J. Fract.*, **35**, pp. 295-310.

Zhou, M., Rosakis, A. J. and Ravichandran, G., (1996a), Dynamically Propagating Shear Bands in Prenotched Plates: I – Experimental Investigations of Temperature Signatures and Propagation Speed, *J. Mech. Phys. Solids*, **44**.

Zhou, M., Ravichandran, G. and Rosakis, A. J., (1996b), Dynamically Propagating Shear Bands in Prenotched Plates: II – Finite Element Simulations, *J. Mech. Phys. Solids*, **44**.

Zhou, M., Rosakis, A. J. and Ravichandran, G., (1996), On the Growth of Shear Bands and Failure-mode Transition in Prenotched Plates: A Comparison of Singly and Doubly Notched Specimens, to appear in *Int. J. Plasticity*.

THE TWO-DIMENSIONAL STRUCTURE OF DYNAMIC BOUNDARY LAYERS AND SHEAR BANDS IN THERMOVISCOPLASTIC SOLIDS

G. GIOIA
Division of Engineering
Brown University
Providence, RI 02912, USA

AND

M. ORTIZ
Graduate Aeronautical Laboratories
California Institute of Technology
Pasadena, CA 91125, USA

1. Introduction

Solids deforming at high rates often develop narrow layers of intense shearing. The realistic modeling of these problems requires consideration of large plastic deformations, rate sensitivity, hardening, heat convection and conduction, thermal softening and inertia effects. Fully nonlinear multidimensional solutions to problems of this nature are rare (see Wright and Walter, 1994, for a notable exception). However, the thinness of the shear layers of interest here makes possible certain approximations in the governing equations which facilitate the analytical characterization of the flow. The systematic use of these approximations results in a much simplified set of *boundary layer equations* which, in some cases, lend themselves to analytical treatment.

Gioia and Ortiz (1996) have applied boundary layer theory to the determination of the two-dimensional structure of dynamic shear bands in thermoviscoplastic solids. They specifically consider the case of a plate which is impacted upon by a flat-ended, rigid projectile, but other geometries can also be treated within the theory. When the impact velocity is sufficiently high, a sharp shear band is often observed to propagate deep into the plate from the edge of the impactor. For instance, Wingrove (1973) studied the

J. R. Willis (ed.), IUTAM Symposium on Nonlinear Analysis of Fracture, 151–162.

penetration of 2014-T6 aluminum alloy plates by flat-ended projectiles. Intense shear bands were observed to be punched through the thickness of the target from the corner of the penetrator, causing a plug to form. Shear bands of a similar character were reported in 70–30 brass by Craig and Stock (1970) at impact velocities of 300 m/s. The same essential geometry arises in pre-notched plates dynamically loaded in shear by an impactor which strikes near the notch (Kalthoff, 1987; Kalthoff and Winkler, 1987; Mason *et al.*, 1994; Zhou *et al.*, 1996).

In the steady-state case, the boundary layer equations can be reduced to a system of ordinary differential equations by the introduction of a similarity variable. The reduced governing equations can then be integrated numerically by a standard shooting method, leading to a full characterization of the the velocity, stress, temperature and plastic work fields. Interestingly, the existence of steady-state solutions is found to place restrictions on the material parameters which can be interpreted as stability conditions. Transient solutions can also be formulated by a natural extension of the similarity methods developed for the steady case. The region where a critical value of the plastic work is exceeded is found to define a narrow shear band whose 'tip' moves away from the origin at constant speed. This shear band tip speed is found to be greatly in excess of the impact velocity, in agreement with the observations of Zhou *et al.* (1996). The ratio of the tip speed to the impact velocity rises steeply as a function of the latter at low impact velocities, also in keeping with the observations of Zhou *et al.* (1996), and saturates at high impact velocities.

The results presented in the present paper are extracted from the article of Gioia and Ortiz (1996). The reader is referred to the original publication for further details.

2. Boundary layer equations

Under certain conditions, the plastic flow of solids may be expected to be confined to layers which are thin relative to all other geometrical dimensions of the problem. The thinness of the layer makes possible certain approximations in the governing equations which facilitate the characterization of the flow within the layer. For fluids, the hypothesis that viscosity effects are significant only in narrow layers, the thicknesses of which approach zero as the Reynolds number increases to infinity, was advanced by Prandtl (1904), who also proceeded to compute the simplified boundary layer equations of motion.

Gioia and Ortiz (1996) have formulated similar approximations for thermoviscoplastic solids. As in Prandtl's original work, the aim is to characterize solutions which are *rapidly varying* in one direction, while being *slowly varying* in the remaining orthogonal directions. As an illustration of how such deformation fields may arise in solids, consider a plate impacted on one side by a flat-ended impactor traveling at speed U, Fig. 1a. Let the x_1 axis point into the plate from the edge of the impactor, and x_2 be the orthogonal direction within the plane of the plate. Imagine cutting the plate along the x_1 axis so that its top and bottom parts can slide freely relative to each other. If the material is further idealized as rigid-plastic in the manner outlined in the foregoing, the impacted portion of the plate will move rigidly with velocity U, Fig. 1b. Evidently, the resulting velocity field is incompatible along the x_1 axis. If U is sufficiently large, we expect compatibility to be restored through the development of a thin layer of shearing deformation, or boundary layer, Fig. 1c.

The equations which determine the structure of boundary layers in thermoviscoplastic solids can be obtained by recourse to a scaling argument. The result is (Gioia and Ortiz, 1996):

$$\tilde{v}_{1,t} + \tilde{v}_1\, \tilde{v}_{1,1} + \tilde{v}_2\, \tilde{v}_{1,2} = \mathcal{R}^{-1} \tilde{s}_{12,2} + \tilde{p}_{,1} \tag{1}$$

$$\tilde{p}_{,2} = 0 \tag{2}$$

$$\tilde{v}_{1,1} + \tilde{v}_{2,2} = 0 \tag{3}$$

$$\tilde{s}_{12} = |\tilde{v}_{1,2}|^{m-1} \tilde{v}_{1,2}\, \tilde{w}^{\tilde{n}}\, \tilde{\theta}^{l} \tag{4}$$

$$\tilde{w}_{,t} + \tilde{v}_1\, \tilde{w}_{,1} + \tilde{v}_2\, \tilde{w}_{,2} = \mathcal{P}\, \tilde{s}_{12}\, \tilde{v}_{1,2} \tag{5}$$

$$-\tilde{q}_{2,2} = \mathcal{T}_d\, \tilde{s}_{12}\, \tilde{v}_{1,2} \tag{6}$$

$$\tilde{q}_2 = \tilde{\theta}^{p}\, |\tilde{\theta}_{,2}|^{k-1} \tilde{\theta}_{,2} \qquad \text{(conduction)} \tag{7}$$

$$\tilde{\theta}^{q}\, (\tilde{\theta}_{,t} + \tilde{v}_1\, \tilde{\theta}_{,1} + \tilde{v}_2\, \tilde{\theta}_{,2}) = \mathcal{T}_v\, \tilde{s}_{12} \tilde{v}_{1,2} \qquad \text{(convection)} \tag{8}$$

In accordance with the flow character of the solutions to be sought, an eulerian description of the motion is adopted. Elastic strains are assumed to be negligible compared to plastic deformations, and the plastic flow to be volume preserving. Asymptotically, either conduction or convection must dominate asymptotically to the exclusion of the other. Which mechanism dominates depends sensitively on material parameters. In the above equations, v_i is the eulerian velocity field; s_{ij} the stress deviator; p the hydrostatic pressure, which we shall take to be positive in tension; w is the plastic work, which is identified as the sole state variable describing the state of hardening of the solid; θ is the absolute temperature; q_i the heat flux vector; q is an exponent characteristic of the variation of the heat capacity with temperature over the range of interest; l, n and m are the thermal softening, strain hardening and rate sensitivity exponents, respectively; the exponent p characterizes the variation of the thermal conductivity with temperature over the range of interest, and k is a thermal conductivity exponent. Commas are used to denote partial differentiation. In writing the

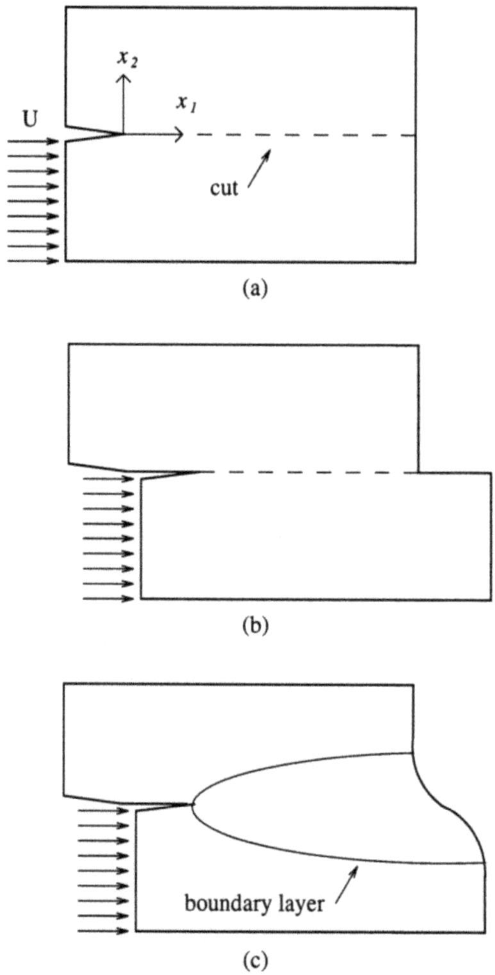

Figure 1. Thought experiment illustrating the formation of an internal boundary layer in a solid. a) Pre-notched plate impacted by a flat-ended projectile; b) incompatible free flow obtained by cutting the solid; and c) internal boundary layer across which continuity of velocities is restored.

boundary layer equations, the following normalization is adopted:

$$\tilde{x}_1 = \frac{x_1}{L}, \qquad \tilde{x}_2 = \frac{x_2}{L}, \qquad \tilde{t} = \frac{Ut}{L}, \qquad \tilde{\psi} = \frac{\psi}{UL}, \qquad \tilde{v}_1 = \frac{v_1}{U}, \tag{9}$$

$$\tilde{v}_2 = \frac{v_2}{U}, \qquad \tilde{w} = \frac{w}{W}, \qquad \tilde{\theta} = \frac{\theta}{T}, \qquad \tilde{s}_{12} = \frac{s_{12}}{S}, \qquad \tilde{p} = \frac{p}{\rho U^2}. \tag{10}$$

whence the dimensionless $\mathcal{R}$, $\mathcal{P}$, $\mathcal{T}_{\rm d}$ and $\mathcal{T}_{\rm v}$ numbers arise. In particular, the dimensionless number

$$\mathcal{R} \equiv \frac{\rho U^2}{S}, \tag{11}$$

plays the role of a generalized Reynolds number. Indeed, in the newtonian case, which is recovered by setting $m = 1$, $n = l = 0$, $\mathcal{R}$ reduces to UL/ν, where ν is the kinematic viscosity. This is the conventional definition of the Reynolds number for newtonian fluids. The asymptotic regime in which the boundary layer equations apply is that of very high Reynolds numbers.

3. Steady similarity solutions

Similarity methods (Rosenhead, 1963) constitute a powerful tool for obtaining semi-analytical solutions of the boundary layer equations stated in the foregoing. An appealing feature of similarity solutions is that their determination requires the solution of a system of *ordinary* differential equations. This system can conveniently be solved by numerical integration, and the complete two-dimensional fields recovered. Gioia and Ortiz (1996) have considered free flows in which the pressure is uniform, i.e., $\tilde{p}_{,\tilde{1}} = 0$. Free flows of the Falkner-Skan type (Falkner and Skan, 1930, 1931), in which the pressure gradient varies as a power of $\tilde{x}_1$, can be treated similarly.

A semi-infinite plate occupying the half-plane $\tilde{x}_1 \geq 0$ is imparted a velocity U on the lower half of the boundary, i. e., on $\tilde{x}_1 = 0$, $\tilde{x}_2 < 0$, Fig. 1. Gioia and Ortiz (1996) have obtained solutions to the steady boundary layer equations by introducing the similarity variable

$$\zeta = \tilde{x}_2 \tilde{x}_1^{-a}. \tag{12}$$

The normalized stream function, temperature and plastic work fields are then expressed as

$$\tilde{\psi} = \tilde{x}_1^a \, f(\zeta), \qquad \tilde{\theta} = \tilde{x}_1^b \, g(\zeta), \qquad \tilde{w} = \tilde{x}_1^c \, h(\zeta), \tag{13}$$

where the functions f, g and h and the exponents a, b and c are to be determined. Inserting representation (13) into the boundary layer equations and balancing out powers of $\tilde{x}_1$, the following system of ordinary differential equations is obtained:

$$-aff'' = \frac{1}{\mathcal{R}} \tau \left(m \frac{f'''}{|f''|} + \bar{n} \frac{h'}{h} + l \frac{g'}{g} \right), \tag{14}$$

$$chf' - afh' = \mathcal{P} \tau |f''|, \tag{15}$$

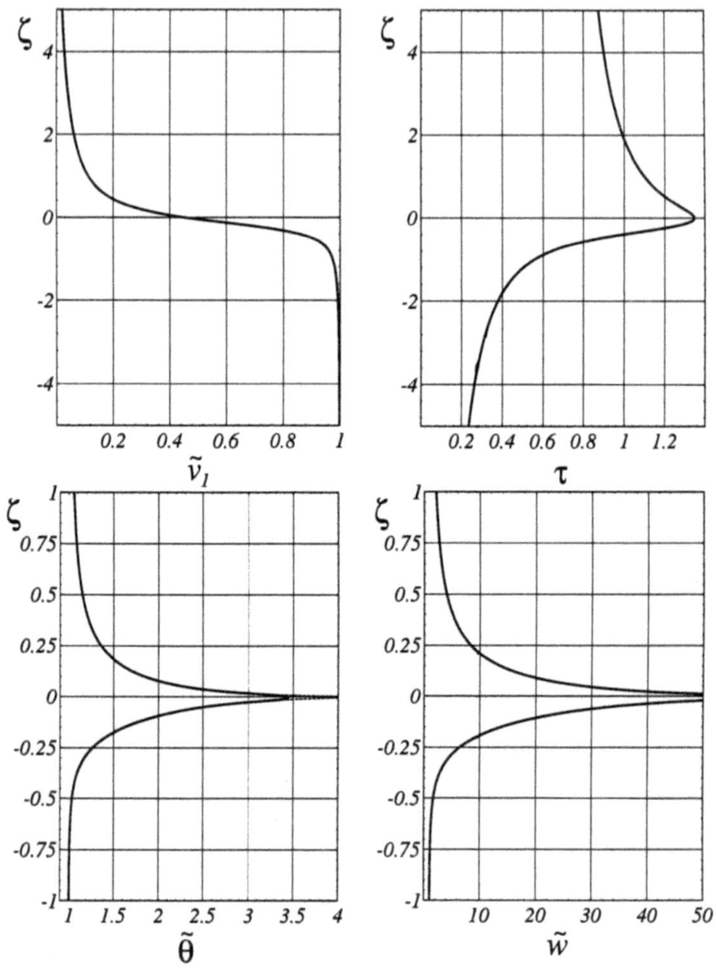

Figure 2. Steady boundary layer in a copper plate impacted by a flat-ended rigid projectile. Profiles of a) velocity; b) shear stress; c) temperature, convection-dominated solution (dotted line) and conduction correction (solid line); d) plastic work.

where $\tau = |f''|^m g^l h^{\bar{n}}$ and $(\cdot)'$ denotes differentiation with respect to ζ. In the convection-dominated case, the energy equation reduces to

$$b\, g^{1+q} f' - a f g^q g' \;=\; T_v\, \tau |f''|, \tag{16}$$

while the characteristic exponents take the values $a = 1/(m+1), b = c = 0$. Consequently, both the temperature and plastic work fields are similar in this case. By contrast, the shear stress field is not similar in general. The convection-dominated case can be treated similarly.

Consider, by way of example, the case of a copper plate which develops a steady internal boundary layer such as described in the foregoing. Gioia and Ortiz (1996) employ a fourth-order Runge-Kutta method to integrate (14) from boundary conditions at $\zeta = 0$. Integration into the positive and negative ζ-directions is performed separately based on assumed boundary values of f' and f''. Simultaneously, (15) and (16) are integrated by the forward-Euler method to determine g and h. Since the remote values of these functions are known, the integration of (15) and (16) can conveniently be effected from the free flow, i.e. from $\zeta = \pm\infty$ towards the origin. The unknown values $f''(0^{\pm})$ are determined iteratively so as to match the remote boundary conditions $f'(-\infty) = 1$ and $f'(\infty) = 0$. Finally, continuity of tractions at the origin is achieved by iteration on $f'(0)$.

The impact velocity U is set at 544 m/s. Lengths are measured in units of the characteristic dimension $L = U/\dot{\gamma}_0 = 1.1$ mm, which corresponds to the gage length over which the reference strain rate is attained for the prescribed impact velocity. We take $m = 0.2$, $n = 0.1$ and $l = -0.1$ as representative of the early stages of deformation of copper in which the material may be expected to behave stably. In addition, we set the Reynolds number $\mathcal{R} = \rho U^2/\sigma_0 = 10$, and the dimensionless numbers $\mathcal{P} = 5$, $\mathcal{T}_\mathrm{v} = 0.25$, and $\mathcal{T}_\mathrm{d} = 1305$. We verify that the Stanton number $\mathcal{S} = \mathcal{T}_\mathrm{v}/\mathcal{T}_\mathrm{d} = 0.000192$ is very small.

The profiles of the similarity solution computed by numerical integration at intervals of $\triangle\zeta = 0.005$ are shown in Fig. 2. As expected, the velocity profile effects a smooth transition from its limiting value of 1 at large and negative ζ to its limiting value of 0 at large and positive ζ, Fig. 2a. The shear stress profile attains a maximum $\tau = 1.34$ at $\zeta = 0$, and decays monotonically but asymmetrically to zero away from the layer, Fig. 2b. The convection temperature profile, shown as the dotted line in Fig. 2c, diverges to infinity at $\zeta = 0$, as expected. However, the conduction sublayer renders the temperatures bounded everywhere. The thinness of the conduction sublayer relative to the convection sublayer is particularly noteworthy. The plastic work distribution also diverges to infinity at $\zeta = 0$, Fig. 2d. This may be regarded as an artifact of the steady solution. The transient solutions presented in Section 4 are devoid of this unphysical behavior. The shear stress and conduction temperature fields predicted by the theory are not similar and, consequently, they vary with $\tilde{x}_1$ along the centerline $\tilde{x}_2 = 0$. The shear stress decreases downstream monotonically as $\tilde{x}_1^{-0.33}$. The temperature exhibits a moderate rise from a value of 733°C near the origin to a value of 910°C at 5.5 cm from the root.

Conditions for the existence of steady boundary layers can be inferred by examining the asymptotic form of the solutions near the boundary (Gioia and Ortiz, 1996). In the convection-dominated case, existence necessitates

$m + \bar{n} + \bar{l} > 0$. Since this inequality involves material parameters only, it can be interpreted as a material stability condition. Remarkably, this result coincides with the stability condition derived by Molinari and Clifton (1987) in a one-dimensional setting. In the conduction-dominated case, by contrast, it is found that the existence of steady boundary layers requires $m+\bar{n} > 0$. The effect of nonlocal hardening on stability can be ascertained likewise. In the convection-dominated case, existence requires $m+\bar{l} > 0$. In the conduction-dominated case the nonlocal material is always stable.

4. Transient boundary layers

Gioia and Ortiz (1996) have constructed a class of unsteady boundary layer solutions which characterizes the mechanical and thermal fields attendant to dynamically propagating shear bands such as develop in impact tests. The transient character of the solutions necessitates consideration of the full boundary layer equations, with rate terms included. Solutions to these equations are obtained by introducing the similarity variables

$$\zeta = \tilde{x}_2 \tilde{x}_1^{-a}, \qquad \xi = \tilde{t}\,\tilde{x}_1^{-1}, \tag{17}$$

and adopting the representation

$$\tilde{\psi} = \tilde{x}_1^a\, f(\zeta,\xi), \qquad \tilde{\theta} = \tilde{x}_1^b\, g(\zeta,\xi), \qquad \tilde{w} = \tilde{x}_1^c\, h(\zeta,\xi), \tag{18}$$

which generalizes (13). Inserting representation (18) into the boundary layer equations and balancing powers of $\tilde{x}_1$ yields the system of partial differential equations

$$\begin{aligned}
(1-\xi f_{,\zeta})\, f_{,\zeta\xi} + \xi f_{,\xi}\, f_{,\zeta\zeta} - a f f_{,\zeta\zeta} &= \frac{1}{\mathcal{R}} s_{,\zeta} && (19)\\
g^q\left[(1-\xi f_{,\zeta})\, g_{,\xi} + (-af + \xi f_{,\xi})\, g_{,\zeta}\right] &= \mathcal{T}_{\mathrm{v}} s\, f_{,\zeta\zeta}\,, && (20)\\
(1-\xi f_{,\zeta})\, h_{,\xi} + (-af + \xi f_{,\xi})\, h_{,\zeta} &= \mathcal{P} s\, f_{,\zeta\zeta}\,, && (21)\\
s = |f_{,\zeta\zeta}|^m g^l h^{\bar{n}} \operatorname{sgn}(f_{,\zeta\zeta}), & && (22)
\end{aligned}$$

together with exponents $a = 1/(m+1)$, $b = c = 0$. Remarkably, the similarity of the velocity, temperature and plastic work fields is preserved in the transient case. Evidently, Eqs. (19), (20) and (21) reduce to their steady counterparts when all derivatives with respect to ξ vanish. To the boundary conditions now one needs to append suitable initial conditions at $t = 0$.

The similar transient equations (19–22) provide a convenient framework for investigating the transition from stable to unstable boundary layers, leading to the formation of a shear band. As a simple model of this transition, we assume that the material initially exhibits stable behavior characterized by a high hardening exponent n, or low thermal softening exponent

l, or both, followed by an unstable regime of low—possibly negative—n, or high l, or both. The stress varies continuously across this transition. Loss of material stability is assumed to occur at a critical accumulated plastic work $\tilde{w} = \tilde{w}_c$. The resulting constitutive law is

$$\tilde{s}_{12} = A\,|\tilde{v}_{1,2}|^m\,\tilde{w}^{\bar{n}}\,\tilde{\theta}^{l}\,\mathrm{sgn}(\tilde{v}_{1,2}) \begin{cases} m+\bar{n}+\bar{l} > 0, & \tilde{w} \le \tilde{w}_c \\ m+\bar{n}+\bar{l} < 0, & \tilde{w} \ge \tilde{w}_c \end{cases} \tag{23}$$

Conveniently, by virtue of the similarity of the transient solutions the critical condition reduces simply to $\tilde{w}(\zeta,\xi) = \tilde{w}_c$, and all fields retain similarity through the transition. At any given time, the locus of points at which the material is critical is defined parametrically by the equation

$$\tilde{w}(\tilde{x}_2\tilde{x}_1^{-a}, \tilde{t}\tilde{x}_1^{-1}) = \tilde{w}_c. \tag{24}$$

This curve encloses a region of highly deformed unstable material, which may therefore be regarded as a shear band. The position of the tip of this region, or 'shear band tip,' follows by particularizing (24) to $\tilde{x}_2 = 0$, which yields the condition $\tilde{w}(0, \tilde{t}\tilde{x}_1^{-1}) = \tilde{w}_c$. This in turn requires $\tilde{t}\tilde{x}_1^{-1} = \xi_c =$ constant, and, consequently, the shear band tip proceeds at the constant speed $\tilde{V} = \xi_c^{-1}$. It bears emphasis that the constancy of the shear band tip speed is a direct consequence of the similarity of the solution and the form (17b) of the time-like similarity variable ξ, which represents a reciprocal speed. The shear band tip speed can be computed simply by integrating the transient similar equations until such time ξ_c as the critical plastic work $\tilde{w}_c$ is attained at $\zeta = 0$. It is noteworthy that the shear band tip speed does not depend on the form of the constitutive relation in the unstable regime.

By way of example, consider a copper plate such as described in Section 3. The material parameters are chosen such that $m + \bar{n} + \bar{l} = 0.04 > 0$ for $\tilde{w} < \tilde{w}_c = 5$, and $m + \bar{n} + \bar{l} = -0.06 < 0$ for $\tilde{w} > \tilde{w}_c = 5$. The Reynolds number is set to 10. The remaining dimensionless numbers of the flow are $\mathcal{P} = 5$ and $\mathcal{T}_v = 0.25$. The transient boundary layer equations are integrated in ξ by the forward-Euler method.

The time evolution of the plastic work field is displayed in Fig. 3. It should be noted that in all plots the $\tilde{x}_2$-axis scale has been magnified by a factor of 2 to aid visualization. The figures clearly bring forth the fully two-dimensional and time-dependent nature of the solution. With increasing time, the level contours of plastic work appear to emanate from the origin and to broaden as they shoot downstream, Fig. 3. This type of growth has been observed by Needleman (1989) and by Chou *et al.* (1992) in finite element simulations of dynamic shear banding. The plastic work at $\zeta = 0$ attains its critical value $\tilde{w}_c$ at $\xi_c = 0.113$, resulting in a normalized shear band tip speed $\tilde{V} = \xi_c^{-1} = 8.84$. The boundary of the shear band in Fig. 3 coincides with the level contour $\tilde{w} = \tilde{w}_c = 5$. As may be seen, the

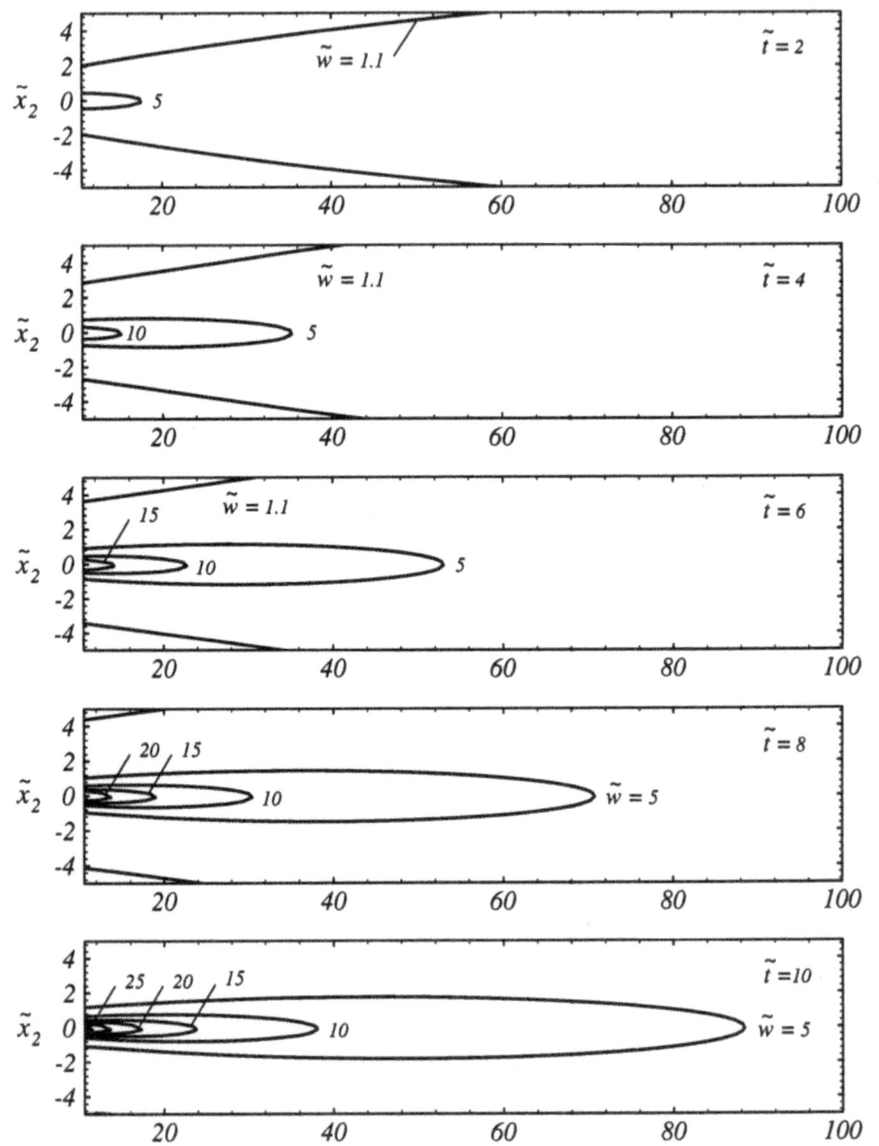

Figure 3. Time evolution of the plastic work field.

shear band emanates from the origin at $\tilde{t} = 0$ and subsequently propagates downstream at the theoretical speed. The remarkable thinness of the band is also noteworthy.

Fig. 4 shows the dependence of the normalized shear band tip speed on the Reynolds number and the critical plastic work. It should be carefully

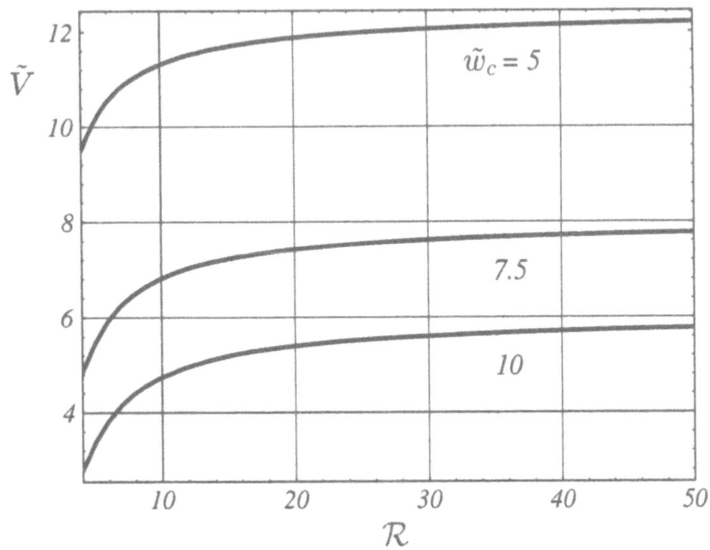

Figure 4. Dependence of the normalized shear band tip speed on the Reynolds number and the critical plastic work.

noted that the shear band tip speeds depicted in Fig. 4 are normalized by the impact velocity. The shear band tip speed is seen to be greatly in excess of the impact velocity and to increase sharply with the Reynolds number at low impact velocities, in agreement with the observations of Zhou *et al.* (1996). At sufficiently high impact velocities, the normalized shear band tip speed saturates and appears to tend asymptotically to a constant value. As expected, high values of $\tilde{w}_c$ have the effect of retarding the propagation of the band. It is remarkable that the speed of propagation of the shear band tip, defined as the leading edge of the unstable region, follows from the theory and is dictated solely by constitutive behavior. In particular, no shear band tip equation of motion needs to be supplied.

5. Acknowledgements

This work has been funded by the National Science Foundation through Brown University's Materials Research Group on "Micro-Mechanics of Failure Resistant Materials." We are grateful to Profs. A. J. Rosakis and G. Ravichandran of Caltech and Prof. M. Zhou of Georgia Tech for helpful discussions and suggestions and for granting us access to as yet unpublished experimental data.

References

Chou, P. C., Flis, W. J., and Konopatski, K. L., *Shock-Wave and High-Strain-Rate Phenomena in Materials*, M. A. Meyers, L. E. Murr and K. P. Staudhammer (eds.), 657–667, 1992.

Craig, J. V., and Stock, T. A., *J. Aust. Inst. Met.*, **15**, 1–5, 1970.

Falkner, V. M., and Skan, S. W., *Rep. Memor. Aero. Res. Coun. London*, No. 1314, 1930.

Falkner, V. M., and Skan, S. W., *Phil. Mag.*, **12** (7), 865–896, 1931.

Gioia, G., and Ortiz, M., *J. Mech. Phys. Solids*, **44** (2) 251–292, 1996.

Kalthoff, J. F., *Photomechanics and Speckle Metrology*, **SPIE 814**, 1987.

Kalthoff, J. F., and Winkler S., *Impact Loading and Dynamic Behavior of Materials*, Chiem, C. Y., Kunze, H. D., and Meyer, L. W., (eds.), **1**, Verlag, 1987.

Mason, J. J., Rosakis, A. J., and Ravichandran, G., *J. Mech. Phys. Solids*, **42** (11) 1679–1697, 1994.

Molinari, A., and Clifton, R. J., *J. Appl. Mech.*, **54**, 806–812, 1987.

Needleman, A., *J. Appl. Mech.*, **56**, 1-9, 1989.

Prandtl, L., *Verh. III Int. Math. Kongr.*, Heildelberg, 484–491, 1904.

Rosenhead, L., *Laminar Boundary Layers*, Oxford, 1963.

Wingrove, A. L. (1973) *Metall. Trans.*, **4**, 1829–1833, 1973.

Wright, T. W., and Walter, T. W., "On Mode III Propagation of Adiabatic Shear Bands," preprint, 1994.

Zhou, M., Rosakis, A. J., and Ravichandran, G., "Dynamically Propagating Shear Bands in Prenotched Plates," *J. Mech. Phys. Solids*, 1996 (in press).

EARTHQUAKES, FRACTURE, COMPLEXITY

DONALD L. TURCOTTE
Department of Geological Sciences
Cornell University, Ithaca, NY 14853 USA

Abstract. Earthquakes generally occur on preexisting faults; big earthquakes occur on big faults, little earthquakes occur on little faults. Several important general observations are: (1) No systematic observable precursory seismic or aseismic slip prior to an earthquake on or near the rupture zone, (2) low stress levels (coefficients of friction less than 0.1), (3) earthquakes nucleate on the deepest part of the rupture zone, (4) rupture propagates at about 60% of the relevant sound speed, (5) rupture occurs in self-healing (Heaton) pulses, (6) background earthquakes in a region satisfy power-law (fractal) frequency-size statistics, and (7) the rate of occurrence of background earthquakes is essentially constant. The classic approach to understanding earthquakes is to apply laboratory derived rate and state dependent friction laws and to model the stick-slip cycle on a planar fault embedded in an elastic media. These models give results that are generally inconsistent with the observations given above. The fundamental differences between theory and observation have led to several new approaches to the earthquake problem:

(1) That high-speed slip on faults results in nonfrictional behavior. One possibility is that the fault gouge fluidizes.

(2) That fault-zone complexity plays an essential role in determining how ruptures propagate. This has led to stranded-cable analogies to the role of fault-zone asperities. Stochastic time-to-failure models are used for a hierarchical array of elements.

(3) The earth's crust is considered to be a complex medium with a fractal distribution of faults. When stressed the medium is subject to fluctuations which result in fractal frequency-size statistics for earthquakes. The generic model for this behavior is an array of slider blocks connected to each other and to a constant velocity driver by springs.

J. R. Willis (ed.), IUTAM Symposium on Nonlinear Analysis of Fracture, 163–175.

1. Introduction

The classical approach to the understanding of an earthquake rupture is to assume that a single preexisting crack exists in an elastic medium. Generally the crack is assumed to be planar and it is assumed that rupture on the crack is controlled by friction.

If the displacement on the rupture surface is prescribed as a function of time and space the entire spectrum of radiated seismic waves can be determined along with the amount of radiated energy. Usually relatively simple ramp models for displacements are specified. Many authors have inverted seismograms to obtain slip histories. These studies have shown (Sholz, 1990) that earthquake ruptures propagate at velocities significantly less than the relevant sound speed (~60%) and that slip velocities are near the maximum allowed values (~1-2 m/s) in large earthquakes. At the same time the rupture motion appears to be relatively strongly damped, oscillations do not occur and there is little evidence that frictional stresses change sign.

In the classic analysis of faults the rupture will initiate at one point on the fault and will spread across the fault surface. Slip occurs across the entire fault zone until the rupture propagation is arrested. However, Heaton (1990) proposed, based on seismic inversions, that self-healing pulses of slip occur in earthquake ruptures. This mode of rupture was strongly confirmed by displacement inversions for the 1992 Landers earthquake (Wald and Heaton, 1994). These pulses are now known as Heaton pulses and the two alternative models for rupture displacement are illustrated in Figure 1.

A variety of observations are relevant to the state of stress prior to, during, and after an earthquake rupture. Inversions of seismograms, specifically the "corner frequency", give stress drops during rupture. Stress drops can also be inferred from surface geodetic observations, before and after an earthquake. These studies indicate a stress drop of less than 10 MPa, even in the largest earthquakes (with a few exceptions). Absolute stress levels during rupture can be inferred from heat flow measurements. Heat flow measurements on the San Andreas fault (Lachenbruch and Sass, 1992) indicate a very small heat-flow anomaly associated with friction on the fault and a dynamic frictional stress of less than 10 MPa. Direct stress measurements were an important component of the Cajon Pass Scientific Drilling Project. The Cajon Pass borehole was drilled to a depth of 3.5 km, about 4 km from the surface trace of the San Andreas fault

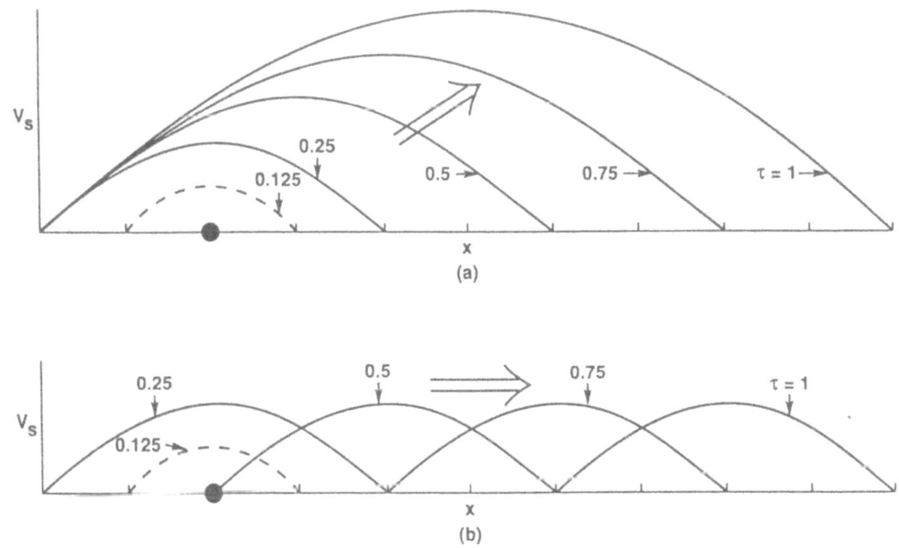

Figure 1. The slip velocity v_s is given as a function of position on the fault x at increasing times τ. The rupture initiates at the black circle. Illustration of two models for the propagation of rupture across a fault. (a) The classical picture that rupture expands across the entire fault. (b) Illustration of a healing (Heaton) pulse propagating along a fault.

near San Bernardino, CA. In situ stress orientations and magnitudes were obtained from depths of 0.9 to 3.5 km using hydraulic fracturing and well-bore breakout analyses. Zoback and Healy (1992) summarized the results of these studies and concluded that the San Andreas fault must be quite weak as there was a complete absence of any right-lateral shear stress on planes parallel to the fault. One of the reasons for selecting this site was that the last major earthquake on this section of the San Andreas was in 1812 and it would be expected that the shear stress would be near its maximum value.

2. Laboratory Friction

Since it is generally accepted that essentially all earthquakes occur on preexisting faults, it is certainly reasonable to conclude that the "frictional" behavior of faults will control the rupture process. Extensive laboratory studies of the frictional behavior of rock surfaces and granular rock (fault gauge) have been carried out. These studies have provided a general understanding of the dependence of friction on slip velocity and other surface characteristics. Velocity weakening is a necessary condition for stick-slip rather than stable sliding on a frictional surface. There is general agreement on frictional behavior under well-constrained laboratory conditions.

The simplest approach to the friction associated with earthquakes is the static-dynamic model. If the slip velocity is zero, $v = 0$, the static coefficient f_s is applicable; if the slip velocity is non zero, $v > 0$, the dynamic coefficient of friction f_d is applicable. As long as $f_s > f_d$ stick-slip behavior is expected. But this model is clearly an oversimplification. The next degree of complexity is to have a functional dependence of friction on slip velocity; an example is

$$f = \frac{f_s}{\left(1 + \frac{v}{v_c}\right)^n} \tag{1}$$

where v_c is a characteristic velocity. However, laboratory experiments have shown that friction does not immediately adjust to a new slip velocity. This led to the introduction of one or more state variables (Ruina, 1983). A typical example with a single state variable is (Okubo, 1989).

$$f = f_o + b \log\left(b_1 \vartheta + 1\right) - a \log\left(\frac{a_1}{v} + 1\right) \tag{2}$$

$$\frac{d\vartheta}{dt} = 1 - \frac{\vartheta v}{D_c} \tag{3}$$

where ϑ is the state variable and f_o, b, b_1, a, a_1, and D_c are empirical constants.

The state variable is generally associated with the time dependence of surface adhesion. Dieterich and Conrad (1984) describe the physics as follows: "A change in slip velocity produces an immediate change in frictional resistance that is of the same sign as the change in velocity. However, as slip progresses at the new velocity, the age of the load bearing contacts begins to evolve to a new population with an average age that is characteristic of the new velocity".

The essential question that must be addressed is whether the empirical friction laws based on laboratory studies are applicable to actual faults (Scholz, 1990, pp 91-96). Of necessity, the laboratory experiments must be carried out at much lower velocities (typically 10^{-6} m/s) than slip velocities on real faults (up to 1 m/s). However, it has also been questioned whether the results are applicable in terms of static friction. Based on their own experimental results, Beeler et al. (1994) argue that the presently accepted friction laws are inadequate to predict real fault behavior. They suggest a much stronger time dependence with a "healing" of a ruptured fault to prevent subsequent slip.

One immediate objection to the rate and state friction law given in (2) and (3) is that the friction coefficient is infinite when $v = 0$, i.e. the static coefficient of friction is infinite. This has led some authors to suggest that faults are always slipping at very low velocities, a very doubtful suggestion. However, modifications of the friction law can be made relatively easily to overcome this objection.

A much more serious objection to the applications of the laboratory derived friction laws to actual faults is the predicted high stress levels and low stress drops. Earthquakes often nucleate at a depth of 10 km where the lithostatic pressure is 250 MPa, a typical static failure stress would be 10 MPa, giving a coefficient of static friction $f = 0.04$ while typical laboratory values are $f = 0.6$. Recognizing this major discrepancy a number of authors have proposed that the low stress is due to a high fluid pressure (Byerlee, 1990; Sleep and Blanpied, 1994). This requires an impermeable fault zone and several mechanisms have been proposed; but at best, this proposal must be considered ad hoc. The stress paradox reinforces other objections to currently accepted concepts of friction on faults.

The rate and state equations predict measurable slip on faults in the interseismic period between major earthquakes. Since the lithostatic normal force increases linearly with depth, it is also expected that the frictional resistance to slip also increases with depth. As the stress on a fault increases during an earthquake cycle it would be expected that slip would occur on the upper portion of the fault while it remains locked at depth (Lorenzetti and Tullis, 1989; Rice, 1993). While there are exceptions, observations, either directly or geodetically, of such fault offsets are quite rare. The locked northern and southern sections of the San Andreas fault are not slipping. On the basis of high resolution strain and tilt data Johnston et al. (1987) found no precursory fault displacements prior to several earthquakes in California. These authors suggest: "that rupture initiation occurs at smaller regions of higher strength which, when broken, allow runaway catastrophic failure".

3. Alternative Friction Models

Many authors have recognized that classic laboratory frictional behavior may not be applicable to real faults. And a variety of alternative hypotheses for rupture mechanics have been proposed.

It has been proposed that the granular fault gauge becomes fluidized during rupture (Melosh, 1979, 1996). Under rather restricted conditions sufficient energy is available to fluidize a narrow zone of fault gauge. Heaton (1990) pointed out that self healing "Heaton " pulses are inconsistent with classic friction experiments and suggested that fault gauge may become fluidized when a critical slip velocity is exceeded. Fluidization could certainly reduce the dynamic friction on a fault but would not be expected to reduce the static friction.

A second approach to dynamic rupture on faults invokes interface waves. Brune et al. (1993) and Anooshelpoor and Brune (1994) have proposed interface waves involving fault surface separation during slip (Mora and Place, 1994). They also suggest that normal interface vibrations associated with these waves can explain the high corner frequency. A number of authors have considered the role of normal vibrations on friction (Tolstoi, 1967; Comninou and Dundurs, 1977, 1978; Freund, 1978). The excitation of Rayleigh waves on a rupture surface can lead to periodic pulses of separation. However, this mechanism is controversial and has not been demonstrated experimentally in a conclusive manner.

Several authors have recently included a linear viscous resistance on the slip surface (Nakanishi, 1994; Morgan et al., 1996). A Barenblatt cohesive zone is introduced at the crack tip and this cohesion models the static friction. With a viscous resistance to slip on the crack surface, the tip singularity is reduced below the value 1/2 associated with the classic stress intensity factor and the velocity of crack propagation is a function of the crack "viscosity". Rupture always initiates at the same value of the cohesive force independent of the viscosity and the rupture velocity increases towards the relevant sound speed as the viscosity is decreased. These authors suggest that there are two slip-mode regimes during earthquake rupture. In the immediate vicinity of the crack tip, slip velocities are small and cohesive forces dominate. This is the regime that has been studied in the laboratory; plastic deformation of the surfaces and gauge dominate and the "frictional" stress is relatively high. At higher slip velocities, away from the crack tip, there is a second frictional mode with low frictional stresses. This second mode may be due to acoustic fluidization or separation waves. A linear viscous rheology is consistent with the fluidization of fault gauge (Savage, 1984; Campbell, 1990). As the driving stress drops, the slip velocity decreases, there is a return to the cohesive mode, and the fault heals.

4. Failure Models

The approaches given above modify the classic approach to friction, but retain the concept of elastic half-spaces on either side of the fault. In fact the crustal rock adjacent to faults is fragmented by secondary faults over a wide range of scales. One consequence of this complexity is the introduction by seismologists of the concepts of asperities and barriers. Asperities often arrest seismic rupture resulting in earthquakes appearing to be a sequence of ruptures. Inversions of seismograms indicate there may be sections of faults, barriers, on which there is little or no coseismic displacement. A conclusion is that faults are rough and complex and stress and strain patterns can be extremely heterogeneous.

In order to model fault heterogeneity Smalley et al. (1985) introduced a hierarchical model for the initiation and propagation of failure on a fault. The asperities on the fault were treated as individual elements with a probabilistic distribution of strengths. If one element failed the stress was transferred to the adjacent element on which an induced failure could occur. If two elements failed the stresses were transferred to the two adjacent elements, and so forth. A cascade of failures resulted.

The behavior of this hierarchical model resembles the failure of a stranded cable or a composite material. These applications raise several interesting questions that are relevant to the behavior of faults. The first is the extent of load sharing. One extreme is the global load-sharing model. If one element fails, the stress on this element is transferred to all other elements. This approach is applicable to "stiff" systems. A laboratory friction apparatus is an example of a stiff system. The second extreme is the nearest neighbor stress transfer. In this case a localized cascade of failures can lead to a catastrophic failure. Localized stress transfer can lead to considerably lower failure stresses than global stress transfer.

When an element is placed under a stress σ there are two approaches to its failure. One is to provide the probability than an element will fail under the prescribed stress. However, studies of the failure of engineering materials indicate that it is preferable to utilize the concept of "time to failure" (Coleman, 1958). The statistical distribution of lifetimes at a prescribed stress is given. The occurrence of aftershocks supports the application of the time to failure model to earthquakes. The cumulative distribution of failure times t_f for an individual element is typically given by (Phoenix and Tierney, 1983)

$$\Pr\left(t_f\right) = 1 - \exp\left(- \nu\, t_f\right) \tag{4}$$

where ν (σ) is the hazard rate under stress σ. One-half of a large number of wires under stress σ will have failed when $t_{1/2} = (\ln 2)/\nu$. For a wide variety of materials a Weibull distribution of failures is found to be applicable, thus

$$\nu = \nu_0 \left(\frac{\sigma}{\sigma_0}\right)^{\rho} \tag{5}$$

The hazard rate has a power-law dependence on stress with the power ρ typically in the range 2-5.

For global load sharing the stress on a failed element is redistributed uniformly to all the remaining elements. The stress on the n surviving elements is given by

$$\sigma = \frac{n_o}{n} \sigma_o \tag{6}$$

The rate at which elements fail is assumed to be given by the rate law

$$\frac{dn}{dt} = - \nu n = - \frac{\nu_o n_o^{\rho}}{n^{\rho - 1}} \tag{7}$$

Solving (7) with $n = n_o$ at $t = 0$ gives

$$n = n_o \left(1 - \rho \nu_o t\right)^{1/\rho} \tag{8}$$

And the time to failure t_f is obtained by setting $n = 0$ with the result $t_f = (\rho \nu_o)^{-1}$. With $\rho = 4$ this gives $t_f = (4 \nu_o)^{-1}$ which is about a factor of two reduction compared with the mean failure time of individual elements.

Numerical solutions of the time to failure model with local load sharing have been obtained by Newman et al. (1995). When an element fails, the stress on the element is transferred to a neighboring element, if two adjacent elements fail, the stress on these elements is transferred to two neighboring elements, if four adjacent elements fail, the stress on these elements is transferred to four neighboring elements, and so forth. A seventh-order illustration of the model is given in Figure 2; there are 128 lowest-order elements.

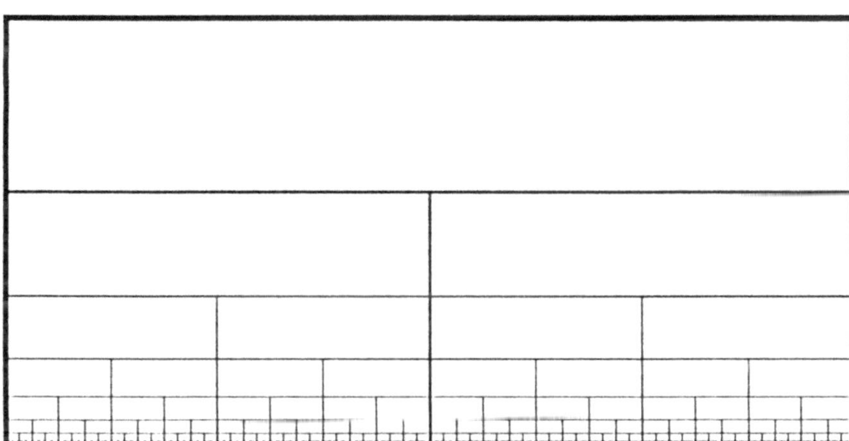

Figure 2. Illustration of a seventh order (N=128) example of the hierarchical model.

In order to accommodate the increase in stress caused by local load sharing from failed elements a reduced time to failure is introduced for each element.

$$t_f = \int_0^{t_{fi}} \left(\frac{s(t)}{s_0} \right)^r dt \tag{9}$$

where t_f is the time to failure under stress σ_0 and t_{fi} is the actual time to failure under stress $\sigma(t)$. Each element i is assigned a random time to failure t_{io} under stress σ_0 based on the distribution given in (4). The actual time to failure of element i, t_{fi}, is reduced below t_{io} if stress is transferred to the element. The time t_{fi} is determined by carrying out the integral in (9) until the condition is satisfied.

Newman et al. (1995) have carried out a sequence of numerical experiments using a 16th order (n = 65,536) realization of this model with $\rho = 4$. An example is given in Figure 3. The total failure sequence is given in Figure 3a. The nondimensional time is taken to be $\tau = \nu_0 t$ and failure in this case occurs at τ = 0.048027. The failure sequence between $\tau = 0.0445$ and failure is illustrated in Figure 3b. It is interesting to note that there is a well defined sequence of partial failures. And for each of these partial failures their is a nested sequence of higher order failures. Further expansions would show higher orders of nesting. The structure is basically self-similar or fractal; there is a scale invariant sequence of precursory failures at all levels. The mean time to failure was found to be

$$\bar{t}_f = \frac{0.048}{\nu_0} \tag{10}$$

Comparing this value with the value obtained for global load sharing shows that local load sharing reduces the mean time to failure by almost an order of magnitude. This could be an explanation for why real faults appear to be weak.

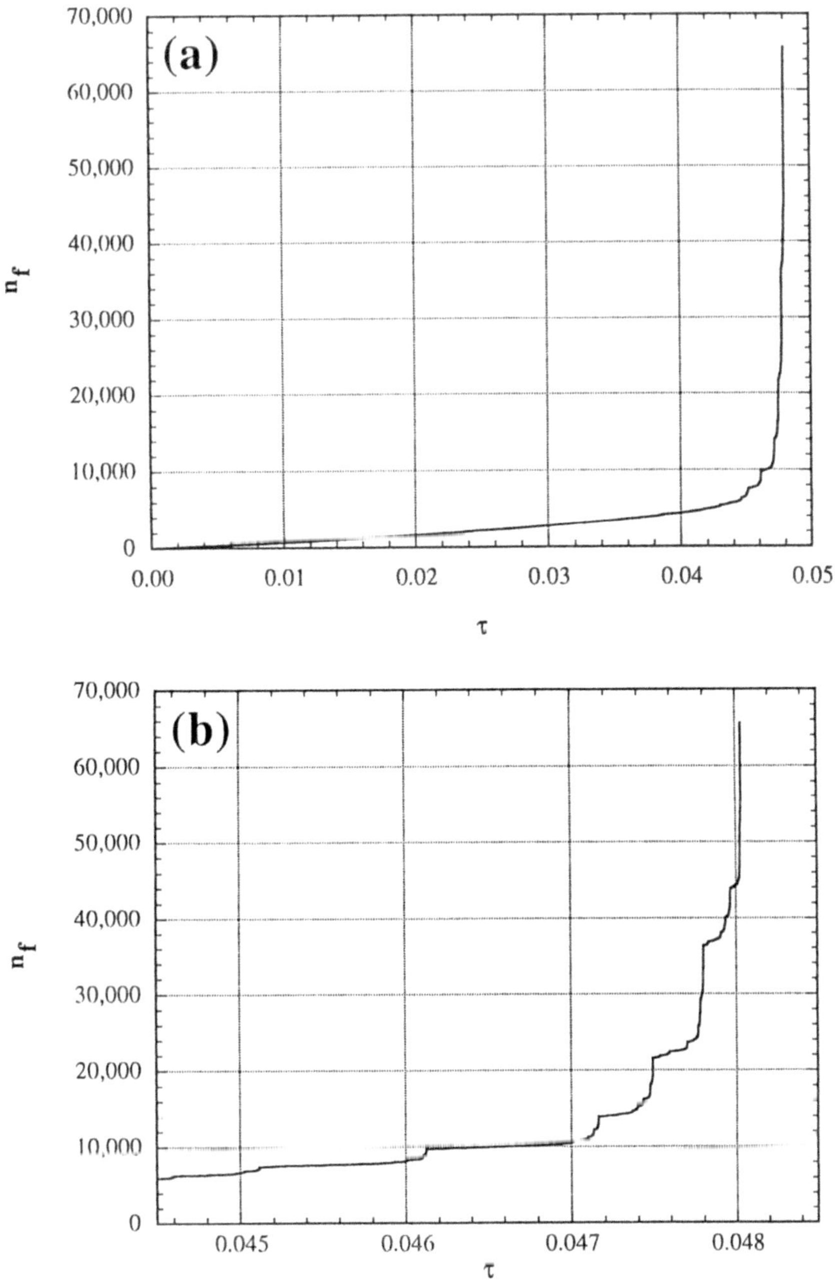

Figure 3. Failure sequence for a 16th order (N=65,536) hierarchical model. (a) Entire failure sequence (failure is complete at $\tau = 0.048031$. (b) Expansion of the final sequence of partial failures.

5. Conclusions

There is a great deal of observational evidence that models for the behavior of faults based on laboratory friction law are not applicable. One explanation is that the laboratory based models are too restrictive and do not simulate the large time scale and relatively high slip velocities on faults. It has been suggested that fault gauge may fluidize, that interfacial waves may be important, and that faults may heal.

An alternative explanation for the discrepancies is the complexity of real faults. The complexity is in both the structure of the fault zone itself and in the fragmentation of the rock in which the fault is embedded. Stochastic time-to-failure models are an alternative approach to the earthquake rupture process.

6. References

Anooshelpoor, A. and Brune, J.N. (1994) Frictional heat generation and seismic radiation in a foam rubber model of earthquakes, *Pure Ap. Geophys.* **142**, 735-747.

Beeler, N.M., Tullis, T.E., and Weeks, J.D. (1994) The roles of time and displacement in the evolution effect in rock friction, *Geophys. Res. Let.* **21**, 1987-1990.

Brune, J.N., Brown, S., and Johnson, P.A. (1993) Rupture mechanism and interface separation in foam rubber models of earthquakes: a possible solution to the heat flow paradox and the paradox of large overthrusts, *Tectonophys.* **218**, 59-67 .

Byerlee, J. (1990) Friction, overpressure and fault normal compression, *Geophys. Res. Let.* **17**, 2109-2112.

Campbell, C.S. (1990) Rapid granular flows, *An. Rev. Fluid Mech.* **22**, 57-92.

Coleman, B.D. (1958) Statistics and time-dependence of mechanical breakdown in fibers, *J. Ap. Phys.* **29**, 968-983.

Comninou, M. and Dundurs, J. (1978) Elastic interface waves and sliding between two solids, *J. Ap. Mech.* **45**, 325-330.

Dieterich, J.H. and Conrad, G. (1984) Effect of humidity on time and velocity-dependent friction of rocks, *J. Geophys. Res.* **89**, 4196-4202.

Heaton, T.H. (1990) Evidence for and implications of self-healing pulses of slip in earthquake rupture, *Phys. Earth Planet. Int.* **64**, 1-20.

Johnston, M.J.S., Linde, M.J.S., Gladwin, M.T., and Borcherdt, R.D. (1987) Fault failure with moderate earthquakes, *Tectonophys.* **144**, 189-206 .

Lachenbruch, A.H. and Sass, J.H. (1992) Heat flow from Cajon Pass, fault strength, and tectonic implications, J. Geophys. Res. **97**, 4995-5015.

Lorenzetti, E. and Tullis, T.E. (1989) Geodetic predictions of a strike-slip fault model: Implications for intermediate- and short-term earthquake prediction, *J. Geophys. Res.* **94**, 12,343-12,361.

Melosh, H.J. (1979) Acoustic fluidization: A new geologic process?, *J. Geophys. Res.* **84**, 7513-7520.

Melosh, H.J. (1996) Dynamical weakening of faults by acoustic fluidization, *Nature,* **379, 601-606.**

Mora, P. and Place, D. (1994) Simulation of the frictional stick-slip instability, *Pure Ap. Geophys.* **143**, 61-87.

Morgan, J.D., Turcotte, D.L., and Ockendon, J.R. (1996) Models for earthquake rupture propagation, *Tectonophys.*,in press.

Nakanishi, H. (1994) Continuum model of mode-III crack propagation with surface friction, *Phys. Rev.* **E49**, 5412-5419.

Newman, W.I., Turcotte, D.L., and Gabrielov, A. (1995) Log-periodic behavior of a hierarchical failure model with applications to precursory seismic activation, *Phys. Rev. E,* **52,** 4827-4835.

Okubo, P.G. (1989) Dynamic rupture modeling with laboratory-derived constitutive relations, *J. Geophys. Res.* **94**, 12,321-12,335.

Phoenix, S.L. and Tierney, L.U. (1983) A statistical model for the time dependent failure of unidirectional composite materials under local elastic load-sharing among fibers, *Eng. Fract. Mech.* **18**, 193-215.

Rice, J.R. (1993) Spatio-temporal complexity of slip on a fault, *J. Geophys. Res.* **98**, 9885-9907.

Ruina, A. (1983) Slip instability and state variable friction laws, *J. Geophys. Res.* **88**, 10,359-10,370.

Savage, S.B. (1984) The mechanics of rapid granular flows, *Adv. Ap. Mech.* **24**, 289-366.

Scholz, C.H. (1990) *The Mechanics of Earthquakes and Faulting* (Cambridge University Press, Cambridge) 439p.

Sleep, N.H. and Blanpied, M.L. (1994) Ductile creep and compaction: A mechanism for transiently increasing fluid pressure in mostly sealed fault zones, *Pure Ap. Geophys.* **143**, 9-40.

Smalley, R.F., Turcotte, D.L., and Solla, S.A. (1985) A renormalization group approach to the stick-slip behavior of faults, *J. Geophys. Res.* **90**, 1894-1900.

Tolstoi, D.M. (1967) Significance of the normal degree of freedom and natural normal vibrations in contact friction, *Wear* **10**, 199-213.

Wald, D.J. and Heaton, T.H. (1994) Spatial and temporal distribution of slip for the 1992 Landers, California, earthquake, *Seis. Soc. Am. Bull.* **84**, 668-691.

Zoback, M.D. and Healy, J.H. (1992) In situ stress measurements to 3.5 km depth in the Cajon Pass scientific research borehole: Implications for the mechanics of crustal faulting, *J. Geophys. Res.* **97**, 5039-5057.

MICRO-BRANCHING AS AN INSTABILITY IN DYNAMIC FRACTURE

J. FINEBERG[†], S. P. GROSS[‡] AND E. SHARON[†]

[†]*The Racah Institute of Physics*
The Hebrew University of Jerusalem
Jerusalem, Israel

[‡]*The Center for Nonlinear Dynamics*
The University of Texas
Austin, TX 78712, U.S.A.

Abstract. Experiments in brittle, amorphous PMMA indicate that the process of dynamic fracture is governed by a micro-branching instability. At a critical velocity, v_c, a single crack undergoes an abrupt, well-defined transition to microscopic crack branching. As a result, the velocity of the crack develops oscillations, the mean acceleration decreases and structure is formed on the fracture surface. Beyond v_c the total fracture surface created is a linear function of the energy release rate. Micro-branch profiles follow a power law and develop into macroscopic crack branching.

1. Introduction

The subject of the rapid propagation of cracks in a brittle material is one whose roots go back to the early work of Mott [1] over half of a century ago. Since that time a great deal of both analytic and experimental work has been dedicated to understanding the phenomenon of dynamic fracture, where the behavior of a moving crack driven by externally imposed stresses is studied.

Much analytical progress has been made in assuming that the medium behaves according to the equations of linear elasticity [2]. To obtain the equation of motion for a single moving crack, one needs to match the fracture energy of the material (defined as the energy needed to create a crack of unit length) to the energy flux into the crack tip which is supplied by the surrounding elastic fields. These theories predict that a single crack should smoothly accelerate until reaching a limiting velocity given by the Rayleigh wave speed, V_R, of the material. The motion of the crack in this picture is dictated by the forms of both the applied stresses and the fracture energy in the material.

J. R. Willis (ed.), IUTAM Symposium on Nonlinear Analysis of Fracture, 177–189.

Experiments fail to confirm this idealized picture. The surface created by the crack is not necessarily smooth and flat. In brittle materials such as glass, ceramics and brittle plastics a characteristic pattern [3] coined "mirror, mist, hackle" is observed where the fracture surface progressively roughens as the velocity of the crack increases. The velocity of the crack, instead of approaching the Rayleigh wave speed, will rarely reach even half of this value [4], with the crack eventually bifurcating (or "crack branching") as the energy flow into the crack tip becomes sufficiently large.

An alternative view of the fracture process was suggested by Ravi-Chandar and Knauss [5]. In a series of experiments on the brittle plastic, Homolite 100, they observed the simultaneous propagation of an ensemble of micro-cracks, instead of a single propagating crack. In light of this, they viewed the fracture process as a coalescence of micro-voids or defects situated in the crack's path. The intense stress field at the crack tip would cause pre-existing defects in the material to propagate. Subsequently, increased energy flux to the tip, in this view, forms an increase in the number of micro-cracks thereby creating a mechanism for enhanced dissipation. As defects exist in most materials, the above picture would suggest that crack propagation via interacting micro-voids should, in general, occur as a randomly activated process.

In this paper we will briefly review a series of recent experiments performed on brittle PMMA (poly-methyl-methacrylate). The results of these experiments offer a different view of the fracture process which complements both of the above approaches. The formation and evolution of micro-cracks result from a *dynamic instability* of a moving crack. These experiments show that a sharp transition [6] from a state of a single propagating crack to one of an ensemble of cracks occurs as the crack velocity increases. Above the critical velocity of $v_c=0.36V_R$, a single crack undergoes a local change in topology and sprouts small microscopic side branches [7]. The dynamics of both the main and local branches (micro-branches) are inter-related. As a function of the mean velocity, v, these branches will increase in length as the mean dynamics of a crack change dramatically; the mean acceleration drops, the crack velocity develops oscillations and structure is formed on the fracture surface [6,8]. As the branches grow in size, they evolve into macroscopic, large scale crack branches. Thus the instability links the behavior of crack branches on both microscopic and macroscopic scales. The general nature of this instability is suggested by the acoustic emissions [9] of moving cracks in both PMMA and glass where large amplitude emissions centered at well defined frequencies appear in both materials as the crack velocity surpasses $0.4V_R$.

Quantitative measurement [10] in PMMA of the energy flux into the tip of a moving crack together with the total amount of surface area created indicates that the micro-cracking instability is the main mechanism for the dissipation of energy by a moving crack. The rate of new surface creation, beyond v_c, is proportional to the energy flux into the tip of the crack. This provides an explanation for why the limiting velocity of a crack, the Rayleigh wave speed, is never realized.

2. Experimental System

Before reviewing these results we will first briefly describe the experimental system. A more detailed description can be found in [6]. The experiments were conducted in thin, quasi - 2D sheets of brittle, cast PMMA [11] having a thickness (Y direction) of either 0.8 mm or 3 mm with vertical (parallel to the direction of applied stress or "Z" direction) and horizontal (parallel to the propagation or "X" direction) dimensions between 50-200 mm and 200-400 mm respectively. Stress was applied to all samples via uniform displacement of the vertical boundaries with the fracture initiated at constant displacement. Applied stresses in the experiment were varied between 10-18 MPa. Prior to loading, a small "seed" crack was introduced at the edge of the sample midway between the vertical boundaries. The sample geometry was varied to provide either steady-state crack propagation [10] at a given energy density within the sample, or a continuously changing velocity throughout the experiment.

Steady-state propagation was achieved by using a thin strip configuration with the ratio of its vertical to horizontal dimensions between 0.25 - 0.5. If the crack tip is sufficiently far from the horizontal boundaries of the system this geometry approximates an infinitely long strip with approximate "translational invariance" in the direction of propagation. This state is realized when the crack reaches a length of about half of the vertical size of the system. At this point, advance of the crack by a unit length frees an amount of energy equal to the (constant) energy per unit length stored in the plate far ahead of the crack.

Under these conditions a crack arrives at a state of constant mean velocity with G, the energy flux into the crack tip per unit extension of the crack, given by $G = \sigma^2 L/(2E)$ where σ is the applied stress at the vertical boundaries, L the vertical size of the system and E the Young's modulus of the material [12]. Using this geometry we can directly measure G with an 8% accuracy. In the experiments described, G was varied between 400-5000 J/m^2.

The crack velocity was measured by first coating the side(s) of the sample with a thin (~ 1000 Å) resistive layer. Upon fracture initiation, a propagating crack will cut the resistive coating thereby changing the sample's resistance. We measure the resistance change by digitizing to 12 bit accuracy at a rate of 10 MHz. Thus, in PMMA, the location of the crack tip can be established to a 0.1mm spatial resolution at 0.1μsec intervals yielding a velocity resolution of better than 25 m/s.

After fracture, the crack profile in the XZ plane was measured optically with a spatial resolution of 1-5μm depending on the magnification used. Although the medium is idealized as 2D, the plates used are of finite thickness. All comparisons between local branches and the crack velocity were made using branches adjacent to the plane where the velocity measurements were performed.

Additional measurements of the fracture surface profiles and surface roughness were performed by means of an X-Y scanning profilometer with a resolution of 0.1μm in the Z direction. Both the optical and profilometer measurements were then correlated with the velocity measurements.

3. Results

A typical example of the dynamics of a moving crack in PMMA is shown in figure 1. After a nearly instantaneous jump to an initial velocity of $0.1\text{-}0.2V_R$ a

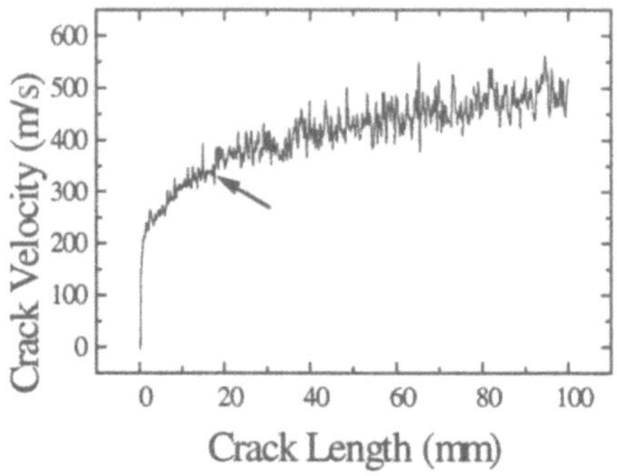

Figure 1. Velocity of the crack tip as a function of its length. At the critical velocity of 335 ± 10 m/s indicated by the arrow, the mean acceleration of the crack slows and the velocity begins to oscillate.

crack will accelerate smoothly and rapidly until reaching a critical propagation velocity, v_c, which in PMMA is 335 m/s (or $0.36V_R$). At that point the mean acceleration of the crack drops and oscillations of the velocity are observed whose amplitude is an increasing function of the mean velocity.

Coincident with the onset of the velocity oscillations in PMMA non-trivial structure is formed on the fracture surface as shown in figure 2. In the upper half of the figure the transition between the featureless "mirror" zone where $v < v_c$ to the onset of the oscillations where $v \sim v_c$ is shown. In the mirror zone there are no features at scales larger than 1μm. The lower half of the figure shows a section of the fracture surface where the oscillations in v are well developed and a pattern with a "wavelength" on the order of 1mm is formed. On this scale PMMA is entirely amorphous.

Although the pattern observed appears roughly periodic as a function of the length of the crack, as the crack tip accelerates the distance between the ribs on the surface increases. The observed pattern on the fracture surface is correlated with the oscillations in velocity [6]. In figure 3 we present a typical time series and power spectrum of the velocity of a crack in steady-state propagation. Beyond v_c the patterns and velocity oscillations are roughly periodic as a function of *time* with a characteristic time scale of 2-3 μsec. Power spectra typically show no single sharp frequency although a well-defined time scale is apparent. The location of the peak is constant as a function of the mean velocity. The oscillation period is an intrinsic time scale, independent of either the lateral dimensions or thickness of the material. Below v_c the spectrum has little power with no apparent time scale.

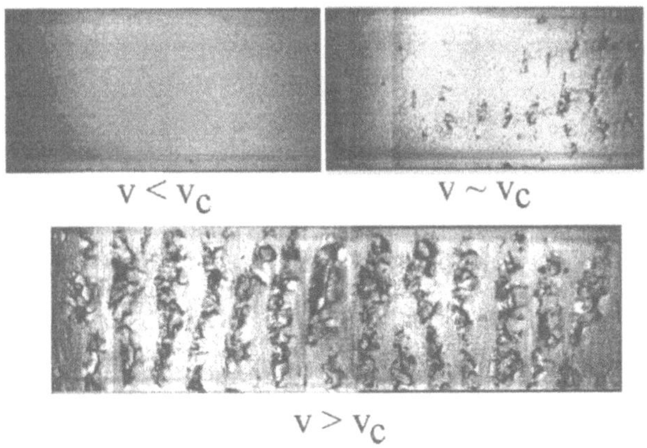

Figure 2. A typical photograph of the fracture surface in PMMA for $v < v_c$ (upper left), $v \sim v_c$ (upper right) and $v > v_c$ (lower figure) where a pattern develops on the surface. In the figure the crack propagates from left to right. The pictures are to scale, where the width of the samples shown is 3 mm.

The acoustic emissions of a running crack in both PMMA and soda-lime glass [9] show similar behavior. Below $0.4V_R$ only very low-level sound emission is observed in both materials with large amplitude, peaked emissions above that point. In both cases the total acoustic power was approximately 3% of

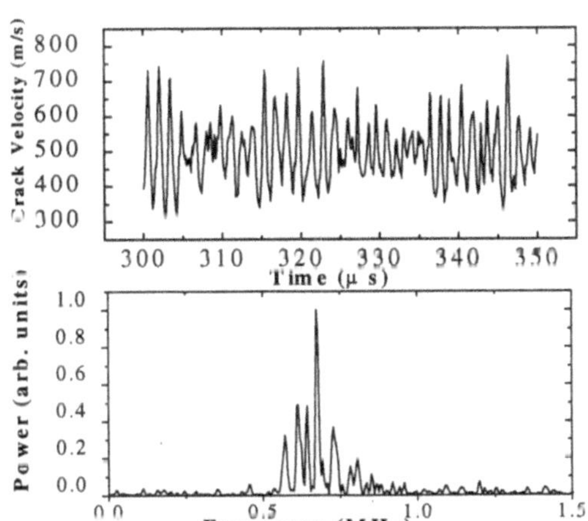

Figure 3. The steady-state velocity of a crack propagating in PMMA beyond v_c as a function of time (upper), with its corresponding power spectrum (lower). Note the existence of a well-defined time scale.

the fracture energy. In PMMA, the characteristic time of the acoustic signal is the same as in the crack velocity oscillations. Subsequent experiments [8] indicate an excellent degree of correlation between the acoustic signals and the velocity fluctuations. The quantitative similarity of the acoustic emission spectra in both glass and PMMA is a strong indication of the generality of this instability as the microscopic structures of the two materials differ substantially.

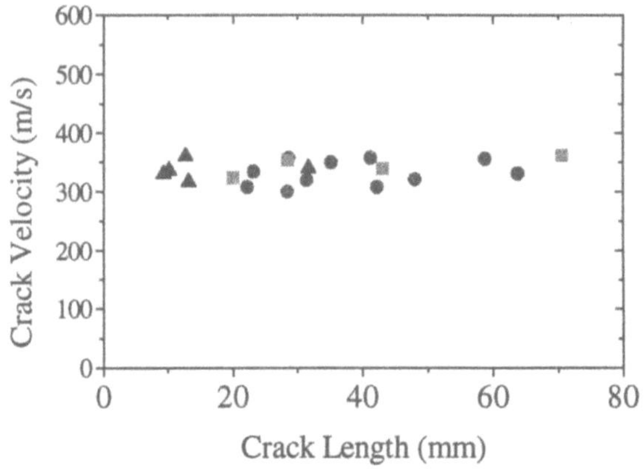

Figure 4. Critical velocity as a function of crack length at the moment of the appearance of surface structure. triangles, 1.6 mm wide extruded PMMA surrounded by air; circles, 3.2 mm wide cell-cast PMMA surrounded by air; squares, 3.2 mm wide cell-cast PMMA surrounded by helium gas.

The critical velocity, as determined from profile measurements of the fracture surface, is the velocity at which the surface height deviations surpass a threshold value (for details see [6]). As figure 4 indicates, the value of v_c is independent of sample geometry, sample thickness, applied stress, surrounding atmosphere and the acceleration rate of the crack; whenever the crack velocity surpasses $0.36V_R$ both oscillations and resultant surface structure develop. Below v_c neither patterns nor velocity oscillations are observed.

We now turn to the mechanism that drives the instability. Optical analysis of the near vicinity of the fracture surface reveals, beyond v_c, considerable structure in the XZ plane. For $v > v_c$ microscopic crack branches, as observed [5] in Homolite 100, appear. Profiles in the XZ plane both below and above v_c are presented in figure 5.

Below v_c (Fig 5a), as in the profilometer measurements, the fracture surface is entirely featureless to the resolution (~ 3μm) of our measurements. Beyond v_c microscopic "daughter" cracks begin to appear. In PMMA micro-branches have a minimum length of approximately 10 μm. At velocities near threshold, as in Fig.

5b, the spacing between branches is relatively large. As the velocity of the crack increases, bunches of micro-branches are seen. Within a bunch, the lengths of

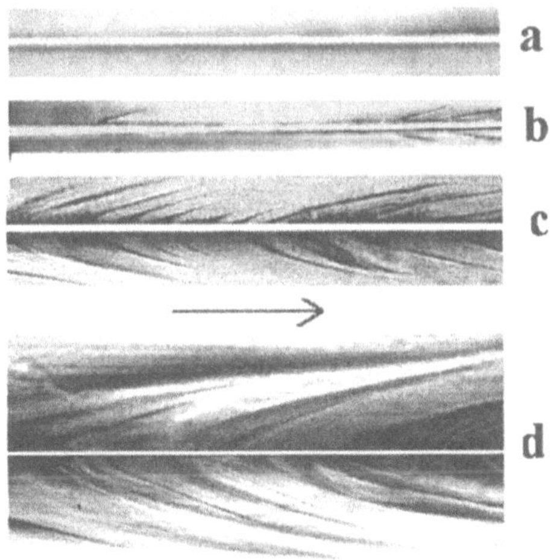

Figure 5. Typical examples of micro-branches imaged optically in the XZ plane. All pictures are to scale and the path of the main crack in each picture is indicated in white. The arrow, indicating the direction of propagation, 0.25 mm in length. (a) $v < v_c = 330$ m/s (b) $v = 340$ m/s (c) $v = 400$ m/s (d) $v = 480$ m/s.

the individual branches in a bunch can vary considerably. The spacing between bunches of cracks corresponds to that between velocity fluctuations. For relatively high crack velocities (Fig. 5d) the structure of the micro-branches becomes increasingly complex.

In figure 6 we demonstrate that the velocity for the onset of micro-branching indeed corresponds to v_c, the critical velocity for the onset of both surface roughness and velocity oscillations. Although the final crack velocity achieved depends on the energy stored in the medium, as the figure indicates, the onset of local crack branching is invariant.

We now turn to the relation between micro-branches and crack dynamics. Upon the onset of the branching instability we can no longer think in terms of a single crack. To characterize the ensemble behavior, we look at the running averages over 5mm of both the crack velocity and micro-branch length. In figure 7a we plot the mean branch length as a function of the mean velocity. The transition to branching behavior is sharp with a well defined (roughly linear) dependence between the mean crack velocity and mean branch length.

In figure 7b we present measurements of the mean amplitude of the fracture surface as a function of the mean crack velocity. As in the case of the branch lengths (fig. 7a) the sharp transition at v_c from a flat, featureless surface to an increasingly rough surface is evident. Note the difference in scale between the

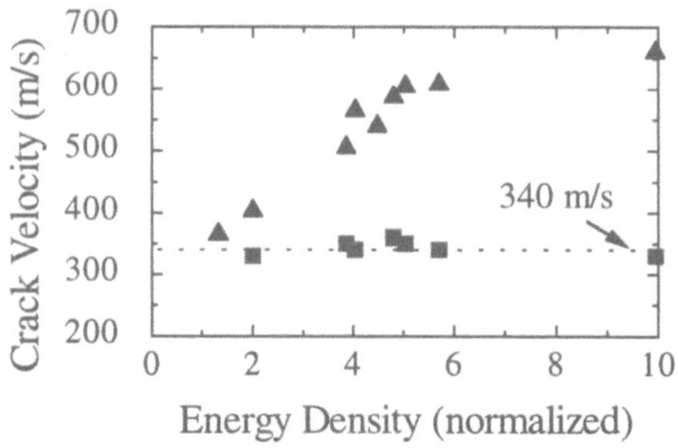

Figure 6. The critical velocity for the onset of local crack branching (squares) and the maximum velocity achieved (triangles) as a function of the density of the energy stored in the plate prior to fracture. The energy density here is normalized by the energy density stored in a plate in which the maximum velocity observed was just below the threshold for crack branching. The data shown were obtained in 3 mm plates of different sizes and aspect ratios indicating that the branching threshold is intrinsic to the material.

branch lengths and the surface amplitudes with the branch length nearly two orders of magnitude larger.

In summary, the onset of local branching occurs as a sharp transition; beyond v_C both the mean dynamics of the crack as well as the formation of non-trivial surface structure are governed by the generation and growth of local crack branching.

We now turn to the functional form of the micro-branches [7]. As can be seen from figure 5, at a given crack velocity the observed lengths of the micro-branches are broadly distributed. In figure 8 we superimpose micro-branches that were measured at the same mean crack velocity. It is evident from the figure that the functional form of these branches is not random but surprisingly well-defined. Although the lengths of micro-branches have a wide distribution, a branch, once formed, follows a distinct trajectory. A log-log plot (figure inset) of the mean profile of the branches shown in the figure indicates that a branch follows a power law trajectory of the form: $y = 0.20 \cdot x^{0.70}$ where y is the normal and x the direction parallel to the propagation direction of the main crack. The data shown were obtained for v = 370m/s = 1.09v_C. For crack velocities within 10% of v_C the superposition of individual trajectories exhibits remarkably little scatter. Although the functional form of the mean branch profile remains the same, the scatter increases with increasing crack velocity reaching several times the measurement uncertainty at velocities 40% over v_C. (The scatter for low velocities is on the order of the ± 3μm uncertainty in the starting point of a given branch.)

How are micro-branches related to macroscopic, large scale crack branches? Consider the branching angle of the crack, defined as the initial angle between a branched crack and the direction of propagation of the main crack. Previous experiments have observed that the macroscopic branching angle [13] falls in the range of 10-15°. Although often observed, a first principles understanding of

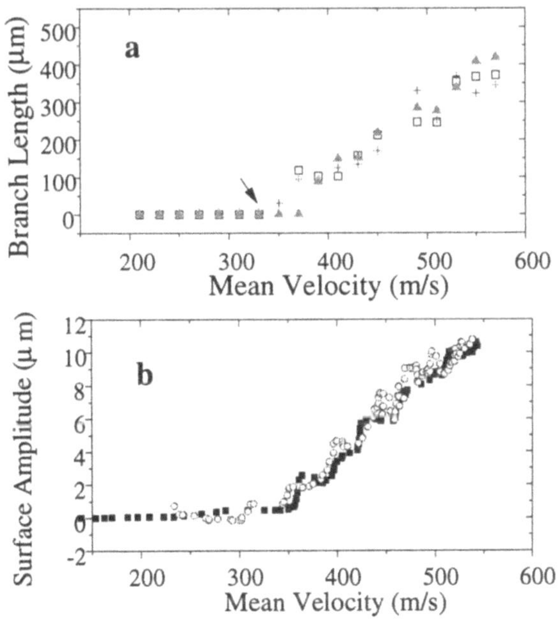

Figure 7. A comparison of the mean length of micro-branches (a) and the mean amplitude of the fracture surface (b) with the mean crack velocity for a number of different experiments. The branch length, surface amplitude, and velocity measurements were averages over 5 mm intervals. The arrow indicates v_c, the transition velocity for the instability. Note both the sharpness of the transition and the large different in the scales of the two figures. As in figure 6, the experiments shown were performed under different stresses and plate geometries.

crack branching including a well-defined criterion for the appearance of crack branches together with a description of the branching angle is lacking. There are a number of predictions for this angle in the literature. The well known Yoffe calculation [14] predicts an angle of 60°, branching angles of 30° were observed in molecular simulations [15], and an angle of 18° was predicted [16] using an energy criterion and considering non-singular terms of the stress field, relevant away from the near vicinity of the crack tip. The functional form of the crack branch profile suggests that the observed "branching angle" is a function of the scale at which an observation is made.

The power-law form of the micro-branches relates the microscopic and macroscopic scales. At scales of 0.1-0.3mm the branching profile would indeed produce apparent branching angles that would fall within the range noted in the literature as theories based on non-singular terms of the stress field predict. At the smallest scales one would expect the singular terms in the stress field to dominate thereby giving rise to large branching angles. The *apparent* branching angle at the smallest scale observed in our experiments (on the order of 5-10 μm) is about 30° but at still smaller scales we might expect to see a larger angle. The basic question of the mechanism *causing* crack branching, in this picture, reduces to the as yet unresolved question of the exact mechanism giving rise to the microscopic instability.

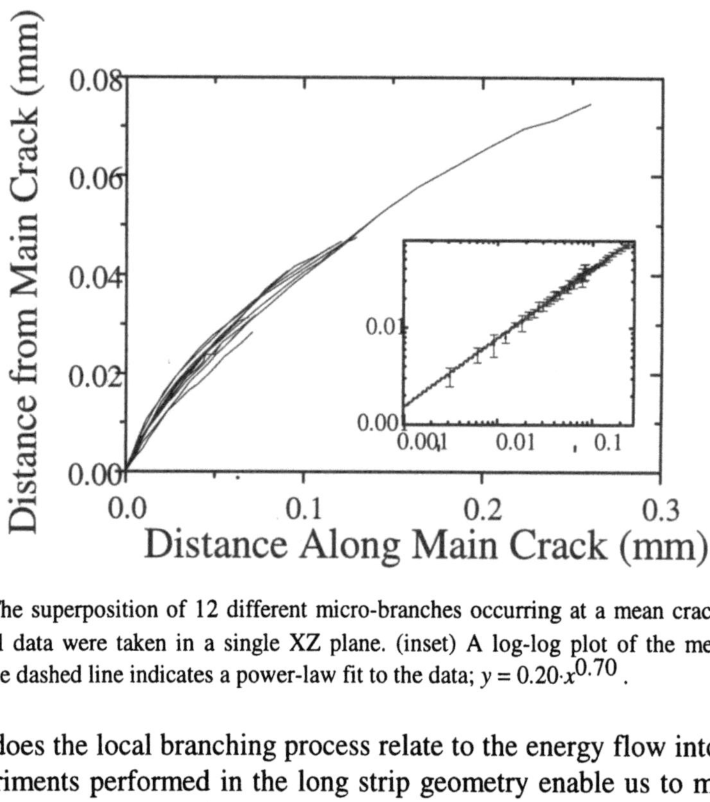

Figure 8. The superposition of 12 different micro-branches occurring at a mean crack velocity of 370 m/s. All data were taken in a single XZ plane. (inset) A log-log plot of the mean trajectory obtained. The dashed line indicates a power-law fit to the data; $y = 0.20 \cdot x^{0.70}$.

How does the local branching process relate to the energy flow into the crack tip? Experiments performed in the long strip geometry enable us to measure the energy flux (or energy release rate), G, into the crack tip in "steady-state" propagation. We can compare this with the total amount of fracture surface formed when the surface area formed by the micro-branches is accounted for. This comparison is shown in figure 9 for fracture in PMMA where the surface area formed is normalized by that which would be formed by a single crack. Normalizing G by the energy release rate immediately preceding the formation of micro-branches (1.0 x 10^6 erg/cm^2) leads, after an initial jump, to a linear dependence between the energy flux into the crack tip and the amount of surface area formed having a slope of 1.0 ± 5%. This indicates that nearly *all* of the energy initially stored in the plate went into creating new surface. The value of the slope yields the *constant* value for the fracture energy of 5 x 10^5 erg/cm^2.

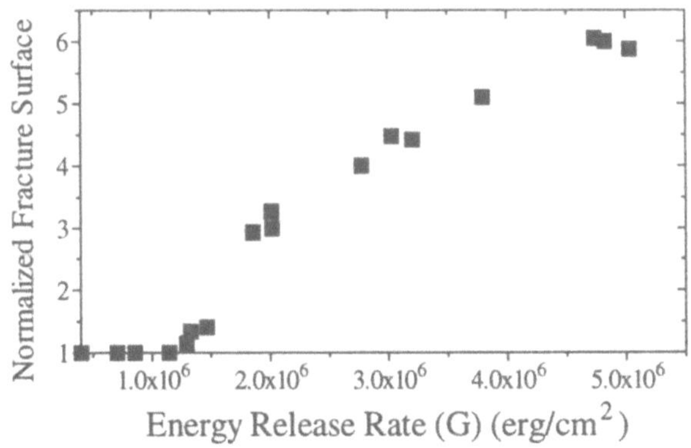

Figure 9. Total fracture surface including the fracture surface formed by micro-cracks normalized by the fracture surface that would be formed by a single crack as a function of the energy flux into the crack tip.

This is surprising in view of the processes that contribute to the fracture energy in PMMA. The value obtained for the fracture energy is 3 orders of magnitude larger than the value of the surface energy (the energy expended in breaking bonds in the material). Most of the energy that goes into the fracture energy [17] is expended in complex, rate dependent, processes such as the shearing of the long molecules comprising PMMA. A constant fracture energy for $v > v_c$ indicates that the system somehow balances these different processes against the creation of surface area.

4. Conclusions

We have found that a micro-branching instability occurs in PMMA where, at a critical velocity, the system undergoes a sharp transition from the propagation of a single crack to the production of microscopic branches. The micro-branches formed are short-lived, possibly screened from the stress field that drives them by the main crack. As a result, the birth and death of these micro-cracks causes oscillations in the velocity of the crack "ensemble".

A possible scenario for these oscillations is as follows. When a micro-branch is formed, the total energy flowing into the process zone (or tip area) now must subdivide between the two competing cracks. As a result, the overall crack slows down. Eventually, the main crack will outrun the daughter crack, screening it from the stress field that drives it. As the daughter crack dies, the total energy is refunneled into the main crack which accelerates until it, once again, bifurcates and the scenario is repeated. The time scale for the oscillations in this picture should be governed by the mean "lifetime" of a branch. We can estimate this lifetime by means of figure 7a which exhibits a linear dependence between the mean length of a branch and the mean propagation velocity of the crack. The slope of this graph yields a time scale of 2μsec, the typical time scale for the oscillations.

The micro-branching picture also links together many hitherto unrelated aspects of dynamic fracture. As we have shown, the onset of local branching gives rise to the formation of structure on the fracture surface. In addition, as the micro-branches increase in size the power law form of their trajectory provides a link between branching on microscopic and macroscopic scales; essentially indicating that the two effects are one and the same. Thus, the onset of the micro-branching instability provides a criterion for crack branching. The instability also provides an explanation for the long-standing problem of the limiting velocity of a crack. As the energy flux into the tip of a crack is increased, more and more fracture surface is formed. As figure 9 shows, the amount of fracture surface formed by both the main and "daughter" cracks is linearly dependent on the amount of stored energy. Thus, although the fracture energy (generalized to the *total* fracture surface formed) is constant in the material, the steady increase in the formation of surface area by the crack provides an effective "dissipation" if one were to view fracture in a single crack picture. For a given energy flow into the system this effective dissipation then limits the velocity attained by the crack.

The view presented here of the fracture process is similar to that of the micro-crack picture proposed in [5], but differs from it in one essential aspect. In the micro-crack picture the propagation of a crack is a noise-dominated process; a single crack never really exists. A propagating crack is viewed as an ensemble of micro-cracks that nucleate from defects or micro-voids that are randomly dispersed in the material and are activated by the strong stress field in the vicinity of the crack's tip. In the picture presented here, a sharp, well defined transition to a branching *instability* occurs when a single crack dynamically becomes unstable and bifurcates. Although, because of pre-existing defects in the material, there is certainly a randomness involved, we view the fracture process, as a whole, to be a deterministic one.

The picture presented here is, of course, just a beginning. Much additional work remains to be done. Some of the many questions which arise are, for example, what is the basic mechanism giving rise to microscopic branching at a critical velocity? What determines v_c and how general are the effects observed in PMMA? Acoustic emission data provide an indication that the instability generally occurs in brittle, amorphous materials but what, if any, effects persist in more plastic materials or materials that posses a basic structure?

The authors are grateful to M. Marder for numerous enlightening conversations over the course of this work. We also would like to acknowledge the United States - Israel Binational Science Foundation (grant no. 92-148) for their support of this work.

1. N. F. Mott, Engineering **165**, 16, (1948).

2. A good review of this work is presented in the book *Dynamic Fracture Mechanics*, L. B. Freund (Cambridge University Press, New York, 1990).

3. J. J. Mecholsky in *Strength of Inorganic Glass*, edited by C. R. Kurkijan (Plenum, New York, 1985).

4. K. Ravi Chandar and W. G. Knauss, Int. J. Fract. **26**, 141 (1984).

5. K. Ravi-Chandar and W. G. Knauss, Int. J. Fract. **26**, 65 (1984).

6. J. Fineberg, S. P. Gross, M. Marder, and H. L. Swinney, Phys. Rev. Lett. **67**, 457 (1991); *ibid.* Phys. Rev. **B45**, 5146 (1992).

7. E. Sharon., S. P. Gross, and J. Fineberg, Phys. Rev. Lett. **74**, 5096 (1995).

8. J. F. Boudet, S. Ciliberto, and V. Steinberg, Europhys. Lett. **30**, 337 (1995).

9. S. P. Gross, J. Fineberg, W. D. McCormick, M. Marder, and H. L. Swinney, Phys. Rev. Lett. **71**, 3162 (1993).

10. E. Sharon, S. Gross, and J. Fineberg, Phys. Rev. Lett. **76**, 2117 (1996).

11. The PMMA used has the following static properties: Young's modulus = 3.1 x 10^3MPa; Poisson ratio = 0.35; V_R=926 m/s.

12. We used the value of 2.8 x 10^{10} erg/cm^2 for E which was obtained by direct measurement using a plate of the dimensions and manufacture used in the experiments.

13. For PMMA see B. Cotterell, App. Mat. Res., 227 (1965);In homalite-100 and polycarbonate see M. Ramulu and A. S. Kobayashi, Int. J. of Fracture **27**, 187 (1985); In glass see J.W. Johnson and D. G. Holloway, Phil. Mag. **17**, 899 (1968).

14. E. H. Yoffe, Philos. Mag. **42**, 739 (1951).

15. F. F. Abraham, D. Brodbeck, R. A. Rafey, and W. E. Rudge, Phys. Rev. Lett. **73**, 272 (1994). Branching at 30° was observed but this may correspond to the direction of symmetry of the hexagonal lattice used in the simulation

16. P. S. Theocaris and H. Georgiadis, Int. J. Fract **29**, 181 (1985); J. G. Michopoulos and P. S. Theocaris, Int. J. Engng. Sci **29**, 13 (1991).

17. R. P. Kusy and M. J. Katz, Polymer **19**, 1345 (1978).

INSTABILITIES IN DYNAMIC FRACTURE

J.S. LANGER
Institute for Theoretical Physics
University of California
Santa Barbara, CA 93106-4030, USA

My purpose in this presentation is to summarize and interpret recent theoretical work by Emily Ching, Hiizu Nakanishi and myself on dynamic stability of mode I fracture [1]. Our investigation has been motivated by the experiments of Fineberg *et al.* [5, 6] in which it appears that mode I cracks in a variety of materials encounter some kind of oscillatory instability at velocities appreciably below the Rayleigh speed. The experiments indicate that the instability involves out-of-plane deformations; that is, the cracks are unstable against deflection or branching away from their directions of motion. We find that a broad class of models of dynamic fracture exhibits this kind of instability. Several features of our results appear to contradict conventional assumptions in this field.

Our basic premise has been that we must solve mathematically well posed free-boundary problems in order to understand fracture dynamics. This means that we must start with physical models that are complete enough that the configuration of the system and its state of motion at one time are, at least in principle, sufficient to determine the configuration and state of motion at all later times. One of our main conclusions is that, in carrying out such analyses, it is necessary to include detailed aspects of the cohesive mechanism that is acting at the crack tip. Simple far-field assumptions, such as the assumption that the mode II stress-intensity factor must vanish ($K_{II} = 0$), cannot generally be correct. Their range of validity includes, for example, the quasi-static situations considered by Cotterell and Rice [3], but it does not include the dynamic instabilities observed in the recent experiments [5, 6].

In all of our work, we have used conventional, cohesive-zone models of fracture in continuous, isotropic solids [8, 9]. (Results similar to ours most

J. R. Willis (ed.), IUTAM Symposium on Nonlinear Analysis of Fracture, 191–199.

likely can be obtained with lattice models of the kind studied recently by Marder [11], but such calculations almost certainly will be even more difficult than those reported here.) In our ideally brittle continuum models, the material that is breaking is linearly elastic everywhere outside of sharply defined fracture surfaces, and a finite-ranged cohesive stress opposes the separation of these surfaces at the crack tip. These models are highly simplified pictures of the complex, nonlinear processes taking place within the process zone near the tip of a real crack; but their simplicity allows us to obtain exact analytic results and, from these, to learn what ingredients will be needed in more realistic theories.

The cohesive stress serves two related purposes; it provides the fracture energy and it regularizes the stress singularities. A more precise way of saying this is that the shape of the cohesive zone is a dynamic entity that moves in response to the stresses in its neighborhood. It automatically must adjust in such a way that the stresses are nonsingular and continuous at all times. Note in particular that the nonsingular stress exactly at the crack tip must be the yield stress appropriate for whatever fracture mode, or combination of modes, is occurring.

To specify this situation more precisely, we must define some terms. Consider a two-dimensional material — the x, y plane, in either plane stress or plane strain — in which a crack is opening along some some trajectory whose centerline is $y = Y_{cen}(x)$. Let the functions $U_N^{[\pm]}$ be the normal displacements, relative to the local orientation of the centerline, of the "upper" [+] and "lower" [−] fracture surfaces. Similarly, let the $U_T^{[\pm]}$ be the corresponding tangential displacements. Then $2U_N \equiv U_N^{[+]} - U_N^{[-]}$ is the crack-opening normal (mode I) displacement, and $2U_S \equiv U_T^{[+]} - U_T^{[-]}$ is the crack-opening shear (mode II) displacement. If the crack is bending, then U_S is a dynamic variable that cannot arbitrarily be assumed to vanish.

Denote the cohesive traction on the fracture surface by the vector quantity $\vec{\Sigma}_c(U_N, U_S)$. (Throughout this presentation, all stresses are expressed in units 2μ, where μ is the elastic shear modulus.) The natural choice for $\vec{\Sigma}_c$ is a central force (per unit area) acting between material points on opposite fracture surfaces:

$$|\vec{\Sigma}_c(U_N, U_S)| = \Sigma_c(|U|); \quad |U| = \sqrt{U_N^2 + U_S^2}, \tag{1}$$

so that the normal and shear components are

$$\Sigma_{cN} = \Sigma_c(|U|)\,\frac{U_N}{|U|}; \quad \Sigma_{cS} = \Sigma_c(|U|)\,\frac{U_S}{|U|}. \tag{2}$$

We have used a model [9] for the magnitude of this central force in which $\Sigma_c(|U|)$ is a constant for all opening displacements $|U|$ less than some range of the forces, and vanishes for larger values of $|U|$. In this case, we can characterize the cohesive zone by just a single parameter, say, the length ℓ of the region in which $\Sigma_c(|U|)$ is nonzero.

There is a simple way to see that detailed properties of the cohesive zone, such as the magnitude of the cohesive shear stress Σ_{cS}, must be relevant in fracture dynamics. Suppose, for the sake of argument, that the magnitude of Σ_{cS} were appreciably less than the central-force value shown in (2). In other words, suppose that deformation mechanisms or dissipative forces cause the work of fracture to be different in mode II than in mode I and, for the moment, consider the possibility that the fracture energy is smallest in mode II. Then the crack would be unstable against bending at least partially into mode II because it would encounter less resistance in that mode. The total energy release, when we account for the fracture energy, would increase even if the flow of elastic energy to the tip decreased. A further complication is that, as the crack bent toward mode II and lowered its fracture energy, it would be further from the Griffith threshold and would accelerate. It is not clear what a quasi-static approximation would mean in such a situation. It does seem clear, however, that the cohesive shear stress must play an essential role in a dynamic stability theory.

We can push this argument further before considering actual calculations. If the crack bends sharply so that the ratio $U_S/|U|$ in (2) is nonzero at the crack tip, then Σ_{cS} must be nonzero there and must be equal to the nonsingular shear stress (related to axes parallel and perpendicular to the direction of crack extension) at that point in the elastic material. In principle, it is possible that U_S vanishes faster than $|U|$ at the tip; but there is no reason to expect that behavior to be a general rule. (It is not.) If the shear stress at the tip of a bending crack is not necessarily zero, then it seems highly unlikely that any simple far-field prescription such as $K_{II} = 0$ can be generally correct for dynamic fracture. It might be true for very nearly straight cracks, but it cannot be the starting point for a stability analysis.

To amplify these remarks, I shall devote the rest of this presentation to a summary of two new results in fracture dynamics [1]. The first of these is neither entirely new nor fully dynamic, but it adds plausibility to the second, which is an explicit demonstration that fracture in this class of isotropic models is unstable at all nonzero speeds.

The first result is an analog of the Yoffe analysis [7, 4]. Here, however, instead of considering the singular stress fields away from the crack tip, we have used the fact that all stresses are nonsingular in a cohesive-zone model, and we have looked to see whether the stresses right at the fracture surface are acting in such a way as to oppose or amplify deviations in the trajectory. Consider a mode I crack moving at steady-state speed v along the centerline — the x axis — of a strip of width $2W$. For an isotropic, linearly elastic material, we find a relation between the normal (yy) and tangential (xx) stresses:

$$\Sigma_{xx}(x,0)-\Sigma_{yy}(x,0)=\Sigma_{xx}(\infty)-\Sigma_{yy}(\infty)+A(v)\,[\Sigma_{yy}(x,0)-\Sigma_{yy}(\infty)]. \quad (3),$$

where $\Sigma_{xx}(\infty)$ and $\Sigma_{yy}(\infty)$ are the normal and tangential stresses in the unbroken material far ahead of the crack tip, and

$$A(v)=\frac{(2-v^2/\kappa)\,(1-v^2/2)-2\sqrt{(1-v^2)\,(1-v^2/\kappa)}}{\sqrt{(1-v^2)\,(1-v^2/\kappa)}-(1-v^2/2)^2}. \qquad (4)$$

The quantity κ is the square of the ratio of the longitudinal to transverse wave speeds, and the crack speed v is measured in units of the transverse wave speed. The function $A(v)$ vanishes like v^2 near $v=0$, diverges at the Rayleigh speed v_R, and is positive and monotonically increasing everywhere in $0<v<v_R$.

The relation (3) is valid at all points along the x axis, that is, everywhere on the centerline ahead of the crack and everywhere on the fracture surfaces including at the tip. Remember that $\Sigma_{xx}(x,0)$ is the stress that would open a crack moving perpendicular to the x axis and $\Sigma_{yy}(x,0)$ is the stress that drives the crack in its original x direction. At the crack tip, $\Sigma_{yy}(x_{tip},0)$ is the yield stress which, under ordinary conditions, is large of order $\sqrt{W}$ in comparison to the external driving stress $\Sigma_{yy}(\infty)$. Thus, for any moving crack, the actual tractions on the fracture surface near the tip strongly favor motion away from the original direction of propagation. This result, by itself, is an indication but not a proof of instability in this class of models. A formula very similar to (3) was obtained by Rice in 1968 [10], and important elements of this interpretation of it were included in a recent paper by Ching [2]. So far as I know, however, its generality and importance have not been emphasized before now, certainly not in connection with a fully dynamic stability analysis of the kind to be reported next.

The derivation of (3) requires only simple, well-known techniques in the theory of elasticity. A dynamic, linear stability analysis, however, is a very much more ambitious project. To make this project feasible, we have attempted to compute only the steady-state response of our system to a small (*i.e.* first order) external force that produces a static, spatially oscillating shear stress along the x-axis:

$$\Sigma_{xy}^{(ext)}(x,0)=\hat{\varepsilon}_m\,e^{imx}, \qquad (5)$$

where $\hat{\varepsilon}_m$ is the amplitude of the perturbation and m is its wavenumber. The goal of the calculation is to compute the perturbed centerline $y=Y_{cen}(x)$ of the resulting fracture to first order in $\hat{\varepsilon}_m$, that is:

$$Y_{cen}(x)=\hat{Y}_m\,e^{imx}\equiv\hat{\chi}_Y(m,v)\,\hat{\varepsilon}_m\,e^{imx}. \qquad (6)$$

Here, $\hat{\chi}_Y$ is a complex steady-state response coefficient that depends on the wave number m and the average crack speed v. If $\hat{\chi}_Y$ diverges at some v and some real value of m, then we would conclude that the system undergoes a change in dynamic stability at that wavenumber and speed. More generally, poles of $\hat{\chi}_Y$ in the complex m-plane are stability eigenvalues. We have chosen a convention in which the unperturbed crack is moving at speed v from right to left, that is, in the negative x direction. Accordingly, poles in the lower half m-plane correspond to stable modes. Changes in stability occur when poles cross the real m-axis.

To compute $\hat{\chi}_Y$, we first have derived and solved a set of Wiener-Hopf equations for the stresses and crack-opening displacements on the x axis. To first order in $\hat{\varepsilon}_m$, the relevant one of these equations involves only the mode II displacement U_S and the corresponding shear stress Σ_S. The crucial step, then, is to note that the condition that Σ_S be nonsingular at the crack tip is sufficient to determine $\hat{Y}_m$ and thus $\hat{\chi}_Y$. This step is directly analogous to using the Barenblatt condition [8] that the normal stress be nonsingular at the tip to determine the size of the cohesive zone in the zero'th order, steady-state problem.

Our final result has the form:

$$\hat{\chi}_Y^{-1}(m,v) = -im\left[\Delta\Sigma_\infty + \tilde{K}_I(-im)^{1/2}\,\mathcal{D}(m\ell,v)\right]. \tag{7}$$

Here, $\Delta\Sigma_\infty = \Sigma_{yy}(\infty) - \Sigma_{xx}(\infty)$; $\tilde{K}_I$ (apart from an uninteresting multiplicative constant) is the static, mode I stress-intensity factor, proportional to $\Sigma_{yy}(\infty)\sqrt{W}$; and $\mathcal{D}(m\ell,v)$ is a function of v and the product of the wavenumber m times ℓ, the length of the cohesive zone. We have defined $\tilde{K}_I$ so that $\mathcal{D}(0,0) = 1$.

As a first step in interpreting (7), note that the limit $m\ell \to 0$ must reproduce the conventional far-field theory. By definition, the far-field stresses are those at distances from the tip much larger than ℓ but still much smaller than macroscopic lengths such as W. It is ℓ that sets the scale of the process zone and the size of the region in which the stresses are concentrated, thus the assumption that $m\ell$ is negligibly small is equivalent to the assumption that we are looking only at stress fields far from the tip. In fact, by setting $m\ell = 0$ in (7), we obtain precisely the theory of Cotterell and Rice (CR) [3] expressed in terms of the response coefficient $\hat{\chi}_Y$ instead of specific crack trajectories. Setting $m\ell = 0$ is exactly the same as omitting the cohesive shear stress (thereby assuming a geometrically sharp crack with no shear cohesion in the process zone) and using the far-field condition $K_{II} = 0$ instead of the condition of nonsingular shear stress at the tip to determine $\hat{Y}_m$.

To emphasize the relationship between (7) and CR, we recast it in the form:

$$-im\hat{\chi}_{YCR} = \frac{1}{\tilde{K}_I(-im)^{1/2} - T}, \tag{8}$$

where $T = -\Delta\Sigma_\infty$ is the CR "T stress". To be consistent with CR, we have taken the quasi-static limit, $v \to 0$ and have set $\mathcal{D} = 1$. For positive T, $\hat{\chi}_{YCR}$ has a pole at $m = m_s = i(T/\tilde{K}_I)^2$, which would correspond to an unstable trajectory of the form

$$Y_{cen} \propto e^{im_s x} \propto \exp[-(T/\tilde{K}_I)^2 x]. \tag{9}$$

This is a growing exponential because the crack is moving in the $-x$ direction. For negative T, on the other hand, the system appears to be stable; the only singularity on the physical sheet of the complex m- plane is the branch point at the origin. The associated branch cut should be drawn along the negative imaginary m axis. In this situation, CR compute the trajectory of a crack tip that initially points away from from the original direction of fracture. We obtain their result by computing the response of our crack to a localized patch of shear stress. That is, we write the perturbing stress as a linear superposition of our Fourier modes $\hat{\varepsilon}_m \exp(imx)$ and, for simplicity, let $\hat{\varepsilon}_m = \bar{\varepsilon}$ be a constant so that

$$\varepsilon_{shear}(x) = \int \frac{dm}{2\pi}\, \hat{\varepsilon}_m\, e^{imx} = \bar{\varepsilon}\, \delta(x). \tag{10}$$

We then find, for $x = -|x| < 0$,

$$\frac{dY_{cen}}{dx} = -\bar{\varepsilon} \int \frac{dm}{2\pi} \frac{e^{-im|x|}}{|T| + \tilde{K}_I(-im)^{1/2}} = -\frac{\bar{\varepsilon}}{\pi \tilde{K}_I} \int_{-\infty}^{+\infty} dw\, \frac{w^2\, e^{-w^2|x|}}{w^2 + \frac{T^2}{\tilde{K}_I^2}}. \tag{11}$$

This formula can be rewritten in terms of an error function, but it is easier to see what is happening in this integral representation. The perturbation is at $x = 0$. As the crack moves to the left past this point, its initial trajectory is $Y_{cen} \approx (2\bar{\varepsilon}/\tilde{K}_I)\sqrt{|x|/\pi}$. After moving a distance of order W, this trajectory is again parallel to the x-axis but is displaced by an amount $\bar{\varepsilon}/|T|$.

Our earlier discussion implies that this result cannot be entirely correct. We have obtained it, in effect, by using a model in which the cohesive shear stress is zero, but we have argued that cracks in such a model must be manifestly unstable. To see what is happening, we return to (7) and look more closely at the behavior of the function $\mathcal{D}(m\ell, v)$ for small but nonzero values of $m\ell$. As described in Ref. [1], this calculation is quite difficult; it requires a careful analysis of how the shear displacement U_S is induced self-consistently by the cohesive shear stress in the elastic field of the moving crack. The result, however, is simple and very interesting. To leading order for small values of both $m\ell$ and v, we find:

$$\mathcal{D}(m\ell, v) \approx 1 + \frac{\frac{3im\ell}{2}}{(1-\frac{1}{\kappa})\frac{v^2}{2} - im\ell}. \tag{12}$$

The first term on the right-hand side — unity — is just what we had before; but the new term — the contribution of the cohesive shear stress — is singular at $m\ell = v = 0$.

For the moment, let us drop the quantity $\Delta\Sigma_\infty$ in (7) on grounds that it ordinarily is small of order $W^{-1/2}$ in comparison to $\tilde{K}_I$. Then, a zero of $\mathcal{D}$ in the complex m- plane is a pole of $\hat{\chi}_Y(m,v)$ and, from (12), we see that $\mathcal{D}$ vanishes at $m = m_s = i(1-\frac{1}{\kappa})\frac{v^2}{\ell}$. For any nonzero v, m_s is in the upper half plane and therefore indicates instability. Unlike the weakly unstable CR trajectory in (9), however, this instability is very strong:

$$Y_{cen} \propto e^{im_s x} \propto \exp\left[-\left(1-\frac{1}{\kappa}\right)v^2\left(\frac{x}{\ell}\right)\right]. \tag{13}$$

Thus, the very small length ℓ sets the scale for unstable motion away from a straight trajectory. This analysis also tells us what happened when, in the far-field theory, we let ℓ become small at fixed v; the pole at m_s moved indefinitely far up the positive imaginary m-axis, where it became mathematically invisible but nevertheless implied strong physical instability. We conclude that the cohesive-zone model is marginally stable at the zero-velocity threshold but becomes unstable at any nonzero velocity. This conclusion is consistent with the tip-stress analysis in Eq. (3).

An obvious question is whether there is any modification of this model that can produce stability in isotropic materials, at least at small speeds. That question is large and complicated, far too large to be addressed here in any generality. But there is one simple modification that is suggested by physical considerations and which, when implemented, helps us to understand the significance of our results. Once one accepts the implications of the tip-stress analysis, it seems clear that any degree of stability against deflection must require a large enough cohesive shear stress at the crack tip to suppress the growth of mode-II components of fracture. Our natural choice of Σ_{cS} in (2), based on our picture of central forces acting between the newly opened fracture surfaces, puts us exactly at the marginal point where the fracture energy is the same in mode-II as it is in mode-I.

It is easy to see what happens if we move just slightly away from this point. We can change the central- force assumption by multiplying the right-hand side of the equation for Σ_{cS} in (2) by a factor, say, $1+\rho$. Let us also say that ρ is a small quantity. (If the difference between fracture energies in modes I and II is caused by dissipative effects, as in principle it must, then ρ must depend on v. But that is a level of detail that is beyond the scope of the present discussion.) The introduction of ρ changes (12) by shifting the singularity in the cohesive shear term away from $v = m\ell = 0$:

$$\mathcal{D}(m\ell, v; \rho) \approx 1 + \frac{\frac{3im\ell}{2}}{(1-\frac{1}{\kappa})\frac{v^2}{2} - im\ell - \rho}. \tag{14}$$

The pole in $\hat{\chi}_Y(m,v)$ is now at $m_s = -2i\frac{\rho}{\ell} + i(1-\frac{1}{\kappa})\frac{v^2}{\ell}$. If ρ is negative, that is, if the mode II cohesive stress is less than the mode I stress, then the only change is that m_s remains in the upper half plane at $v = 0$. As expected, the crack with less than marginal resistance against bending into mode II is unstable even in the quasi-static limit.

When ρ is positive, on the other hand, the pole is at $m_s = -2i\rho/\ell$ for $v = 0$, and it moves into the unstable upper half plane at a nonzero critical velocity $v_c = \sqrt{2\rho/(1-1/\kappa)}$. For $v < v_c$, we can compute the response to perturbations that are slowly varying on the scale of ℓ by taking the limit $m\ell \to 0$. In this way, we recover the CR theory in all respects. In particular, (8) is correct. But such a calculation is entirely incapable of telling us whether or not the crack is stable and, thus, whether or not a CR calculation of this kind is physically sensible. To determine stability, we must look in detail at the cohesive forces acting at the crack tip and try to understand the physical mechanisms that might cause the cohesive shear stress to be larger than its central force value.

In conclusion, we can indeed construct an isotropic model in which there is a transition from stability to instability at a nonzero crack speed, but so far only by using a large cohesive shear stress. The search for an explanation of why the cohesive shear stress might be larger than its central-force value, and more generally the search for other possible stabilizing mechanisms, are important projects for further research. These projects inevitably must involve more realistic pictures of what is happening at crack tips.

Acknowledgements

This research has been supported by U.S. DOE Grant DE-FG03-84ER45108 and NSF Grant PHY- 9407194.

References

1. E.S.C. Ching, J.S. Langer, and H. Nakanishi, preprint.
2. E.S.C. Ching, Phys. Rev. E **49**, 3382 (1994).
3. B. Cotterell and J.R. Rice, Int. J. Fracture **16**, 155 (1980).
4. L. B. Freund, *Dynamic Fracture Mechanics* (Cambridge University Press, New York, 1990).

5. J. Fineberg, S.P. Gross, M. Marder, and H.L. Swinney, Phys. Rev. Lett. **67**, 457, (1991); Phys. Rev. B **45**, 5146 (1992).
6. S.P. Gross, J. Fineberg, M. Marder, W.D. McCormick, and H. Swinney, Phys. Rev. Lett. **71**, 3162 (1993).
7. E. Yoffe, *Phil. Mag.* **42**, 739 (1951).
8. G.I. Barenblatt, Adv. Appl. Mech. **7**, 56 (1962).
9. D.S. Dugdale, J. Mech. Phys. Solids **8**, 100 (1960).
10. J.R. Rice, "Mathematical Analysis in the Mechanics of Fracture", in *Fracture: An Advanced Treatise*, Vol. 2, Chap. 3, pp. 191-311, ed. H. Liebowitz (Academic Press, 1968).
11. M. Marder, Phys. Rev. Lett. **74**, 4547 (1995).

RADIAL CRACK DYNAMICS WITH CLOSURE

L.I. SLEPYAN
Department of Solid Mechanics, Materials and Structures
Tel Aviv University, Ramat Aviv 69978 Tel Aviv, Israel

AND

J.P. DEMPSEY
Department of Civil and Environmental Engineering
Clarkson University, Potsdam, NY 13699-5710, USA

Abstract. An infinite elastic-brittle plate subjected to a transverse impact is considered. Many cracks, each of which extends through the full thickness of the plate, are assumed to grow radially from the impact point. Due to circumferential bending, the faces of these radial cracks tend to close over a portion of the plate thickness. In this paper, the dynamics of the radial crack system within the influence of a tangential unilateral constraint which prevents crack face interpenetration is examined. The analysis is simplified by assuming that the faces interact only over the lines where these faces meet a surface of the plate, rather than over some unknown area of the fracture surfaces. In other words, the local contact strain is neglected. The radial cracks are assumed to be large in number and evenly distributed so that the deformation retains axial symmetry. The governing equations are derived, and a transient, self-similar solution of these equations is constructed, relating the plate response to the impact force. The extension of the radial cracks according to the principles of fracture mechanics is explored.

1. Introduction

The rapid formation of a system of radial cracks typically accompanies the dynamic concentrated impact of a brittle plate. In all likelihood, the quasi-continuum approach adopted in this paper is more suitable for dynamic radial cracking than for the elastostatic studies published to date (Hellan, 1984; Dempsey *et al.*, 1995b). The first quasi-continuum 'multiple crack' analytical solution was derived by Hellan (1984) for an elastic Kirchhoff-Poisson plate with line contact of the crack faces. The last restric-

J. R. Willis (ed.), IUTAM Symposium on Nonlinear Analysis of Fracture, 201–210.

tion was removed by Dempsey *et al.* (1995a&b), and an analytical solution was obtained for a nonzero closure contact width, based on both the corresponding 'inner' plane contact problem of elasticity as well as the 'outer' coupled plane-bending problem of thin plate theory. The local and global expressions for the energy release rate, and the bearing capacity of such a cracked plate were also determined. Further, an accurate evaluation and approximate expression for the asymptotic closure contact strip width and stress distribution were obtained. The contact strip width was found to depend on the ratio of the contact bending moment to the in-plane force.

In the paper by Dempsey *et al.* (1995b) as well as in the paper by Slepyan *et al.* (1995), it was shown that the contact strip is accurately modeled as line contact once the crack length has increased sufficiently, relative to the plate thickness. The initial formulation by Hellan (1984) is therefore adequate for relatively long cracks. In the present paper, a similar simplified formulation is used. In addition, the contact closure width and stress distributions are assumed to be quasi-static but to be expressed in terms of the contact force and moment distributions that are determined from the dynamic problem under consideration. Such a decomposition of the problem into dynamic and quasi-static (in-plane) deformation fields concerns both the global and local actions: only the out-of-plane inertia forces are taken into account. In other words, the relatively small in-plane motion is assumed to be quasi-static. Under these conditions, as is shown below, a self-similar solution in the form $t^k f(\zeta)$, $\zeta = r/\sqrt{2hct}$, exists not only for an intact plate but also for the extensively radially cracked plate in spite of the coupled bending–stretching deformation that occurs in the crack closure area (CCA) . In the above representation, $2h, c, t, r, k$ and f are the plate thickness, the plane wave velocity, time, the radial coordinate, a number (it may not be integer), and an unspecified function, respectively; $c = \sqrt{E/\rho}$, while E, ν and ρ are the Young's modulus, Poisson's ratio and density of the plate material, respectively. Note that the functions of ζ depend on k as well.

The plate is assumed to be loaded by a concentrated force or impulse at the central point, $r = 0$, only. Thus, an axisymmetric homogeneous problem is considered for $r > 0$. Note that a number of circumferential cracks usually appear in addition to the radial cracks. While the governing equations derived below are valid for the case of circumferential cracking, the self-similar solution developed in the paper is valid solely for radial cracking.

The problem under consideration displays one important feature which cannot be obtained via conventional fracture mechanics, *viz.*, the determination of the number of cracks. In addition, the total energy release by the axisymmetric quasi-continuum crack array (by the moving interface) may differ from that in a system comprised of a finite number of cracks. This difference, the magnitude of which is an unknown function of the speed of the crack zone boundary, may be produced by additional elastic waves generated by such a star crack system, these waves carrying energy away. One can avoid these difficulties using the principle of maximum energy dissipation rate directly in the axisymmetric formulation of the problem.

2. Formulation

The key to the formulation is an adequate description of the coupling between the in-plane (u, S_r, S_θ) and out-of-plane (w, M_r, M_θ, Q_r) quantities. The quantities $w(r,t)$ and $u(r,t)$ denote the vertical and radial displacements of the plate in the central plane $(z = 0)$, respectively $(w' \equiv \partial w/\partial r)$; S_r and S_θ denote the in-plane radial and tangential forces, respectively, per unit length and M_r and M_θ denote the radial and tangential bending moments. In addition, let Q_r denote the radial shear force per unit length.

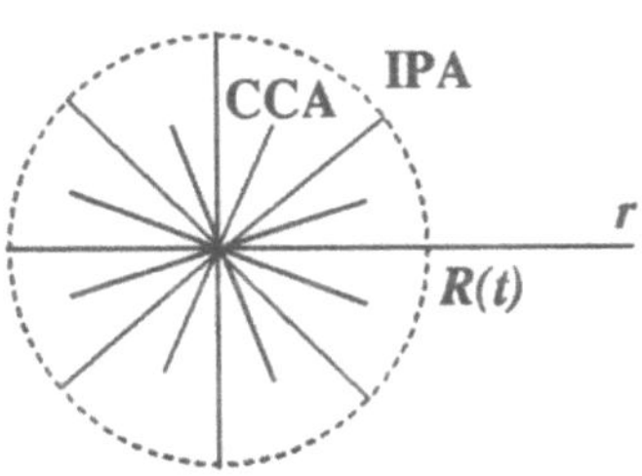

Figure 1. The crack system.

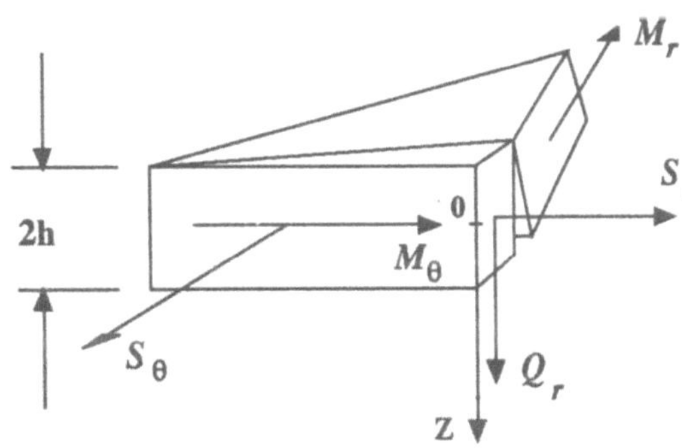

Figure 2. The forces and moments.

The radially cracked plate configuration (Fig. 1) is assumed to be separated into the following two regimes: the radial crack closure area (the inner region $0 < r < R$, $R = R(t)$) which is hereafter called the CCA, and the unbroken or intact plate area (the outer region, $r > R$), which is hereafter called the IPA. The in-plane interaction force S_θ is compressive in the CCA $(S_\theta \leq 0)$. The formulation here prescribes that S_θ acts at $z = 0$. In addition to this in-plane force, a moment M_θ acts at the same radial location on $\theta = \text{const}$. This force and moment are statically equivalent to the force S_θ acting at $z = -h$ (Fig. 2).

The occurrence of this in-plane force S_θ thus couples the planar and bending deformations. Note that the plate becomes wavy under the considered dynamic action, and if the intensity of the action is large enough the CCA can include alternate sub-areas with the crack surfaces contacting at $z = \pm h$. In all cases with crack closure, the solution must satisfy the inequality $S_\theta \leq 0$, and this condition serves to determine the extent of the sub-areas. In the case when the inequality cannot be satisfied at both $z = h$ and $z = -h$, an open crack area (OCA) arises with $S_\theta = 0$. In the general case, sub-areas with the contact at $z = \pm h$ and $S_\theta = 0$ may exist. However to be brief, only the case $z = -h$ is considered here. The general solution may be derived in the same way.

The governing equations for dynamic out-of-plane deformations but quasi-static in-plane deformations are given by ($I = 2h^3/3$):

$$EIw'' = M_r - \nu M_\theta, \quad M_\theta = (rM_r)' + rQ_r, \quad (rQ_r)' = 2\rho h r \ddot{w}. \tag{1}$$

$$2Ehu' = S_r - \nu S_\theta, \quad S_\theta = (rS_r)'. \tag{2}$$

Crack closure in the CCA with line contact at $z = -h$ gives

$$M_\theta = S_\theta h. \tag{3}$$

The coupling between the in-plane and out-of-plane quantities occurs solely through the expressions for the strain, ϵ_θ, in the continuous circular material line $z = -h$ in the CCA:

$$\epsilon_\theta = (u + hw')/r = (S_\theta - \nu S_r)/2Eh + h(M_\theta - \nu M_r)/EI. \tag{4}$$

The following definitions for the moments, shear force and tangential strain are applicable in the IPA only [$D = EI/(1-\nu^2)$]:

$$\begin{aligned} M_r &= D(w'' + \nu w'/r), \quad M_\theta = D(w'/r + \nu w''), \\ Q_r &= -D(w'' + w'/r)', \quad 2Eh\, u/r = S_\theta - \nu S_r. \end{aligned} \tag{5}$$

In the OCA, the relations for the CCA are valid with (4) replaced by the equality $S_\theta = 0$.

Under the condition of a central impact transverse force,

$$\lim_{r\to 0} 2\pi r Q_r = -P(t), \quad \lim_{r\to 0} rM_r = 0, \quad \lim_{r\to 0} rS_r = 0. \tag{6}$$

while at infinity:

$$w = u = 0. \tag{7}$$

Continuity conditions at the moving interface, $r = R(t)$ require that

$$[S_r] = [M_r] = [Q_r] = [u] = [w] = [w'] = 0. \tag{8}$$

The general expression for the energy release per crack in the quasi-continuum axisymmetric problem is derived in Dempsey *et al.* (1995b). In the present simplified case of line contact, in which the contact force S_θ is assumed to be applied at $z = -h$, this expression coincides with that used by Hellan (1984), and is given by

$$G = \frac{2\pi R}{nh}\frac{[M_\theta]^2}{EI}, \quad [M_\theta] \equiv M_\theta(R^+) - M_\theta(R^-). \tag{9}$$

3. Governing Equations

In the IPA, the deflection of the plate must satisfy

$$D\Delta^2 w + 2\rho h \ddot{w} = 0, \quad r(ru')' - u = 0, \quad \Delta = \frac{1}{r}\frac{\partial}{\partial r}\left(r\frac{\partial}{\partial r}\right). \tag{10}$$

In the OCA, where $S_\theta \equiv 0$,

$$EI(rw'')'' + 2\rho h r \ddot{w} = 0, \quad (ru')' = 0. \tag{11}$$

In the CCA, the system of equations (1)-(4) may be reduced to two coupled equations in terms of $w(r,t)$ and $S_r(r,t)$:

$$\begin{aligned} &\frac{2}{3}Eh^2(rw'')'' + 2\rho r\ddot{w} - (rS_r)'' + \nu(r(rS_r)')'' = 0, \\ &2Eh^2[w'' + \nu(rw'')'] + S_r - (4-3\nu^2)[r(rS_r)']' = 0. \end{aligned} \tag{12}$$

4. Self-Similar Solution

In the CCA the self-similar solution is posed as

$$\begin{aligned} w(r,t) &= \left(\frac{ct}{h}\right)^k hW(\zeta), \quad u(r,t) = \left(\frac{ct}{h}\right)^{k-1/2} hU(\zeta), \\ S_\theta(r,t) &= \left(\frac{ct}{h}\right)^{k-1} Eh\sigma_\theta(\zeta), \quad S_r(r,t) = \left(\frac{ct}{h}\right)^{k-1} EhS(\zeta), \\ M_\theta(r,t) &= \left(\frac{ct}{h}\right)^{k-1} Eh^2 m_\theta(\zeta), \quad M_r(r,t) = \left(\frac{ct}{h}\right)^{k-1} Eh^2 m_r(\zeta), \\ Q_r(r,t) &= \left(\frac{ct}{h}\right)^{k-3/2} Ehq_r(\zeta), \quad \zeta = \frac{r}{\sqrt{2hct}}, \end{aligned} \tag{13}$$

where the functions of ζ are different for different k, and this is noted below by the corresponding subscript: $w(r,t) = w_k(r,t)$, $W(\zeta) = W_k(\zeta)$.

In the IPA, the plane and bending problems are not coupled by the equations but they are coupled by the CCA – IPA interface conditions (8).

For the self-similar representation to be valid here, in both the CCA and the IPA, one has to assume that the interface (the phase transition front) corresponds to a constant ζ, defined here as Z. In other words, the radius $R(t)$ must be proportional to $\sqrt{t}$. In turn, the corresponding functions in the CCA have to have the same time-multipliers (the same k) as in the IPA. In the terms of (13),

$$r\ddot{w} = \frac{h\sqrt{2hct}}{t^2}\left(\frac{ct}{h}\right)^k \{k\zeta[kW_k(\zeta) - (\zeta W_k(\zeta))'] + \frac{1}{4}(\zeta^3 W_k'(\zeta))'\}, \tag{14}$$

where now primes denote differentiation with respect to ζ, and from (12) one has

$$\frac{1}{3}(\zeta W_k'')'' + (\zeta^3 W_k')' + 4k\zeta[kW_k - (\zeta W_k)'] - (\zeta S_k)'' + \nu[\zeta(\zeta S_k)']'' = 0,$$
$$W_k'' + \nu(\zeta W_k'')' + S_k - (4 - 3\nu^2)[\zeta(\zeta S_k)']' = 0. \tag{15}$$

For the case $k = 0$, which corresponds to an impulse, p, one has $P = 0$ for $t > 0$ $(P = p\delta(t), p = \text{const})$. With this in mind, based on the conditions in (6), one can see that these equations reduce to

$$\frac{1}{3}(\zeta W_0'')' + \zeta^3 W_0' - (\zeta S_0)' + \nu[\zeta(\zeta S_0)']' = 0,$$
$$W_0'' + \nu(\zeta W_0'')' + S_0 - (4 - 3\nu^2)[\zeta(\zeta S_0)']' = 0. \tag{16}$$

For $k = 0$, the central point is regular $(t > 0)$, and one can find

$$W_0''(0) = 3(1 - \nu)S_0(0) \quad (\zeta = 0). \tag{17}$$

The solution to (16) can be expressed in series-form as

$$W_0 = C_1 + C_2\zeta + \sum_{n=0}^{\infty} \frac{a_n}{(n+1)(n+2)}\zeta^{n+2}, \quad S_0 = \sum_{n=0}^{\infty} b_n\zeta^n,$$
$$U_0 = -\frac{1}{\sqrt{2}}W_0'(0) + \frac{1}{\sqrt{2}}\sum_{n=0}^{\infty}\left(\frac{1}{n+1} - \nu\right) b_n\zeta^{n+1}, \tag{18}$$

with the coefficients being represented by the following recurrence relations:

$$a_n = -\frac{3}{4}\,\frac{1 - (4 - 3\nu^2)(n+1)^2}{n(n+1)(n+2)(n-3)}\,a_{n-4},$$
$$b_n = -\frac{3}{4}\,\frac{1 + \nu(n+1)}{n(n+1)(n+2)(n-3)}\,a_{n-4}, \quad n = 4 + m, \quad m = 0, 1, 2, \ldots;$$
$$a_0 = 3(1 - \nu)b_0, \quad b_0 = S_0(0), \quad a_1 = a_2 = b_1 = b_2 = 0,$$
$$a_3 = \frac{1 - 16(4 - 3\nu^2)}{80}\,C_2, \quad b_3 = -\frac{1 + 4\nu}{80}\,C_2. \tag{19}$$

The equality $U_0(0) = W_0'(0)/\sqrt{2}$ follows from (4), (6) and (13).

The dynamic bending problem for the IPA was originally considered by Sneddon (1945). For $k = 0$, the general solutions can be represented as

$$W_0(\zeta) = I_1 \operatorname{si}(\kappa) + I_2 \operatorname{ci}(\kappa), \quad \kappa = \frac{\sqrt{3}}{2}\sqrt{1-\nu^2}\zeta^2, \tag{20}$$

where I_1 and I_2 are arbitrary constants, and

$$\operatorname{ci}(x) = -\int_x^\infty \frac{\cos\xi}{\xi}d\xi, \quad \operatorname{si}(x) = -\int_x^\infty \frac{\sin\xi}{\xi}d\xi. \tag{21}$$

For the OCA with no crack surface interaction (if this region exists or with crack face interpenetration being ignored), $S_\theta = 0$, and the first equation in (16) reduces to

$$(\zeta W_0'')' + 3\zeta^3 W_0' = 0. \tag{22}$$

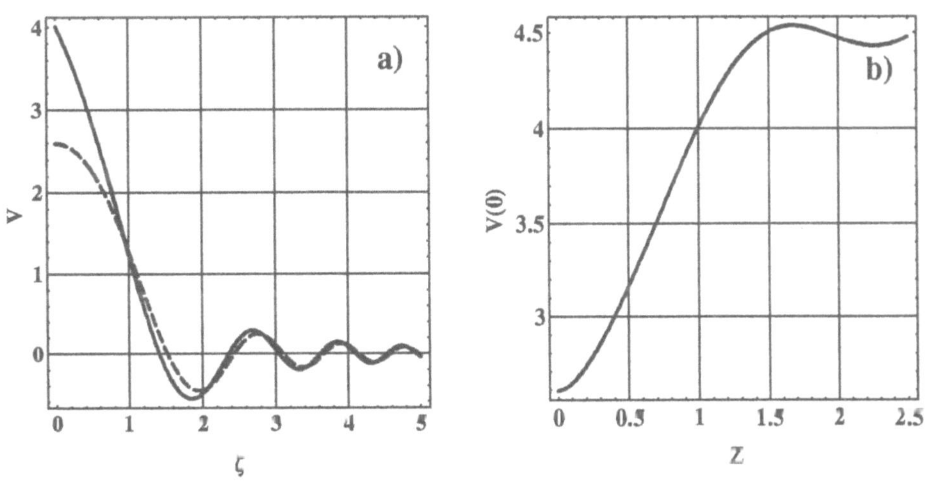

Figure 3. The plate under a concentrated impulse: a) The lateral displacement of the intact plate, $Z = 0$ (dashed curve), and in the CCA – IPA problem, $Z = 1$. b) The dependence of the central displacement on the interface motion.

The solution to this equation that is regular at $\zeta = 0$ looks as

$$W_0'(\zeta) = C_0 J_0(\beta), \quad \beta = \frac{\sqrt{3}}{2}\zeta^2 = \frac{\kappa}{\sqrt{1-\nu^2}}, \quad C_0 = \text{ const.} \tag{23}$$

The other Bessel functions of the same index and argument satisfy this equation as well.

In the CCA – IPA problem under consideration, there are six arbitrary constants in the solution, three in the CCA: C_1, C_2 and $S_0(0)$, and three in the IPA: I_1, I_2 and $U(Z)$. These constants are defined by the interface conditions in (8). Note, however, that in the considered case, $k = 0$, the continuity conditions in (8) $[Q_r] = 0$ and $[w'] = 0$ are equivalent to each other, and the relation for the total momentum, p, has to be used as an additional equality required for the determination of these constants. The total momentum can be expressed as follows

$$p = 4\pi h^2 \rho \frac{d}{dt} \int_0^\infty w\, r\, dr = p^0 \int_0^\infty W_0(\zeta)\zeta\, d\zeta, \quad p^0 = \pi(2h)^3 \rho c. \tag{24}$$

Now based on (18) and (20) one can find ($\kappa_* = \kappa(Z)$)

$$\frac{p}{p^0} = \frac{1}{2} Z^2 C_1 + \frac{1}{3} Z^3 C_2 + \sum_{n=0}^{\infty} \frac{a_n Z^{n+4}}{(n+1)(n+2)(n+4)} -$$
$$- \frac{1}{\sqrt{3(1-\nu^2)}} [(\cos\kappa_* + \kappa_* \mathrm{si}(\kappa_*))I_1 - (\sin\kappa_* - \kappa_* \mathrm{ci}(\kappa_*)I_2] \tag{25}$$

Some results of calculations are shown in Figure 3 where the normalized displacement $V(\zeta) = p^0 W_0(\zeta)/p$ is used, and $\nu = 0.3$.

The self-similar solution for any k can be found by the superposition

$$w(r,t) = w_k(r,t) = h \int_0^{ct/h} W_0\left(\frac{r}{\sqrt{2hc\tau}}\right) \left(\frac{ct}{h} - \frac{c\tau}{h}\right)^{k-1} d\left(\frac{c\tau}{h}\right)$$
$$= h \left(\frac{ct}{h}\right)^k \int_0^1 W_0\left(\frac{1}{\sqrt{\alpha}}\zeta\right)(1-\alpha)^{k-1} d\alpha \equiv \left(\frac{ct}{h}\right)^k h W_k(\zeta). \tag{26}$$

5. Energy Criterion And Crack Number

To determine the specific form of the interface motion one can appeal to the energy criterion (9). In terms of the notations in (13), with $k = 0$,

$$G = \frac{3\pi E h^2 Z [s_\theta]^2}{ctn} \sqrt{\frac{2h}{ct}}. \tag{27}$$

Assuming $G = G_c$ and Z to be constant during crack growth one has no choice but to consider the number of cracks as a function of r. In fact, to satisfy the latter condition it is necessary that n be proportional $1/r^3$. In this case, and only in this case, G is constant, and the relation (27) is given by

$$G = G_c = \frac{3\pi E h^2 Z^4 [s_\theta]^2}{2hn_*}, \quad n = \left(\frac{2h}{r}\right)^3 n_*. \tag{28}$$

The above criterion has made it possible to determine *the dependence* $n(r)$ but not the arbitrary constant n_*. The number of cracks, under the chosen quasi-continuum formulation, remains indeterminate within the conventional fracture mechanics framework.

An alternative approach, one which does not require that the number of cracks be accentuated, is to assume that the interface motion obeys the principle of maximum energy dissipation rate (Slepyan, 1992&1993). This criterion requires that the total energy flux into the propagating interface be maximal:

$$N = (2hnG - G^{tot}_{min})\dot{R} = 6\pi E h^2 c \left(\frac{h}{ct}\right)^2 [s_\theta]^2 Z^2 - G^{tot}_{min} Z \sqrt{\frac{h}{2ct}}. \tag{29}$$

where G^{tot}_{min} be the minimal energy release rate required for the crack system motion. The system of cracks can be assumed to arrest, however, when the total energy release rate falls below this minimum which, in turn, corresponds to the final number of radial macro-cracks.

6. Concluding Remarks

The above self-similar solution corresponds to an impulsive force. An arbitrary solution for $k > 0$, and a superposition of a set of k-modes can be used. However, for the latter case, the interface motion must be the same for each k-term. The question arises about when such a set of the self-similar solutions is meaningful. The answer is as follows: the self-similar solutions are valid as long as the crack zone frontal velocity is proportional to $\sqrt{t}$. In a more general case, the quasi-continuum approach for the crack area description and the equations (10)-(12) can be used.

The solution presented in this paper is related to the problem of high-speed impact or perforation of a brittle plate. In this sense, it is only a part of the general problem. The associated reactive deformations of the impactor are not considered here. In the case of a high-speed-impact, the problem can be separated, to first order, into two problems. The first concerns the force-central displacement relation, which can be considered assuming that $R = \infty$. In this case, the superposition is valid, and one has:

$$\dot{w}(0,t) = hW_0(0)P(t), \quad W_0(0) \approx 4.5p^0. \tag{30}$$

The second problem concerns the final radius of the cracked zone. This final radius can be deduced via the above self-similar problem ($k = 0$) with the impulse, p, derived from the first problem.

Finally note that the quasi-continuum approach adopted here is suitable for determination of the radial crack zone propagation and arrest. In fact, circumferential cracking, which usually arises in such a process, can be taken into consideration based on the above formulation as well as that in the solution derived by Dempsey *et al.* (1995b).

7. Acknowledgement

This study was supported by the U.S Office of Naval Research through its Sea Ice Mechanics Accelerated Research Initiative [Grant No. N00014-90-J-1360].

References

Dempsey, J.P., Adamson, R.M. and DeFranco, S.J. (1995a) Fracture analysis of base-edge-cracked reverse-tapered plates, *Int. J. Fract.* **69**, 281–294.

Dempsey, J. P., Slepyan, L. I. and Shekhtman, I. I. (1995b) Radial cracking with closure, *Int. J. Fract.* **73**, 233–261.

Hellan, K. (1984) An asymptotic study of slow radial cracking, *Int. J. Fract.* **26**, 17–30.

Slepyan, L. I. (1992) The criterion of maximum dissipation rate in crack dynamics, *Sov. Phys. Dokl.* **37**, 259–261.

Slepyan, L. I. (1993) Principle of maximum energy dissipation rate in crack dynamics, *J. Mech. Phys. Solids* **41**, 1019–1034.

Slepyan, L. I., Dempsey, J. P. and Shekhtman, I. I. (1995) Asymptotic solutions for crack closure in an elastic plate under combined extension and bending, *J. Mech. Phys. Solids* **43**, 1727–1749.

Sneddon, I. N. (1945) The symmetrical vibrations of a thin elastic plate, *Proc. Camb. Phil. Soc.* **41**, 27–43.

Zener, C. (1941) The intrinsic inelasticity of large plates, *Physical Review* **59**, 669–673.

NUMERICAL STUDIES OF FAST CRACK GROWTH IN ELASTIC-PLASTIC SOLIDS

T. SIEGMUND AND A. NEEDLEMAN
Division of Engineering
Brown University, Providence, RI, 02912 USA

Abstract. Dynamic crack growth is analyzed numerically for a plane strain block with an initial central crack. The material on each side of the bond line is characterized as an isotropically hardening elastic-viscoplastic solid. A cohesive surface constitutive relation is also specified that relates the tractions and displacement jumps across the crack plane. Crack initiation, crack growth and crack arrest emerge naturally as outcomes of the imposed loading, without any ad hoc assumptions concerning crack growth criteria.

1. Introduction

Analyses of fast crack growth in structural metals have generally been based on an approach where a material and crack speed dependent value of a characterizing parameter, such as the energy release rate or the stress intensity factor, is used in conjunction with a crack tip equation of motion, Freund (1990). Another approach, where physically based models of the micromechanisms of failure are incorporated in the material's constitutive relation, is described in Needleman and Tvergaard (1991) and Tvergaard and Needleman (1993).

The basis for our analyses of dynamic crack growth in elastic-viscoplastic solids is the cohesive surface formulation in Needleman (1987). This cohesive surface framework has been used previously to model quasi-static crack growth in plastically deforming solids, e.g., Needleman (1990a,b), Tvergaard and Hutchinson (1992), and dynamic crack growth in elastic solids, e.g., Xu and Needleman (1994). In this study, material and cohesive parameters are chosen to give deformation behavior and toughness values representative of a structural steel. The focus here is on the evolution of

J. R. Willis (ed.), IUTAM Symposium on Nonlinear Analysis of Fracture, 211–220.

the crack speed for straight ahead crack growth; more detailed results are presented in Siegmund and Needleman (1996).

Full finite strain transient analyses are carried out for a plane strain block with an initial central crack and subject to tensile impact loading. Attention is restricted to a single cohesive surface, which restricts the crack to grow along the initial crack line. For the given loading, crack initiation, crack growth and crack arrest are determined directly in terms of the material properties and of the parameters characterizing the cohesive surface separation law. The cohesive parameters include a strength and the work of separation per unit area so that, from dimensional considerations, a characteristic length enters the formulation.

2. Formulation

The cohesive surface formulation and numerical method follow that in Xu and Needleman (1994). The difference here is that the material is taken to be elastic-viscoplastic and attention is confined to a single cohesive surface. A finite strain Lagrangian formulation is used, with the principle of virtual work written as

$$\int_V \mathbf{s} : \delta\mathbf{F} dV - \int_{S_{int}} \mathbf{T} \cdot \delta\mathbf{\Delta} dS = \int_{S_{ext}} \mathbf{T} \cdot \delta\mathbf{u} dS - \int_V \rho \frac{\partial^2 \mathbf{u}}{\partial t^2} \cdot \delta\mathbf{u} dV \qquad (1)$$

where $\mathbf{s}$ is the nonsymmetric nominal stress tensor, $\mathbf{u}$ is the displacement vector, $\mathbf{F}$ is the deformation gradient, $\mathbf{\Delta}$ is the displacement jump across the cohesive surface, $\mathbf{A} : \mathbf{B}$ denotes $A^{ij}B_{ji}$, V, S_{ext} and S_{int} are the volume, external surface area and internal cohesive surface area, respectively, of the body in the reference configuration. The density of the material in the reference configuration is ρ and the traction vector $\mathbf{T}$ and the reference configuration normal $\mathbf{n}$ are related by $\mathbf{T} = \mathbf{n} \cdot \mathbf{s}$. Also, $\mathbf{s} = \mathbf{F}^{-1} \cdot \boldsymbol{\tau}$, where $\boldsymbol{\tau} = \det(\mathbf{F})\boldsymbol{\sigma}$, with $\boldsymbol{\sigma}$ being the Cauchy stress.

Plane strain conditions are assumed and a Cartesian coordinate system is used as reference, with the $x_1 - x_2$ plane being the plane of deformation. Computations are carried out for center cracked rectangular specimens of dimension $2w \times 2L$ with an initial crack of length $2a_i$ along $x_2 = 0$. At $t = 0$, the body is stress free and at rest. Equal and opposite normal velocities are prescribed on the edges at $x_2 = \pm L$, with the shear traction required to vanish. The edges at $x_1 = \pm w$ are traction free. Symmetry conditions are presumed so that only one quarter of the specimen is analyzed numerically.

The volumetric constitutive law is that of an elastic-viscoplastic isotropically hardening solid. The total rate of deformation, $\mathbf{D} = \mathrm{sym}(\dot{\mathbf{F}} \cdot \mathbf{F}^{-1})$, is written as the sum of an elastic part, $\mathbf{D}^e$, and a plastic part $\mathbf{D}^p$. Small elastic strains and elastic isotropy are presumed. The viscoplastic flow law is

$$\mathbf{D}^p = \frac{3\dot{\bar{\epsilon}}}{2\bar{\sigma}}\boldsymbol{\tau}' \tag{2}$$

where $\dot{\bar{\epsilon}}$ is the effective plastic strain rate and

$$\boldsymbol{\tau}' = \boldsymbol{\tau} - \frac{1}{3}(\boldsymbol{\tau} : \mathbf{I})\mathbf{I} \ , \ \bar{\sigma}^2 = \frac{3}{2}\boldsymbol{\tau}' : \boldsymbol{\tau}' \tag{3}$$

$$\dot{\bar{\epsilon}} = \dot{\epsilon}_0[\bar{\sigma}/g(\bar{\epsilon})]^{1/m} \ , \ g(\bar{\epsilon}) = \sigma_0(\bar{\epsilon}/\epsilon_0 + 1)^N \ , \ \epsilon_0 = \sigma_0/E \tag{4}$$

Here, a superposed dot denotes differentiation with respect to time, $\bar{\epsilon} = \int \dot{\bar{\epsilon}} dt$, σ_0 is a reference strength, and N and m are the strain hardening exponent and strain rate hardening exponent, respectively. Neither thermal softening nor a variation in strain rate sensitivity with strain rate are accounted for in (4).

The material properties are taken to be representative of a high strength steel with $E = 211$ GPa, $\nu = 0.3$, $\sigma_0 = 1000$ MPa, $N = 0.1$, $m = 0.01$, $\dot{\epsilon}_0 = 1/\mathrm{s}$ and $\rho = 7800\ \mathrm{kg/m^3} = 7.8 \times 10^{-3}\mathrm{MPa/(m/s)^2}$. The dilational, shear and Rayleigh wave speeds are 6034 m/s, 3226 m/s and 2987 m/s, respectively.

The constitutive law for the cohesive surface is taken to be elastic so that any dissipation associated with separation is neglected. The traction across the cohesive surface, which lies on $x_2 = 0$ and $|x_1| > a_i$, is given by

$$\mathbf{T} = -\frac{\partial \phi}{\partial \Delta_n} \tag{5}$$

Because the crack is constrained to grow along the initial crack line and symmetry conditions prevail about that line, only normal separation occurs. The specific form used for the potential is, Rose *et al.* (1981),

$$\phi = \phi_n - \phi_n \left(1 + \frac{\Delta_n}{\delta_n}\right) \exp\left(-\frac{\Delta_n}{\delta_n}\right) \tag{6}$$

where ϕ_n is the work of normal separation, $\Delta_n = 2u_2(x_1, 0)$ is the normal displacement along the cohesive surface and δ_n is the cohesive characteristic length. At $\Delta_n = \delta_n$, the magnitude of the traction across the cohesive surface attains a maximum, σ_{max}. The work of separation, the cohesive surface strength and the cohesive surface characteristic length are related by $\phi_n = \exp(1)\sigma_{max}\delta_n$. When δ_n is substantially smaller than all geometric lengths, crack growth predictions are not sensitive to the shape of the potential.

Under dynamic loading conditions the J–integral, Rice (1968), involves an area integral as well as a line integral (Nakamura *et al.*, 1985),

$$J = \int_\Gamma \left[(W + L)n_1 - \mathbf{T} \cdot \frac{\partial \mathbf{u}}{\partial x_1}\right] ds + \int_A \left[\rho \frac{\partial \mathbf{v}}{\partial t} \cdot \frac{\partial \mathbf{u}}{\partial x_1} - \rho \mathbf{v} \cdot \frac{\partial \mathbf{v}}{\partial x_1}\right] dA \tag{7}$$

where $\mathbf{v} = \partial \mathbf{u}/\partial t$, $W = \int \boldsymbol{\tau} : \mathbf{D} dt$, $L = \rho \mathbf{v} \cdot \mathbf{v}/2$, Γ is a path surrounding the crack tip and n_1 is the component of the normal to Γ in the x_1−direction.

The finite element discretization is based on linear displacement triangular elements that are arranged in a 'crossed-triangle' quadrilateral pattern. An explicit time integration scheme based on the Newmark β−method, with $\beta = 0$, is used together with a lumped mass matrix. The constitutive relation is integrated using the rate tangent method of Peirce *et al.* (1984).

3. Results

We confine attention to specimens having L = 30 mm, w = 30 mm and an initial crack length of a_i = 10 mm. A finite element mesh of 225 × 60 quadrilaterals is used, with a uniform 200 × 10 mesh in a 4 mm × 0.2 mm region in front of the initial crack tip. The square elements in the uniform mesh region have side lengths of 0.02 mm.

The cohesive surface characteristic length is fixed at $\delta_n = 2\ \mu$m and, unless stated otherwise, $\sigma_{max} = 3\sigma_0 = 3000$ MPa so that $\phi_n = 16.2$ KJ/m^2.

The first loading wave arrives at the crack tip at 4.98 μs after impact and the next loading wave at 14.94 μs. Assuming the impact itself does not cause plastic deformation, the stress carried by the loading wave is the density times the dilational wave speed times the impact velocity. Impact velocities of 5 m/s, 10 m/s 15 m/s, 20 m/s and 30 m/s then correspond to stress levels of 235 MPa, 470 MPa, 705 MPa, 940 MPa and 1410 MPa, respectively. Because there is impact at both ends, the stress level on the crack plane at 4.98 μs is twice that carried by each of the two loading waves.

In the following, to facilitate comparison with the quasi-static results of Tvergaard and Hutchinson (1992), the reference quantities

$$K_0 = \sqrt{\frac{E\phi_n}{(1-\nu^2)}} \qquad R_0 = \frac{1}{3\pi}\left(\frac{K_0}{\sigma_0}\right)^2 \tag{8}$$

are used for normalization. The quantity K_0 defines a reference stress intensity factor and R_0 is a reference plastic zone size. With σ_{max} = 3000 MPa, $K_0 = 61$ MPa$\sqrt{\text{m}}$ and $R_0 = 0.4$ mm.

Although the small scale yielding normalization, (8), is used, this is only to make contact with previous results; small scale yielding conditions do not prevail in all cases. For impact velocities of 5 m/s and 10 m/s, plastic deformation is pretty much confined to the vicinity of the crack tip region. On the other hand, for impact velocities of 20 m/s and 30 m/s, plastic deformation extends from the crack tip to the block boundary well before the crack has reached the end of the uniform mesh region. The case with an impact velocity of 15 m/s is an intermediate one, where plastic deformation is confined to the crack tip in the earlier stages of crack growth, but some

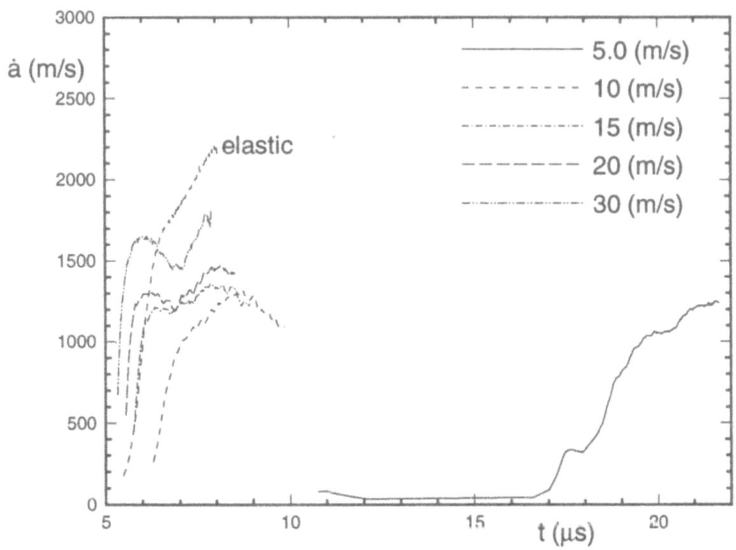

Figure 1. Crack speed versus time for $\sigma_{max} = 3000$ MPa and for various values of the impact velocity.

plastic deformation is present over much of the ligament region toward the end of the calculation.

Curves of crack speed, $\dot{a}$, versus time for various values of impact velocity are shown in Fig. 1. The crack location is identified with $\Delta_n \geq 5\delta_n$ and the curve of crack location versus time is differentiated as described in Xu and Needleman (1994) to obtain the crack speed. For impact velocities of 10 m/s and greater, crack growth begins shortly after arrival of the first loading wave. However, crack growth does not begin until after the second loading wave arrival for an impact velocity of 5 m/s. The calculations are terminated when the crack approaches the end of the uniform mesh region. For comparison purposes, the results of a calculation with $\sigma_{max} = 3000$ MPa, an impact velocity of 10 m/s and with elastic material response are also shown. Comparing the two calculations with an impact velocity of 10 m/s shows that plastic deformation prior to the onset of crack growth gives rise to a delay in crack initiation.

The variation of crack speed with normalized crack extension, $\Delta a/R_0$, is shown in Fig. 2. For the viscoplastic solids in Figs. 1 and 2 the crack speed exhibits a local maximum, except for the case with the lowest impact

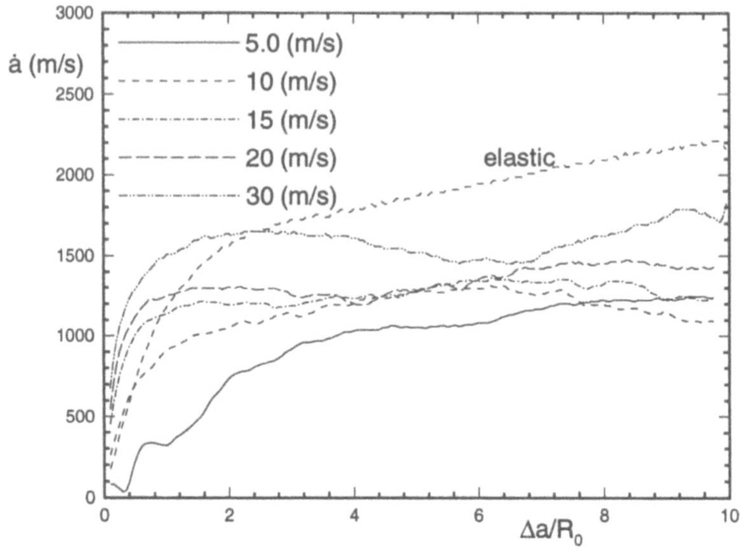

Figure 2. Crack speed versus crack extension for $\sigma_{max} = 3000$ MPa and for various values of the impact velocity.

velocity, 5 m/s. The time at which the maximum crack speed is attained depends on impact velocity, indicating that the crack speed maximum is not a consequence of wave effects. For the elastic solid, the crack speed increases over the entire range of crack growth calculated.

Figure 3 shows curves of the normalized stress intensity factor, K/K_0, versus the normalized crack extension, $\Delta a/R_0$, where K is defined in terms of the J–integral in (7) by

$$K = \sqrt{\frac{EJ}{(1-\nu^2)}} \tag{9}$$

The relation (9) is used here to define K, which is the stress intensity factor in small scale yielding, even when small scale yielding conditions do not prevail.

In our calculations, the value of the work of separation, ϕ_n, is fixed and the elevation of K above K_0 in Fig. 3 is a consequence of material dissipation. Crack growth appears to begin when K is somewhat larger than K_0, but the initiation value of K is sensitive to the value of Δ_n used to define

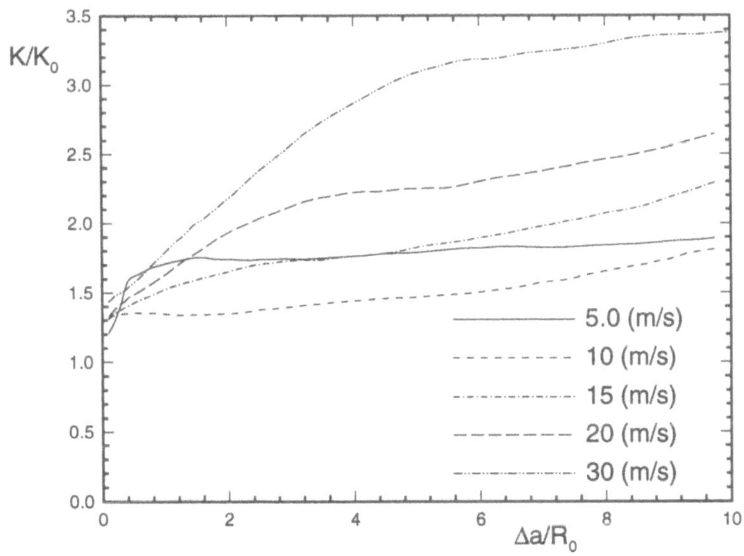

Figure 3. Normalized stress intensity factor versus normalized crack growth for $\sigma_{max} = 3000$ MPa and for various values of the impact velocity.

the crack location. Once significant crack growth takes place this sensitivity disappears. The general trend is for the value of K to increase rapidly in the early stages of crack growth, reach a plateau and then increase again. The K/K_0 level for the case in Fig. 3 with an impact velocity of 10 m/s, in the range $\Delta a/R_0 \leq 2$, is in good agreement with the corresponding quasi-static result in Tvergaard and Hutchinson (1992). Because plastic deformation precedes crack growth with an impact velocity of 5 m/s, this case has a larger plastic zone around the crack tip and a higher apparent toughness than does the case with an impact velocity of 10 m/s. The increase in K occurs in the very early stages of crack growth so that, effectively, the apparent initiation toughness is raised.

The data from Figs. 1 and 3 is used to plot curves of K/K_0 versus crack speed in Fig. 4. These curves show a strong upturn in the stress intensity factor as the crack speed increases. In Xu and Needleman (1994), for an elastic solid, crack branching limits the crack speed. Here, for a viscoplastic solid, the combination of material inertia and dissipation act to limit the attainable crack speed even when the crack is constrained to grow along the initial crack line.

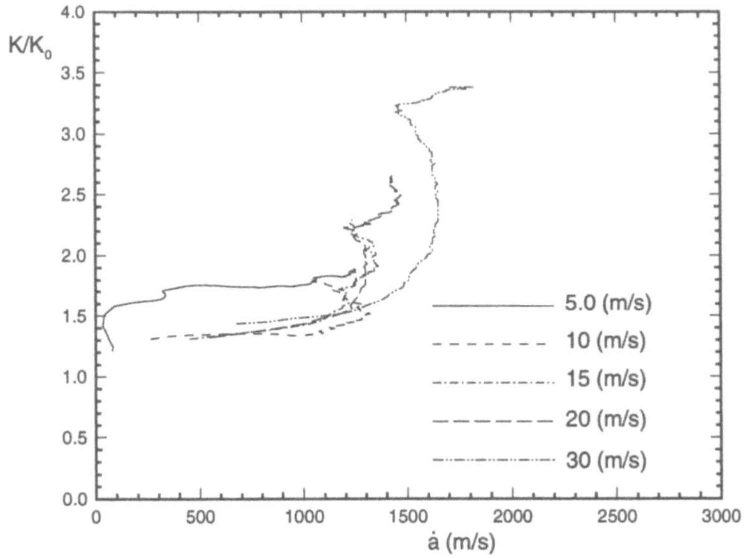

Figure 4. Normalized stress intensity factor versus crack speed for $\sigma_{max} = 3000$ MPa and for various values of the impact velocity.

The dependence of stress intensity factor on crack speed in Fig. 4 is consistent with experimental observations, see e.g. Rosakis and Zehnder (1985) and Zehnder and Rosakis (1990). While there clearly is not a one-to-one correspondence between the value of K and the crack speed, three cases, those with impact velocities of 10 m/s, 15 m/s and 20 m/s, fall within a fairly narrow band. For these cases, the crack speed at which the sharp increase in K takes place, $\approx$ 1200 m/s, is in good agreement with the experimental values in Rosakis and Zehnder (1985) and Zehnder and Rosakis (1990). Lam and Freund (1985) and Freund (1990) found the stress intensity factor to be a strongly increasing function of crack speed in computations where dynamic crack growth in a plastic solid was assumed to occur at a constant speed. In the formulation here, the crack speed history, as well as the history of K, are outcomes of the analysis.

The effect of varying σ_{max} is shown in Fig. 5 where the impact velocity is fixed at 10 m/s, $\sigma_0 = 1000$ MPa and the value of the cohesive strength, σ_{max} is varied. With $\sigma_{max} = 2000$ MPa, the crack speed increases from about 1200 m/s to about 2100 m/s and the decrease in crack speed following the peak is smaller. Thus, the crack speed at which the sharp upturn occurs in

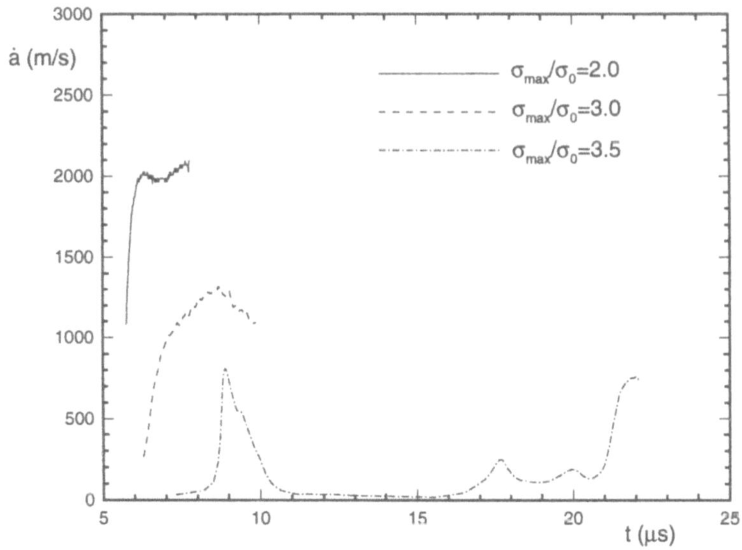

Figure 5. Crack speed versus time for an impact velocity of 10 m/s and various values of the ratio of cohesive surface strength to reference flow strength.

Fig. 4 is dependent on the value assumed for σ_{max} as well as on the impact velocity.

In Siegmund and Needleman (1996) a calculation is carried out for $\sigma_{max}/\sigma_0 = 2.0$, but with $\sigma_{max} = 3000$ MPa and $\sigma_0 = 1500$ MPa. In that case, the crack speed increases monotonically to a steady state value. Thus, the occurrence of the crack speed peak is not a generic feature of the model, but depends on parameter values. In this regard, it is interesting to compare with the results in Needleman and Tvergaard (1991), which were based on a physical model of a fracture mechanism involving two size scales of void nucleating particles, larger particles that nucleate voids at a rather early stage in the deformation history and smaller particles that nucleate voids much later. In that study, the crack speed was found to increase to a constant value once the crack had grown through several of the large particles.

Figure 5 shows that the delay between the arrival of the first loading wave and the onset of crack growth increases with increasing values of the cohesive strength, σ_{max}. For the case in Fig. 5 with the cohesive strength equal to 3500 MPa, crack growth begins after the arrival of the first loading wave, but crack arrest occurs and the crack remains essentially stationary until after the arrival of the second loading wave. There is a significant delay between the arrival of the second loading wave (at 14.94 μs) and the reinitiation of crack growth. It is worth emphasizing that the crack speed histories in Fig. 5 are obtained as outcomes of the initial/boundary value problem formulation.

4. Acknowledgments

We are pleased to acknowledge support from the Air Force Office of Scientific Research under Grant F49620-94-1-0300. T.S. is also grateful for the support provided by Schrödinger Fellowship J00994 – TEC, FWF Vienna, Austria. The computations reported on here were carried out on the Cray C90 at the Pittsburgh Supercomputer Center.

References

Freund, L.B. (1990) *Dynamic Fracture Mechanics.* Cambridge University Press, Cambridge.

Lam, P.S. and Freund, L.B. (1985)Analyses of dynamic growth of a tensile crack in an elastic-plastic material. *J. Mech. Phys. Solids* **33**, 153-167.

Nakamura, T., Shih, C.F. and Freund, L.B. (1985) Computational methods based on an energy integral in dynamic fracture. *Int. J. Fract.* **27**, 229-243.

Needleman, A. (1987) A continuum model for void nucleation by inclusion debonding. *J. Appl. Mech.* **54**, 525-531.

Needleman, A. (1990a) An analysis of decohesion along an imperfect interface. *Int. J. Fract.* **42**, 21-40.

Needleman, A. (1990b) An analysis of tensile decohesion along an interface. *J. Mech. Phys. Solids* **38**, 289-324.

Needleman, A. and Tvergaard, V. (1991) An analysis of dynamic, ductile crack growth in a double edge cracked specimen, *Int. J. Fract.*, **49**, 41-67 (1991).

Peirce, D., Shih, C.F. and Needleman, A. (1984) A tangent modulus method for rate dependent solids. *Comp. Struct.* **18**, 875-887.

Rice, J.R. (1968) A path independent integral and the approximate analysis of strain concentration by notches and cracks. *J. Appl. Mech.* **35**, 379-386.

Rosakis, A.J. and Zehnder, A.T. (1985) On the dynamic fracture of structural metals. *Int. J. Frac* **27**, 169-186.

Rose, J.H., Ferrante, J. and Smith, J.R. (1981) Universal binding energy curves for metals and bimetallic interfaces. *Phys. Rev. Letts.* **47**, 675-678.

Siegmund, T. and Needleman, A. (1996) A numerical study of dynamic crack growth in elastic-viscoplastic solids. *Int. J. Solids Struct.*, to be published.

Tvergaard, V. and Hutchinson, J.W. (1992) The relation between crack growth resistance and fracture process parameters in elastic-plastic solids. *J. Mech. Phys. Solids* **40**, 1377-1397.

Tvergaard, V. and Needleman, A. (1993) An analysis of the brittle-ductile transition in dynamic crack growth, *Int. J. Fract.*, **59**, 53-67.

Xu, X.-P. and Needleman, A. (1994) Numerical simulations of fast crack growth in brittle solids. *J. Mech. Phys. Solids* **42**, 1397-1434.

Zehnder, A.T. and Rosakis, A.J. (1990) Dynamic fracture initiation and propagation in 4340 steel under impact loading. *Int. J. Frac* **43**, 271-285.

DYNAMIC CRACK GROWTH IN VISCOPLASTIC MATERIALS

D. GROSS, ST. HEIMER AND J. HOHE
Institute of Mechanics, TH Darmstadt
Hochschulstr. 1, D-64289 Darmstadt, Germany

Abstract. Fast propagation of a semi-infinite crack in a Perzyna-type material is investigated by two different approaches. First, the asymptotic crack tip field is considered. The field quantities are determined and the dependence on material parameters and crack speed is discussed. Applying a strain criterion a $K(v)$ curve is derived for small scale yielding. In the second part, a damage yield strip model for stationary crack growth is described. Numerical results are presented including the dependence of the stresses, the crack driving force and the temperature distribution as a function of crack velocity.

1. Introduction

The behavior of fast running cracks under small scale or large scale yielding conditions is far from being fully understood. For example, the predetermination of the crack speed for a given material, geometry and loading in general is not possible. A theory based explanation of the observed crack speed dependence on the driving force is still an open question of basic interest. An overview about different approaches to fast crack growth can be found in Freund [3].

Inelastic material behavior usually is rate dependent especially at high strain rates as appear near running crack tips. Therefore, application of viscoplastic material models might be sensible in the description of such processes. Contrary to the enormous literature about quasistatic creep crack growth, cf. Hui & Riedel [5], Chang et al. [2], Li et al. [7], Stamm & Walz [11], there exist only a few investigations on fast crack growth in viscoplastic materials. For example, Lo [8] using a power law presented some analytical

J. R. Willis (ed.), IUTAM Symposium on Nonlinear Analysis of Fracture, 221–230.

results for the antiplane shear case, while a purely numerical study was given by Östlund [10]. A numerical investigation based on the Gurson model can be found in [9].

In this paper, fast growth of a semi-infinite crack in a Perzyna-type viscoplastic material is investigated. The assumed constitutive model is preferably applicable at high strain rates [12]. First, a classical asymptotic crack tip field analysis is carried out. Analytical expressions for the r-dependence of stresses and strains are found, while the angular dependence is determined numerically. The influence of the material parameters is discussed and conclusions are drawn e.g. about unloading regions and limit cases. Using a critical strain criterion a relation between the applied load and the crack speed is determined.

In the second approach a damage yield strip model is applied. The constitutive behavior of the cohesive zone is described by Gurson's damage model in conjunction with the already mentioned Perzyna-type viscoplastic material law. The advantage is that no external fracture criterion is needed now. The governing system of equations is solved numerically. As results, the dependence of the cohesive zone stresses, the yield strip length and the crack driving force on the crack speed are discussed. Furthermore, conclusions on the temperature distribution on account of inelastic energy dissipation can be drawn.

2. Crack tip fields

2.1. BASIC EQUATIONS

Consider a semi-infinite crack, propagating with crack tip speed $v = \dot{a}$ in a Perzyna-type viscoplastic material (fig. 1). The basic equations describing this problem are given by

$$
\begin{aligned}
\sigma_{ij,j} &= \rho\, \ddot{u}_i \\
\varepsilon_{ij} &= \frac{1}{2}\left(u_{i,j} + u_{j,i}\right) \\
\dot{\varepsilon}_{ij} &= \dot{\varepsilon}^{e}_{ij} + \dot{\varepsilon}^{vp}_{ij} \\
\dot{\varepsilon}^{e}_{ij} &= \frac{1}{2\mu}\dot{s}_{ij} + \frac{1-2\nu}{E}\dot{\sigma}_{kk}\,\delta_{ij}\,, \qquad \dot{\varepsilon}^{vp}_{ij} = \frac{1}{\eta}\left\langle \frac{\sqrt{J_2}}{k} - 1 \right\rangle^{n} \frac{s_{ij}}{\sqrt{J_2}}
\end{aligned}
\tag{1}
$$

and the boundary conditions (traction-free crack faces). The material parameters are Young's modulus E, shear modulus μ, Poisson's ratio ν, yield stress k, a viscosity parameter η, and the hardening exponent $n \geq 1$.

Assuming singular crack tip stresses of the type

$$
\sigma_{ij}(r,\varphi) = r^{\lambda}\tilde{\sigma}_{ij}(\varphi)\;; \qquad -\frac{1}{2} \leq \lambda \leq 0 \tag{2}
$$

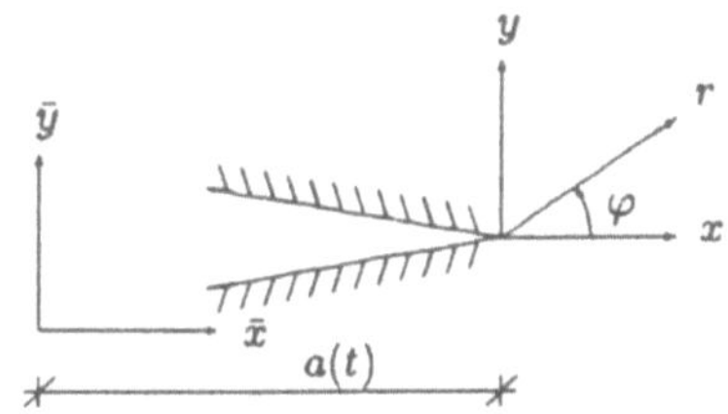

Figure 1. Moving crack tip

the equations can be transformed to a moving coordinate system, fixed at the crack tip by setting $(.)^{\cdot} = -v(.)_{,1}$, $(.)^{\cdot\cdot} = v^2(.)_{,11}$. From this system the strength of the singularity is determined directly by the same arguments as for the Hui-Riedel-field [5]. Since the elastic part of the strain rate dominates the viscoplastic one for $n < 3$ and both parts have the same singularity for $n \geq 3$, λ is found to be the following function of the hardening exponent n:

a) $\dot{\varepsilon}^e$ dominates: $\lambda = -\frac{1}{2}$; $1 \leq n < 3$

b) otherwise: $\lambda = \frac{1}{1-n}$; $n \geq 3$.

For $n < 3$ the well known elastodynamic crack tip field (see [3]) is valid, while for $n \geq 3$ the stresses and displacements can be written as

$$\sigma_{ij}(r,\varphi) = \left(\frac{\eta v k^n}{r\mu^n}\right)^{1/(n-1)} \mu\,\tilde{\sigma}_{ij}(\varphi) \tag{3}$$

$$u_i(r,\varphi) = \left(\frac{\eta v k^n}{r\mu^n}\right)^{1/(n-1)} r\,\tilde{u}_i(\varphi)\ . \tag{4}$$

Substitution of (3, 4) into the governing equations yields a system of nonlinear differential equations for the angular distribution of the field quantities, which has the structure

$$\mathbf{M}\,\frac{\mathrm{d}\mathbf{Y}}{\mathrm{d}\varphi} = \mathbf{f}(\mathbf{Y},\varphi) \tag{5}$$

where $\mathbf{Y}=(\tilde{u}_{r,\varphi},\tilde{u}_{\varphi,\varphi},\tilde{\sigma}_{rr},\tilde{\sigma}_{\varphi\varphi},\tilde{\sigma}_{zz},\tilde{\sigma}_{r\varphi},\tilde{u}_r,\tilde{u}_\varphi)^{\mathrm{T}}$ in plane strain (similar in plane stress). The matrix $\mathbf{M}$ and function $\mathbf{f}$ depend on Poisson's ratio ν, the hardening exponent n and the Mach number $m = v/c_s$, where $c_s^2 = \mu/\rho$ is the shear wave velocity. After transformation into a nonlinear optimization problem and using a 5th order Runge-Kutta integration method the system (5) is solved numerically by a multiple shooting technique in conjunction with a specific optimization routine. By this, all field quantities can be determined in dependence on crack speed and material parameters.

It is remarked that the viscosity parameter η does not appear in (5). Consequently, the dimensionless field quantities $\tilde{\sigma}_{ij}$, $\tilde{u}_i$ are independent of η. Because the dependence on Poisson's ratio is weak, all calculations were carried out with $\nu = 1/4$.

2.2. RESULTS

In figure 2 the angular variation of the stresses is depicted for different n and v/c_s values. The results for $n = 20$ show a strong similarity to the known results for the ideal plastic material. Indeed, from a rigorous analysis of

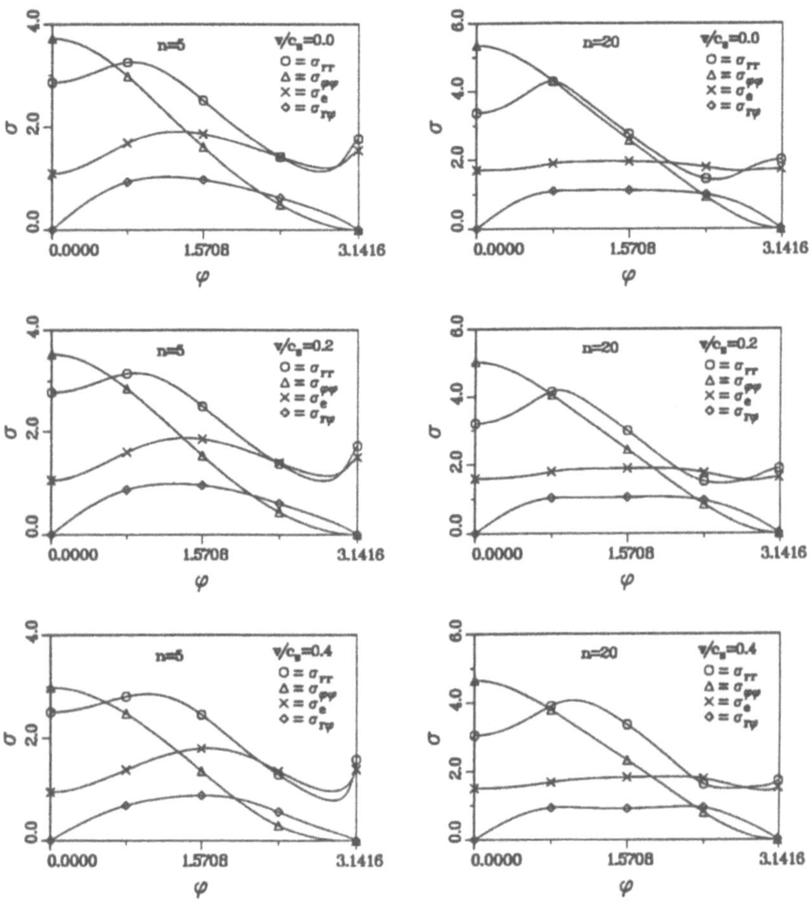

Figure 2. Stress distribution for n=5, n=20 and different v/c_s

the basic equations it can be seen that for $n \to \infty$ the ideal plastic results of Freund [3] are reproduced. In this context it might be interesting to note that for $n = 3$ and $\eta \to \infty$ the linear elastic stress distribution is obtained as well as for $n < 3$ independent on η. Also certain creep crack growth results may be obtained in the limit case $v \to 0$, see e.g. [2].

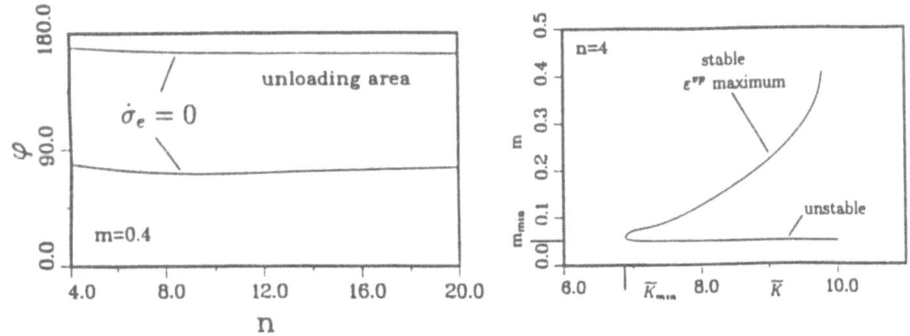

Figure 3. a) Unloading zone b) Small scale yielding

The distribution of the equivalent stress indicates that unloading and reloading zones exist whose sizes depend weakly on n and the crack speed. As an example, the boundary angles between these zones (given by the condition $\dot{\sigma}_e = 0$) are shown for $m = v/c_s = 0.4$ in figure 3a in dependence on n for plane stress. In the unloading zone the material yields because σ_e exceeds $\sqrt{3}\,k$. Contrary to other inelastic material models no pure elastic unloading zones exist at the crack tip in the Perzyna material.

As for the Hui-Riedel-field, no external load parameter appears in (3), (4). Nevertheless, for small scale yielding, in analogy to [5] the crack tip field (3), (4) may be matched with the remote K-determined field by adopting the strain fracture criterion $\varepsilon_e^{vp}(x_{cr}) = \varepsilon_{cr}^{vp}$. In this manner a $K(v)$ curve can be constructed, as plotted in nondimensional form in figure 3b for plane stress and the parameter set $n = 4$, $\mu = 84000$ MPa, $k = 300$ MPa, $\rho = 8000\ kg/m^3$, $\eta = 10^{-2}s$, $\varepsilon_{cr}^{vp} = 0.01$, $x_{cr} = 0.05\ mm$ ($K = \widetilde{K}\varepsilon_{cr}\mu\sqrt{x_{cr}}$). There is a minimum load and a minimum speed below which no crack growth will occur. The upper branch is stable in the sense that the crack will accelerate or decelerate if its velocity is lower or higher than that given for a certain load level. The lower branch is unstable in the same sense. It can be seen that the crack velocity increases with increasing applied load K.

3. Damage yield strip model

3.1. BASIC EQUATIONS

Again, consider a crack tip loaded by an applied dynamic stress intensity factor K_{Id} and running with constant speed $\dot{a}$ through a plate. The local viscoplastic damage behavior now is described by a damage yield strip model or in other words by a specific cohesive model (fig. 4). According to this model the material behavior within the strip is described by the Gurson model [4], [13], [14] in conjunction with the Perzyna-type law. Using the Gurson model, which describes the loss of stress carrying capacity by nucleation, growth and coalescence of microvoids, no external fracture criterion is needed. In onedimensional and modified form its constitutive equations are given by [6], [15]

$$
\begin{aligned}
\sigma &= c_\sigma (1 - q_1 f^*)\, \sigma_M \\
\frac{\partial f}{\partial \varepsilon^{vp}} &= \frac{3}{2}\,\frac{1-f}{1-q_1 f^*}\, q_1 f^* \sinh\left(\frac{c_f}{2}\,(1-q_1 f^*)\right) + \frac{f_N}{s_N\sqrt{2\pi}}\, e^{-\frac{1}{2}\left(\frac{\varepsilon^{vp}-\varepsilon_N}{s_N}\right)^2} \\
\dot{\varepsilon}^{vp} &= \frac{1}{\eta}\left(\frac{\sigma_M}{\sigma_0} - 1\right)^n
\end{aligned}
\qquad (6)
$$

where

$$
c_\sigma = \begin{cases} 1 & \text{for plane stress} \\ \frac{2}{3}\sqrt{3} & \text{for plane strain} \end{cases}, \qquad c_f = \begin{cases} 1 & \text{for plane stress} \\ \sqrt{3} & \text{for plane strain} \end{cases} .
$$

Herein σ_M, f and f^* denote the microscopic stress, actual and effective void fraction respectively; σ_0, q_1, f_N, s_N and ε_N are the initial yield stress and material parameters describing the interaction and nucleation of microvoids. The effective f^* is calculated from f by the common bilinear relation given in [13].

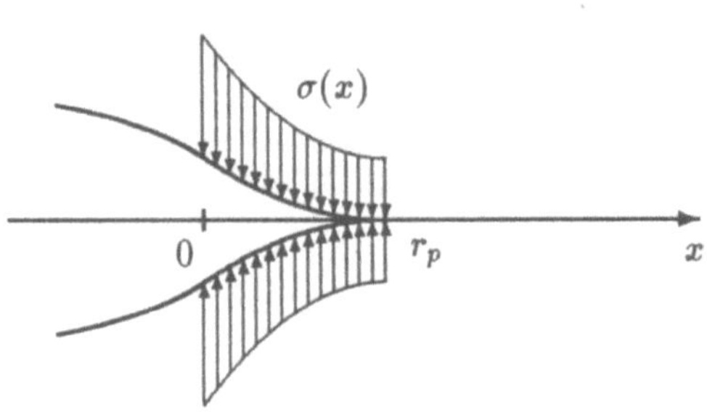

Figure 4. Yield strip model

The relation between the viscoplastic strain and the opening displacement δ within the strip is defined by

$$\varepsilon^{vp} = \frac{\delta}{b_p} \tag{7}$$

where b_p is a finite but small length parameter.

The area surrounding the yield strip is regarded to be linear elastic. Therefore, the mechanical problem can be solved by superposition of the elastodynamic K_{Id}-determined crack tip field and the field induced by the cohesive stresses. The last one is given by integrating the fundamental solution for a moving pair of concentrated forces acting on the crack surfaces at a constant distance from the crack tip [3]. The yield strip length r_p is determined from the condition that the stress singularity at the crack tip ($x = r_p$) must vanish. The following system of equations is obtained

$$\begin{aligned}
\varepsilon^{vp} &= \frac{4V(\dot{a})}{\pi E b_p} \int_0^{r_p} \sigma(x^*) \left(2\frac{\sqrt{r_p - x}}{\sqrt{r_p - x^*}} - \log\frac{\sqrt{r_p - x} + \sqrt{r_p - x^*}}{\left|\sqrt{r_p - x} - \sqrt{r_p - x^*}\right|} \right) dx^* \\
0 &= K_{Id} - \sqrt{\frac{2}{\pi}} \int_0^{r_p} \frac{\sigma(x^*)}{\sqrt{r_p - x^*}} dx^* \\
V(\dot{a}) &= \frac{\dot{a}^2}{c_s^2} \frac{(1+\nu)\,\alpha_d}{4\alpha_d\alpha_s - (1+\alpha_s^2)^2}, \quad \alpha_s = \sqrt{1 - \frac{\dot{a}^2}{c_s^2}}, \quad \alpha_d = \sqrt{1 - \frac{\dot{a}^2}{c_d^2}}
\end{aligned} \tag{8}$$

where c_d is the dilatational wave speed.

The integration in (8) is carried out numerically by a special integration formula taking the singularity of the integrand into account. The obtained system of algebraic equations is solved by the Newton-Raphson-method in conjunction with parameter tracking.

The stress work in the yield zone is mostly converted into heat, which might cause a significant local temperature rise in the vicinity of the crack tip. Using the viscoplastic work rate as the heat source the temperature rise ΔT can be calculated by integrating the fundamental solution of Fourier's equation given by Carslaw and Jaeger [1]

$$\Delta T(x,y) = \int_{A^{vp}} \frac{\sigma_{ij}\dot{\varepsilon}_{ij}^{vp}}{2\pi\lambda_{th}} e^{-\frac{\dot{a}(x-x^*)}{2\kappa_{th}}} \cdot K_0\left(\frac{\dot{a}\, r(x^*,y^*)}{2\kappa_{th}}\right) dx^* dy^* \tag{9}$$

where x and y are coordinates related to the actual crack tip. A^{vp} denotes the plastically deforming region, λ_{th} and κ_{th} are thermal conductivity and diffusivity respectivly, K_0 is the modified 0th-order Bessel function of the second kind and $r = \sqrt{(x-x^*)^2+(y-y^*)^2}$.

3.2. RESULTS

Here, some results for the same pressure vessel steel as in section 2.2 as a material example are presented. In addition to the macroscopic material constants the following Gurson parameter set was chosen: $f_0 = 0.0025$, $f_c = 0.021$, $f_f = 0.19$, $q_1 = 1.5$, $f_N = 0.01$, $\varepsilon_N = 0.3$, $s_N = 0.1$. In fig. 5, the stress σ, the viscoplastic strain ε^{vp}, the microscopic stress σ_M and the void fraction f are plotted along the ligament for different crack tip velocities. It can be seen that with increasing velocity the maximum stress increases due to an increase in deformation rate. On the other hand, the viscoplastic strain at the physical crack is the same for all velocities, which can be explained by the fact that the void volume fraction is a function of the viscoplastic strain only. Thus, the ultimate void volume fraction for a total loss of stress carrying capacity can only be reached at a certain amount of plastic strain. Note that only a part of the yield strip opens while the other part remaines perfectly closed, so that the material behaves as linear elastic in a certain part of the cohesive zone.

In fig. 6, the length r_p^* of that part of the yield strip that opens is depicted as a function of the crack tip velocity. This plastic zone size decreases with increasing velocity due to viscous effects. It vanishes if the Rayleigh wave velocity is reached. Conversely, the crack driving force $\mathcal{G}$, calculated from the remote stress intensity factor K_{Id}, becomes infinite if the crack tip velocity approaches the Rayleigh wave velocity. Neither vanishing yield

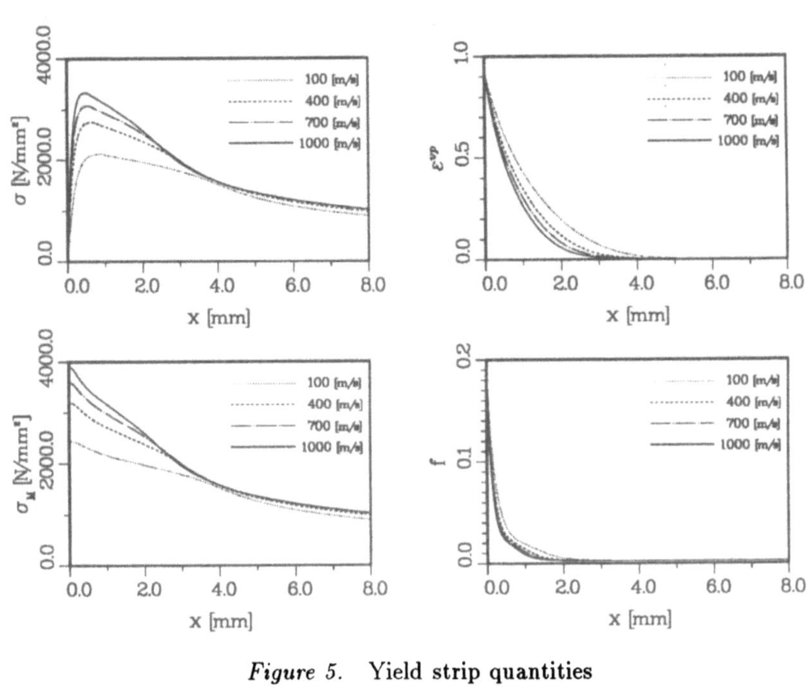

Figure 5. Yield strip quantities

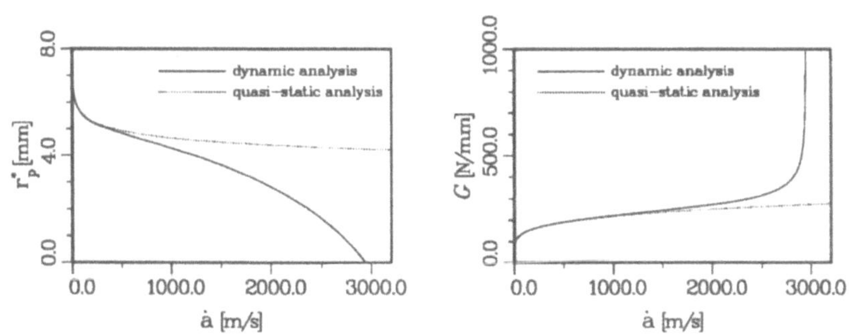

Figure 6. Active yield strip length and crack driving force

strip length nor infinite crack driving force are obtained in a simplified quasistatic analysis, which neglects inertia terms ($V = const.$).

As noted before, the viscoplastic work is converted into heat in the strip ahead of the crack tip. The resulting temperature rise is plotted in fig. 7 along the ligament for different velocities. A significant temperature rise, large enough to affect the fracture process, is observed. The thermal effect increases with crack speed and is strongly localised to the vicinity of the ligament and the crack faces. This can be seen from the contour plot showing the temperature distribution at a velocity of 100 m/s.

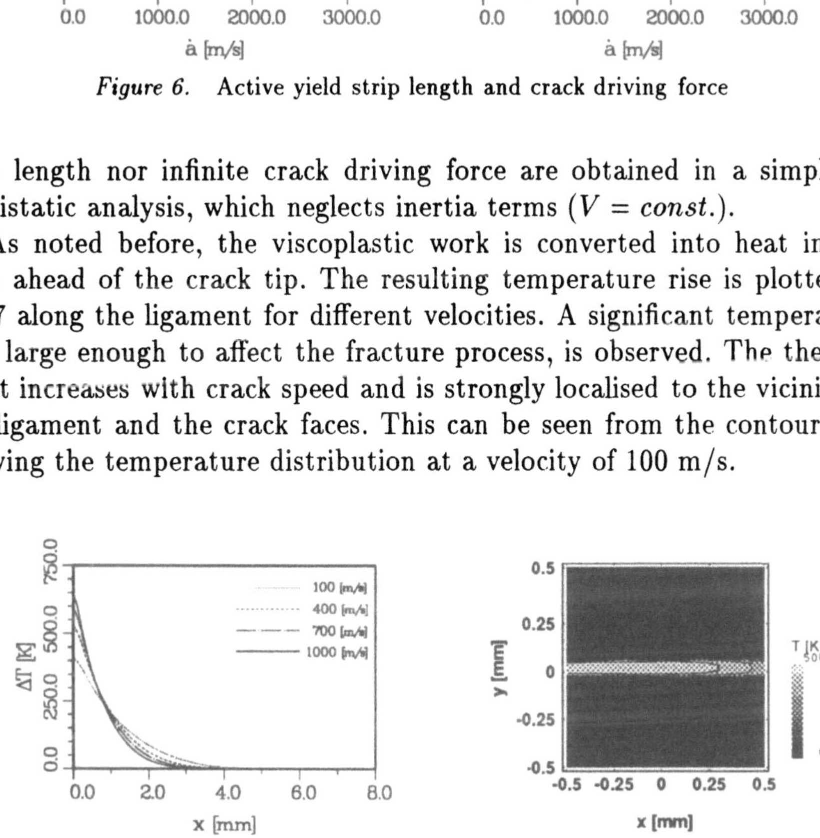

Figure 7. Temperature rise

References

1. Carslaw, H.S. and Jaeger, J.C. (1959) *Conduction of heat in solids*. Oxford University Press, London
2. Chang, T.C., Popelar, C.H. and Staab. G.H. (1987) Creep Crack Growth in an Elastic-Creeping Material, *International Journal of Fracture*, **33**, pp. 31-45
3. Freund, L.B. (1990) *Dynamic fracture mechanics*. Cambridge University Press, Cambridge
4. Gurson, A.L. (1977) Continuum theory of ductile rupture by void nucleation and growth: Part I - yield criteria and flow rules for porous ductile media, *Journal of Engineering Materials and Technology*, **99**, pp. 2–15

5. Hui, C.Y. and Riedel, H. (1981) The Asymptotic Stress and Strain Field near the Tip of a Growing Crack under Creep Conditions, *International Journal of Fracture*, **17**, pp. 409-426
6. Heimer, St., Hohe, J. and Gross, D. (1995) On fast crack propagation in viscoplastic materials, *Archives of Mechanics*, **47**, pp. 899-914
7. Li, F.Z., Needleman, A. and Shih, C.F. (1988) Characterization of Near Tip Stress and Deformation Fields in Creeping Solids, *International Journal of Fracture*, **36**, pp. 163-186
8. Lo, K.K. (1983) Dynamic Crack-Tip Fields in Rate-Sensitive Solids, *Journal of Mechanics and Physics of Solids*, **31**, pp. 287-305
9. Needleman, A. and Tvergaard, V. (1994) Mesh effects in the analysis of dynamic ductile crack growth, *Engineering Fracture Mechanics*, **47**, pp. 75-91
10. Östlund, S. (1990) On Numerical modeling and Fracture Criteria of Dynamic Elastic-Viscoplastic Crack growth, *International Journal of Fracture*, **44**, pp. 283-299
11. Stamm, H. and Walz, G. (1993) Analytical Investigation of Crack Tip Fields in Viscoplastic Materials, *International Journal of Fracture*, **64**, pp. 135-155
12. Perzyna, P. (1966) Fundamental problems in viscoplasticity, *Advances in Applied Mechanics*, **9**, pp. 243–377
13. Tvergaard, V. (1989) Material failure by void growth to coalescence, *Advances in Applied Mechanics*, **27**, pp. 83–151
14. Tvergaard, V. and Needleman, A. (1984) Analysis of the cup-cone fracture in a round tensile bar, *Acta Metallurgica*, **32**, pp. 157–169
15. Zhang, Ch. and Gross, D. (1993) Analysis of ductile crack growth by a simple damage model, *Transactions of the 12th International Conference on Structural Mechanics in Reactor Technology (Stuttgart, 15.8.-20.8.1993, ed: Kussmaul, K.F.)*, **B**, pp. 363–368

NEAR-TIP FIELDS FOR CRACKS IN MATERIALS WITH STRAIN GRADIENT EFFECTS

Y. HUANG
Department of Mechanical Engineering - Engineering Mechanics, Michigan Technological University Houghton, Michigan 49931, USA

L. ZHANG, T.F. GUO and K.C. HWANG
Department of Engineering Mechanics Tsinghua University, Beijing 100084, China

1. Introduction

The linearized couple-stress theory of elastic behavior, originating with the Cosserat brothers (1909), has induced considerable interest. A particularly comprehensive study of this theory is due to Mindlin and Tiersten (1962).

A strain gradient theory of plasticity was recently introduced by Fleck and Hutchinson (1993) in the general framework of couple stress theory. The hardening originates from both statistically stored and geometrically necessary dislocations. The degree of hardening due to statistically stored dislocations is assumed to scale with the von Mises effective strain. Hardening due to geometrically necessary dislocations is taken to scale with an isotropic scalar measure of the strain gradient in the deformed solid, and with a material length parameter ℓ. Two versions, namely, deformation and flow theory versions, are formulated. The theory can explain a size dependence on strength and compares favorably with tension and torsion experiments on thin copper wires (Fleck et al., 1994).

The present paper aims at the investigation of the near-tip fields for opening mode cracks in the couple-stress theory. The outline is as follows. In Section 1.1 is given a recapitulation of the couple stress theory. In Section 2 a general investigation of the near-tip field in elastic material is made. The near-tip fields can be characterized by two independent parameters. In Section 3 is studied crack-tip plasticity of deformation type with strain gradient effects without interrelation between the Cauchy stress-strain and the couple stress-curvature constitutive relations. Section 4 is devoted to crack-tip plasticity according to the theory of Fleck and Hutchinson (1993)

J. R. Willis (ed.), IUTAM Symposium on Nonlinear Analysis of Fracture, 231–243.

with the above mentioned interrelation.

1.1. RECAPITULATION OF COUPLE STRESS THEORY

Throughout the paper the notations of Fleck and Hutchinson (1993) will be followed. The unsymmetric Cauchy stress $\mathbf{t}$ is decomposed into a symmetric part $\boldsymbol{\sigma}$ and an anti-symmetric part $\boldsymbol{\tau}$. The force traction vector $\mathbf{T}$ on a plane with unit normal $\mathbf{n}$ is related to the Cauchy stress by (in dyadic and in Cartesian coordinates forms, respectively)

$$\mathbf{T} = \mathbf{n}\cdot\mathbf{t} = \mathbf{n}\cdot(\boldsymbol{\sigma}+\boldsymbol{\tau}), \qquad T_j = n_i t_{ij} = n_i(\sigma_{ij}+\tau_{ij}). \tag{1.1}$$

Similarly the couple-stress traction vector $\mathbf{q}$ is related to the couple-stress tensor $\mathbf{m}$ by

$$\mathbf{q} = \mathbf{n}\cdot\mathbf{m}, \qquad q_j = n_i m_{ij} \tag{1.2}$$

where $\mathbf{m}$ is assumed to be deviatoric.

Equilibrium of forces within the body gives

$$\nabla\cdot\mathbf{t} = \nabla\cdot(\boldsymbol{\sigma}+\boldsymbol{\tau}) = 0, \qquad t_{ji,j} = \sigma_{ji,j} + \tau_{ji,j} = 0 \tag{1.3}$$

and equilibrium of moments gives

$$\nabla\cdot\mathbf{m} + \mathbf{e}:\boldsymbol{\tau} = 0, \qquad m_{ji,j} + e_{ikl}\tau_{kl} = 0 \tag{1.4}$$

where $\mathbf{e}$ is the permutation tensor, and we have neglected the presence of body forces and body couples.

The rotation $\boldsymbol{\omega}$ and the curvature $\boldsymbol{\chi}$ are defined as, respectively,

$$\boldsymbol{\omega} = \frac{1}{2}\mathbf{e}:(\nabla\mathbf{u}) = \frac{1}{2}\nabla\times\mathbf{u}, \qquad \omega_i = \frac{1}{2}e_{ijk}u_{k,j} \tag{1.5}$$

and

$$\boldsymbol{\chi} = \boldsymbol{\omega}\nabla, \qquad \chi_{ij} = \omega_{i,j}. \tag{1.6}$$

The curvature tensor $\boldsymbol{\chi}$ can be expressed in terms of the strain gradients as

$$\chi_{ij} = e_{ikl}\varepsilon_{jl,k}. \tag{1.7}$$

(1.7) is called the $\chi \sim \varepsilon$ compatibility equations.

The deformation version of the couple stress (or strain gradient) theory can be formulated by means of the strain energy density W :

$$W = W(\varepsilon_e, \ell\chi_e, \varepsilon_m) \tag{1.8}$$

where $\varepsilon_e=(2\varepsilon' : \varepsilon'/3)^{1/2}$– the von Mises strain invariant (ε' denoting the strain deviator), $\chi_e=(2\chi : \chi/3)^{1/2}$– the curvature invariant and $\varepsilon_m = \varepsilon_{kk}$ – the volumetric strain.

The Cauchy stress and couple stress are given by

$$\sigma_{ij} = \frac{\partial W}{\partial \varepsilon_{ij}}, \qquad \ell^{-1} m_{ij} = \frac{\partial W}{\partial (\ell \chi_{ji})}. \tag{1.9}$$

For linearized couple stress theory to be studied in Section 2, the strain energy density (1.8) is taken as

$$W = \frac{3}{2} G(\varepsilon_e^2 + \ell^2 \chi_e^2) + \frac{1}{2} K \varepsilon_m^2 \tag{1.10}$$

where $G \equiv E/2(1+\nu)$ and $K \equiv E/3(1-2\nu)$ are respectively the shear and bulk moduli of the solid.

For strain gradient theory of plasticity in the study of Section 3, we assume "*separate laws of hardening*" for Cauchy stress and for couple stress :

$$\varepsilon_e = \left(\frac{\sigma_e}{\sigma_0}\right)^n, \qquad \ell \chi_e = \left(\frac{\ell^{-1} m_e}{\sigma_0}\right)^n \tag{1.11}$$

where $\sigma_e = (\frac{3}{2}\mathrm{s} : \mathrm{s})^{\frac{1}{2}}$is the von Mises effective stress , and $m_e = (\frac{3}{2}\mathbf{m} : \mathbf{m})^{\frac{1}{2}}$ the analogous effective couple stress. s is the deviator of the symmetric part $\boldsymbol{\sigma}$ of Cauchy stress. The strain energy density (1.8) is taken as

$$W = \frac{n}{n+1}\sigma_0 \left[\varepsilon_e^{\frac{n+1}{n}} + (\ell \chi_e)^{\frac{n+1}{n}}\right] + \frac{1}{2} K \varepsilon_m^2. \tag{1.12}$$

Finally, Section 4 is devoted to the study based on the strain gradient theory of plasticity introduced by Fleck and Hutchinson (1993) with interrelated law of hardening :

$$\mathcal{E} = \left(\frac{\Sigma}{\Sigma_0}\right)^n \tag{1.13}$$

where

$$\mathcal{E}^2 \equiv \varepsilon_e^2 + \ell^2 \chi_e^2, \qquad \Sigma^2 \equiv \sigma_e^2 + \ell^{-2} m_e^2. \tag{1.14}$$

The corresponding expression for strain energy density is

$$W = \frac{n}{n+1} \Sigma_0 \mathcal{E}^{\frac{n+1}{n}} + \frac{1}{2} K \varepsilon_m^2. \tag{1.15}$$

For states of plane strain (as is the case of Sections 2—4) parallel to the plane $x_3 = 0$, we have

$$\begin{gathered} \omega_\alpha = \varepsilon_{3i} = \varepsilon_{i3} = \chi_{i3} = \chi_{\alpha\beta} = 0, \\ t_{3\alpha} = t_{\alpha 3} = m_{33} = m_{\alpha\beta} = 0 \end{gathered} \tag{1.16}$$

where the Greek indices range over 1,2. We introduce the abridged symbols

$$\omega = \omega_3, \qquad \chi_\alpha = \chi_{3\alpha}, \qquad m_\alpha = m_{\alpha 3}. \tag{1.17}$$

We can prove the path independent $\mathcal{J}$-integral for any path Γ surrounding the crack tip (with crack surfaces free of load):

$$\mathcal{J} = \int_\Gamma (W n_1 - T_i u_{i,1} - q_i \omega_{i,1}) ds. \tag{1.18}$$

It should be noted that the strain gradient theory of plasticity with *separate* strain energy density W (1.12) can also predict a size dependence of the strength, which also compares favorably with tension and torsion experiments on thin copper wires (Fleck et al., 1994).

2. Near-Tip Fields in Elastic Material

Substituting the strain energy density expression (1.10) into (1.9) gives the elastic constitutive relations for elastic materials. For plane strain problems ($\varepsilon_{33} = 0$), we have

$$\varepsilon_{\alpha\beta} = \frac{1}{2G}(\sigma_{\alpha\beta} - \nu\delta_{\alpha\beta}\sigma_{\gamma\gamma}), \qquad \chi_\alpha = \frac{1}{2G\ell^2} m_\alpha. \tag{2.1}$$

In polar coordinates, centered at the crack tip, the expressions for strains, rotation and curvatures are

$$\begin{gathered} \varepsilon_{rr} = \frac{\partial u_r}{\partial r}, \qquad \varepsilon_{\theta\theta} = \frac{1}{r}\frac{\partial u_\theta}{\partial \theta} + \frac{u_r}{r} \\ \varepsilon_{r\theta} = \frac{1}{2}\left(\frac{1}{r}\frac{\partial u_r}{\partial \theta} + \frac{\partial u_\theta}{\partial r} - \frac{u_\theta}{r}\right) \end{gathered} \tag{2.2}$$

$$\omega = \frac{1}{2}\left(\frac{\partial u_\theta}{\partial r} - \frac{1}{r}\frac{\partial u_r}{\partial \theta} + \frac{u_\theta}{r}\right) \tag{2.3}$$

$$\chi_r = \frac{\partial \omega}{\partial r}, \qquad \chi_\theta = \frac{1}{r}\frac{\partial \omega}{\partial \theta}. \tag{2.4}$$

Equilibrium of forces and moments, (1.3) and (1.4) gives, in polar coordinate form,

$$\begin{aligned}\frac{\partial t_{rr}}{\partial r}+\frac{1}{r}\frac{\partial t_{\theta r}}{\partial \theta}+\frac{t_{rr}-t_{\theta\theta}}{r} &= 0\\ \frac{\partial t_{r\theta}}{\partial r}+\frac{1}{r}\frac{\partial t_{\theta\theta}}{\partial \theta}+\frac{t_{r\theta}+t_{\theta r}}{r} &= 0 \qquad (2.5)\\ t_{r\theta}-t_{\theta r}+\frac{\partial m_r}{\partial r}+\frac{1}{r}\frac{\partial m_\theta}{\partial \theta}+\frac{m_r}{r} &= 0. \qquad (2.6)\end{aligned}$$

The expression (1.7) of curvature in terms of strain gradients (or $\varepsilon \sim \chi$ compatibility) is simplified to

$$\begin{aligned}\chi_r &= \frac{\partial \varepsilon_{r\theta}}{\partial r}-\frac{1}{r}\frac{\partial \varepsilon_{rr}}{\partial \theta}+2\frac{\varepsilon_{r\theta}}{r}\\ \chi_\theta &= \frac{\partial \varepsilon_{\theta\theta}}{\partial r}-\frac{1}{r}\frac{\partial \varepsilon_{r\theta}}{\partial \theta}+\frac{\varepsilon_{\theta\theta}-\varepsilon_{rr}}{r}.\end{aligned} \qquad (2.7)$$

From (2.4) we have the χ-compatibility equation:

$$\frac{\partial \chi_r}{\partial \theta}-\frac{\partial}{\partial r}(r\chi_\theta)=0. \qquad (2.8)$$

Assume the near-tip asymptotic expansion for stresses (with Greek subscripts ranging over r, θ)

$$\begin{aligned}t_{\alpha\beta}(r,\theta) &= t^{(0)}_{\alpha\beta}(\theta)r^p+t^{(1)}_{\alpha\beta}(\theta)r^{p+1/2}+\cdots\\ \ell^{-1}m_\alpha(r,\theta) &= m^{(0)}_\alpha(\theta)r^p+m^{(1)}_\alpha(\theta)r^{p+1/2}+\cdots\end{aligned} \qquad (2.9)$$

We shall consider the first leading term in asymptotic expansions. Equilibrium of forces (2.5) leads to (hereafter prime means derivative with respect to θ)

$$\begin{aligned}t^{(0)\prime}_{\theta r}(\theta) &= -(p+1)t^{(0)}_{rr}(\theta)+t^{(0)}_{\theta\theta}(\theta)\\ t^{(0)\prime}_{\theta\theta}(\theta) &= -(p+1)t^{(0)}_{r\theta}(\theta)-t^{(0)}_{\theta r}(\theta).\end{aligned} \qquad (2.10)$$

Equilibrium of moments (2.6) dictates

$$m^{(0)\prime}_\theta(\theta) = -(p+1)m^{(0)}_r(\theta). \qquad (2.11)$$

The $\chi \sim \varepsilon$ compatibility equation (2.7) takes the asymptotic forms:

$$\begin{aligned}(p+2)\varepsilon^{(0)}_{r\theta}(\theta)-\varepsilon^{(0)\prime}_{rr}(\theta) &= 0\\ (p+1)\varepsilon^{(0)}_{\theta\theta}(\theta)-\varepsilon^{(0)}_{rr}(\theta)-\varepsilon^{(0)\prime}_{r\theta}(\theta) &= 0.\end{aligned} \qquad (2.12)$$

The χ compatibility (2.8) gives

$$m_r^{(0)\prime}(\theta) = (p+1)m_\theta^{(0)}(\theta). \tag{2.13}$$

The finiteness of the $\mathcal{J}$-integral (1.18) requires $p = -1/2$. We easily obtain the following results:

$$\begin{aligned}
t_{rr} &= r^{-\frac{1}{2}}B^{(0)}\left(\tfrac{3+2\nu}{4}\cos\tfrac{\theta}{2} - \tfrac{7-6\nu}{4}\cos\tfrac{3\theta}{2}\right) + o\left(r^{-1/2}\right)\\
t_{\theta\theta} &= r^{-\frac{1}{2}}B^{(0)}\left(\tfrac{5-2\nu}{4}\cos\tfrac{\theta}{2} + \tfrac{7-6\nu}{4}\cos\tfrac{3\theta}{2}\right) + o\left(r^{-1/2}\right)\\
t_{r\theta} &= r^{-\frac{1}{2}}B^{(0)}\left(-\tfrac{9-10\nu}{4}\sin\tfrac{\theta}{2} + \tfrac{7-6\nu}{4}\sin\tfrac{3\theta}{2}\right) + o\left(r^{-1/2}\right)\\
t_{\theta r} &= r^{-\frac{1}{2}}B^{(0)}\left(\tfrac{7-6\nu}{4}\sin\tfrac{\theta}{2} + \tfrac{7-6\nu}{4}\sin\tfrac{3\theta}{2}\right) + o\left(r^{-1/2}\right) ;
\end{aligned} \tag{2.14}$$

$$\begin{aligned}
\ell^{-1}m_r &= r^{-\frac{1}{2}}A^{(0)}\sin\tfrac{\theta}{2} + o(r^{-1/2})\\
\ell^{-1}m_\theta &= r^{-\frac{1}{2}}A^{(0)}\cos\tfrac{\theta}{2} + o(r^{-1/2}).
\end{aligned} \tag{2.15}$$

The rotation ω and displacements u_α can be found from the known stress field :

$$\omega = r^{\frac{1}{2}}\tfrac{1}{G}\ell^{-1}A^{(0)}\sin\tfrac{\theta}{2} + o\left(r^{1/2}\right) ; \tag{2.16}$$

$$\begin{aligned}
u_r &= r^{\frac{1}{2}}\tfrac{B^{(0)}}{G}\left(\tfrac{3-6\nu}{4}\cos\tfrac{\theta}{2} - \tfrac{7-6\nu}{4}\cos\tfrac{3\theta}{2}\right) + o\left(r^{1/2}\right)\\
u_\theta &= r^{\frac{1}{2}}\tfrac{B^{(0)}}{G}\left(-\tfrac{1-2\nu}{4}\sin\tfrac{\theta}{2} + \tfrac{7-6\nu}{4}\sin\tfrac{3\theta}{2}\right) + o\left(r^{1/2}\right).
\end{aligned} \tag{2.17}$$

Sternberg and Muki (1967) solved the stress distribution in a linear couple stress solid due to a transverse crack of finite length $2a$ in an all-round infinite body that is otherwise in a state of uniform uni-axial tension t_0 at right angles to the plane of the crack. The near-tip stress fields obtained by a limiting process from their equation (4.4) agree completely with our equations (2.14) and (2.15), if the constants $A^{(0)}$ and $B^{(0)}$ are given the values

$$\ell A^{(0)} = \frac{\Gamma_1}{2(1-\nu)}\frac{t_0\sqrt{a}}{\sqrt{2}}a, \qquad B^{(0)} = \frac{\Gamma_2}{1-\nu}\frac{t_0\sqrt{a}}{\sqrt{2}}$$

where Γ_1 and Γ_2 are parameters depending on the ratio ℓ/a.

Substituting the near-tip expressions for quantities occurring in $\mathcal{J}$-integral (1.18), we get after some lengthy calculation

$$\mathcal{J} = \frac{1-\nu}{2G}K_I^2 = \frac{\pi}{G}\left[(1-\nu)(3-2\nu)\left(B^{(0)}\right)^2 + \frac{1}{2}\left(A^{(0)}\right)^2\right]. \tag{2.18}$$

Here K_I is the far field stress intensity factor. Hence the near-tip field is characterized by two parameters $A^{(0)}$ and $B^{(0)}$, whose values cannot be determined from the local asymptotic analysis.

3. Near-Tip Fields Based on Strain-Gradient Theory of Plasticity with Separate Laws of Hardening

The strain energy density (1.12) substituted into (1.9), gives the constitutive relations

$$\begin{aligned}\varepsilon_{ij} &= \frac{3}{2}\left(\frac{\sigma_e}{\sigma_0}\right)^{n-1}\frac{s_{ij}}{\sigma_0} + \frac{1-2\nu}{3E}\sigma_{kk}\delta_{ij} \\ \ell\chi_{ij} &= \frac{3}{2}\left(\frac{\ell^{-1}m_e}{\sigma_0}\right)^{n-1}\frac{\ell^{-1}m_{ji}}{\sigma_0}.\end{aligned} \tag{3.1}$$

Confining ourselves to the first dominating term of near-tip fields, we assume

$$\begin{aligned}t_{ij}/\sigma_0 &= t_{ij}^{(0)}(\theta)r^p \\ \ell^{-1}m_\alpha/\sigma_0 &= m_\alpha^{(0)}(\theta)r^p.\end{aligned} \tag{3.2}$$

The angular distribution functions for the dominating term are labeled with the superscript (0). If stresses have singularity at the crack tip (i.e. $p < 0$), the plane strain condition $\varepsilon_{33} = 0$, from $(3.1)_1$ and noting $n > 1$, leads to $s_{33}^{(0)} = 0$, so the in-plane components of constitutive relations (3.1) are

$$\begin{aligned}\varepsilon_{\alpha\beta} &= \frac{3}{2}\left(\sigma_e^{(0)}(\theta)\right)^{n-1} s_{\alpha\beta}^{(0)}(\theta)r^{np} \\ \ell\chi_\alpha &= \frac{3}{2}\left(m_e^{(0)}(\theta)\right)^{n-1} m_\alpha^{(0)}(\theta)r^{np}\end{aligned} \tag{3.3}$$

where $\sigma_e^{(0)}(\theta)$, $s_{\alpha\beta}^{(0)}(\theta)$ and $m_e^{(0)}(\theta)$ are the angular functions of σ_e, $s_{\alpha\beta}$ and m_e.

3.1. SOLUTION FOR CAUCHY STRESSES

Relations (2.10) are still valid as the dominate equations. Substituting (3.3) into the $\varepsilon \sim \chi$ compatibility equation (2.7) gives

$$\begin{aligned}\frac{d}{d\theta}\left[\left(\sigma_e^{(0)}(\theta)\right)^{n-1} s_{rr}^{(0)}(\theta)\right] &= (np+2)\left(\sigma_e^{(0)}(\theta)\right)^{n-1} s_{r\theta}^{(0)}(\theta) \\ \frac{d}{d\theta}\left[\left(\sigma_e^{(0)}(\theta)\right)^{n-1} s_{r\theta}^{(0)}(\theta)\right] &= -(np+2)\left(\sigma_e^{(0)}(\theta)\right)^{n-1} s_{rr}^{(0)}(\theta).\end{aligned} \tag{3.4}$$

Solving simultaneously (2.10) and (3.4) for 4 unknowns $t_{rr}^{(0)}(\theta)$, $t_{\theta\theta}^{(0)}(\theta)$, $t_{r\theta}^{(0)}(\theta)$, $t_{\theta r}^{(0)}(\theta)$, and noting the symmetry about $\theta = 0$, we obtain

$$\begin{aligned} t_{\theta\theta}^{(0)}(\theta) &= B^{(0)}\cos p\theta + C^{(0)}\frac{2(p+1)}{np+p+2}\cos(np+2)\theta \\ t_{\theta r}^{(0)}(\theta) &= -B^{(0)}\sin p\theta + C^{(0)}\frac{2(p+1)}{np+p+2}\sin(np+2)\theta \end{aligned} \tag{3.5}$$

where $B^{(0)}$ and $C^{(0)}$ are integration constants. The traction-free boundary conditions, $t_{\theta\theta}^{(0)}(\pi) = t_{\theta r}^{(0)}(\pi) = 0$, lead to $\sin(n+1)p\pi = 0$,for the determination of eigenvalue of p . The boundedness of the $\mathcal{J}$-integral (1.18) dictates $p \geq -1/(n+1)$, hence we have

$$p = -\frac{1}{n+1}, \qquad C^{(0)} = \frac{n+1}{2n}B^{(0)}. \tag{3.6}$$

Then the dominant Cauchy stresses are summarized as follows

$$\begin{aligned} t_{rr}/\sigma_0 &= r^{-\frac{1}{n+1}}B^{(0)}\left\{\cos\left(\tfrac{\theta}{n+1}\right) - \tfrac{1}{n}\cos\left(\tfrac{n+2}{n+1}\theta\right)\right\} \\ t_{\theta\theta}/\sigma_0 &= r^{-\frac{1}{n+1}}B^{(0)}\left\{\cos\left(\tfrac{\theta}{n+1}\right) + \cos\left(\tfrac{n+2}{n+1}\theta\right)\right\} \\ t_{r\theta}/\sigma_0 &= r^{-\frac{1}{n+1}}B^{(0)}\left\{-\sin\left(\tfrac{\theta}{n+1}\right) + \tfrac{1}{n}\sin\left(\tfrac{n+2}{n+1}\theta\right)\right\} \\ t_{\theta r}/\sigma_0 &= r^{-\frac{1}{n+1}}B^{(0)}\left\{\sin\left(\tfrac{\theta}{n+1}\right) + \sin\left(\tfrac{n+2}{n+1}\theta\right)\right\}. \end{aligned} \tag{3.7}$$

3.2. SOLUTION FOR COUPLE STRESSES

Equation (2.11) is still valid as the dominant equation of equilibrium of moments. Substituting the curvatures χ_α from $(3.3)_2$ into the χ-compatibility equation (2.8) gives

$$\frac{d}{d\theta}\left[\left(m_e^{(0)}(\theta)\right)^{n-1} m_r^{(0)}(\theta)\right] = (np+1)\left(m_e^{(0)}(\theta)\right)^{n-1} m_\theta^{(0)}(\theta) \tag{3.8}$$

where p is given by (3.6). The solution of (2.11) and (3.8) satisfying symmetry condition $m_r(0) = 0$ and traction-free boundary condition $m_\theta(\pi) = 0$ can be easily found. Then the couple stress fields may be written as

$$\begin{aligned} \ell^{-1}m_r/\sigma_0 &= r^{-1/(n+1)}A^{(0)}\left[\tfrac{1}{2}\left(1+\tfrac{1}{n^2}\right) + \tfrac{1}{2}\left(1-\tfrac{1}{n^2}\right)\cdot\right. \\ &\qquad \left.\cos 2(\theta - \varphi(\theta))\right]^{\frac{1}{2(n+1)}}\sin\varphi(\theta) \\ \ell^{-1}m_\theta/\sigma_0 &= r^{-1/(n+1)}A^{(0)}\left[\tfrac{1}{2}\left(1+\tfrac{1}{n^2}\right) + \tfrac{1}{2}\left(1-\tfrac{1}{n^2}\right)\cdot\right. \\ &\qquad \left.\cos 2(\theta - \varphi(\theta))\right]^{\frac{1}{2(n+1)}}\cos\varphi(\theta) \end{aligned} \tag{3.9}$$

where $A^{(0)}$ is an integration constant, and

$$\varphi(\theta) = \frac{1}{2}\left[\theta - \sin^{-1}\left(\frac{n-1}{n+1}\sin\theta\right)\right]. \tag{3.10}$$

The rotation ω and displacement u_α can be easily obtained.

$$\begin{aligned} u_r &= -r^{1/(n+1)}(n+1)\left(\tfrac{\sqrt{3}}{2}\right)^{n+1}\left(\tfrac{n+1}{n}\right)^{n} \mid B^{(0)} \mid^{n} \cdot \\ &\quad \left(\mathrm{sign}B^{(0)}\right)\cos\left(\tfrac{n+2}{n+1}\theta\right) \\ u_\theta &= r^{1/(n+1)}(n+1)\left(\tfrac{\sqrt{3}}{2}\right)^{n+1}\left(\tfrac{n+1}{n}\right)^{n} \mid B^{(0)} \mid^{n} \cdot \\ &\quad \left(\mathrm{sign}B^{(0)}\right)\sin\left(\tfrac{n+2}{n+1}\theta\right) \end{aligned} \tag{3.11}$$

$$\begin{aligned} \omega &= r^{1/(n+1)}(n+1)\left(\tfrac{3}{2}\right)^{(n+1)/2} \mid A^{(0)} \mid^{n} \left(\mathrm{sign}A^{(0)}\right) \cdot \\ &\quad \left[\tfrac{1}{2}\left(1+\tfrac{1}{n^2}\right) + \tfrac{1}{2}\left(1-\tfrac{1}{n^2}\right)\cos 2(\theta - \varphi(\theta))\right]^{\frac{n}{2(n+1)}} \sin\varphi(\theta). \end{aligned} \tag{3.12}$$

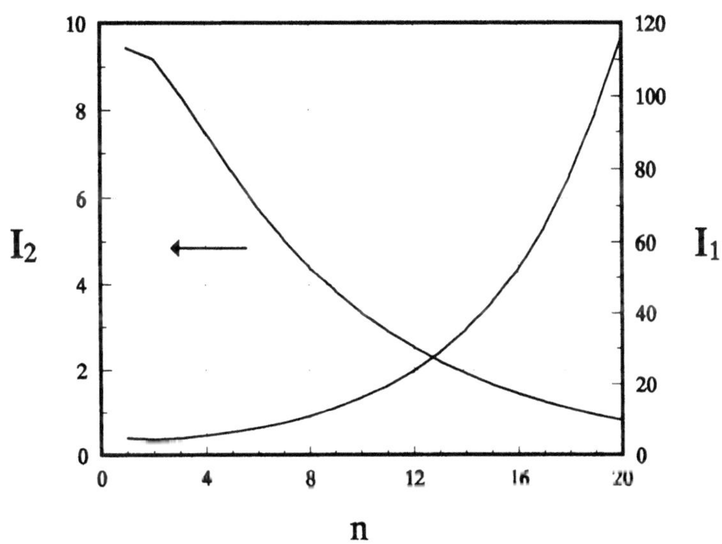

Figure 1: Curves of $I_1(n)$ and $I_2(n)$.

The value of $\mathcal{J}$-integral can be found from (1.18) :

$$\mathcal{J} = \sigma_0\left[\mid A^{(0)} \mid^{n+1} I_1(n) + \mid B^{(0)} \mid^{n+1} I_2(n)\right] \tag{3.13}$$

where

$$
\begin{aligned}
I_1(n) &= 2\left(\frac{3}{2}\right)^{(n-1)/2}\int_0^{\pi}\left[\frac{1}{2}\left(1+\frac{1}{n^2}\right)+\frac{1}{2}\left(1-\frac{1}{n^2}\right)\cos 2(\theta-\varphi(\theta))\right]^{1/2}\cdot \\
&\quad \left[\frac{n}{n+1}\cos\theta-\sin(\varphi(\theta)-\theta)\sin\varphi(\theta)\right]d\theta \\
I_2(n) &= 2\pi\left(\frac{\sqrt{3}}{2}\right)^{n+1}\left(\frac{n+1}{n}\right)^n .
\end{aligned}
\tag{3.14}
$$

$I_1(n)$ and $I_2(n)$ are shown in Figure 1.

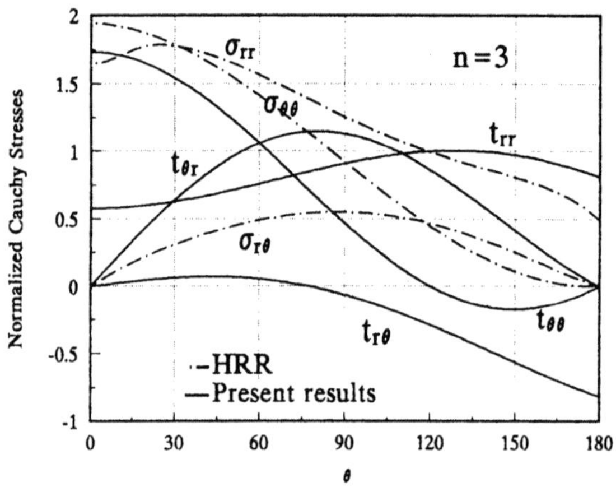

Figure 2: Angular distribution of Cauchy stress ($n = 3$).

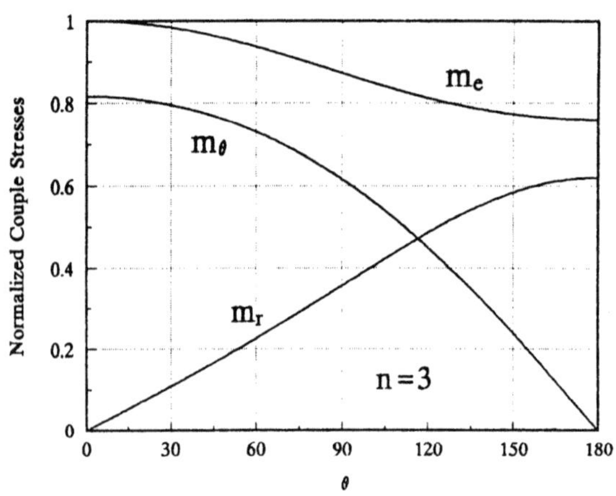

Figure 3: Angular distribution of couple stress ($n = 3$).

The angular distribution functions for Cauchy stress components $t^{(0)}_{\alpha\beta}(\theta)$ and couple stress $m^{(0)}_{\alpha}(\theta)$, normalized by means of $\max_\theta \sigma^{(0)}_e(\theta)$ and $\max_\theta m^{(0)}_e(\theta)$, respectively, are shown in Figures 2 and 3 ($n = 3$).

Remark 1. *It should be noted that according to the dominant near-tip asymptotic solution in the form of products of functions of r and θ, the loading history is basically proportional. So the above results, obtained based on the deformation theory version, remain valid for the flow theory version of gradient theory of plasticity (Fleck and Hutchinson, 1993).*

4. Near-Tip Fields Based on Strain-Gradient Theory of Plasticity with Interrelated Law of Hardening (Fleck and Hutchinson, 1993)

According to Fleck and Hutchinson (1993), the strain energy density W in the form (1.15) leads to the constitutive relations

$$\begin{aligned} \varepsilon_{ij} &= \frac{3}{2}\left(\frac{\Sigma}{\Sigma_0}\right)^{n-1}\frac{s_{ij}}{\Sigma_0} + \frac{1-2\nu}{3E}\sigma_{kk}\delta_{ij} \\ \ell\chi_{ij} &= \frac{3}{2}\left(\frac{\Sigma}{\Sigma_0}\right)^{n-1}\frac{\ell^{-1}m_{ji}}{\Sigma_0} \end{aligned} \tag{4.1}$$

in which Σ is given by (1.14).

Similar to (3.2), we assume the asymptotic expansion with equal dominance (i.e., same singularity) of Cauchy and couple stresses

$$\begin{aligned} t_{ij}/\Sigma_0 &= t^{(0)}_{ij}(\theta)r^p \\ \ell^{-1}m_\alpha/\Sigma_0 &= m^{(0)}_\alpha(\theta)r^p. \end{aligned} \tag{4.2}$$

If stress has singularity at crack tip ($p < 0$), the plane strain condition $\varepsilon_{33} = 0$, from $(4.1)_1$ and noting $n > 1$, leads to $s^{(0)}_{33} = 0$, and the material behaves asymptotically as incompressible.

By the same way as in deriving (3.4), we get the leading term of the $\varepsilon \sim \chi$ compatibility equation (2.7)

$$\begin{aligned} \frac{d}{d\theta}\left[\left(\Sigma^{(0)}(\theta)\right)^{n-1} s^{(0)}_{rr}(\theta)\right] &= (np+2)\left(\Sigma^{(0)}(\theta)\right)^{n-1} s^{(0)}_{r\theta}(\theta) \\ \frac{d}{d\theta}\left[\left(\Sigma^{(0)}(\theta)\right)^{n-1} s^{(0)}_{r\theta}(\theta)\right] &= -(np+2)\left(\Sigma^{(0)}(\theta)\right)^{n-1} s^{(0)}_{rr}(\theta) \end{aligned} \tag{4.3}$$

where

$$\left(\Sigma^{(0)}(\theta)\right)^2 = \left(\sigma^{(0)}_e(\theta)\right)^2 + \left(m^{(0)}_e(\theta)\right)^2. \tag{4.4}$$

As the counterpart of (3.8), the χ-compatibility equation (2.8) takes the form

$$\frac{d}{d\theta}\left[\left(\Sigma^{(0)}(\theta)\right)^{n-1} m_r^{(0)}(\theta)\right] = (np+1)\left(\Sigma^{(0)}(\theta)\right)^{n-1} m_\theta^{(0)}(\theta). \tag{4.5}$$

Equations (2.11), (4.3) and (4.5) furnish 4 simultaneous differential equations for 4 unknowns. Their nontrivial solutions can be expressed in terms of two functions $c(\theta)$ $(0 \leq c(\theta) \leq \pi/2)$ and $m^{(0)}(\theta)$ $(0 \leq m(\theta) \leq 2\pi)$:

$$\begin{aligned} s_{rr}^{(0)}(\theta) &= (\mathrm{sign}A^*)\tfrac{1}{\sqrt{3}}\Sigma^{(0)}(\theta)\sin c(\theta)\cos(np+2)\theta \\ s_{r\theta}^{(0)}(\theta) &= -(\mathrm{sign}A^*)\tfrac{1}{\sqrt{3}}\Sigma^{(0)}(\theta)\sin c(\theta)\sin(np+2)\theta \end{aligned} \tag{4.6}$$

$$\begin{aligned} m_r^{(0)}(\theta) &= \sqrt{\tfrac{2}{3}}\Sigma^{(0)}(\theta)\cos c(\theta)\sin m(\theta) \\ m_\theta^{(0)}(\theta) &= \sqrt{\tfrac{2}{3}}\Sigma^{(0)}(\theta)\cos c(\theta)\cos m(\theta) \end{aligned} \tag{4.7}$$

where A^* is an arbitrary constant, and

$$\Sigma(\theta) = \mid A^* \mid \left(\sqrt{3}/\sin c(\theta)\right)^{1/n}. \tag{4.8}$$

The two functions $c(\theta)$ and $m(\theta)$ satisfy the equations

$$\begin{aligned} &\frac{d}{d\theta}\left[\Sigma^{(0)}(\theta)\cos c(\theta)\cos m(\theta)\right] = -(p+1)\Sigma^{(0)}(\theta)\cos c(\theta)\sin m(\theta) \\ &\frac{d}{d\theta}\left[\left(\Sigma^{(0)}(\theta)\right)^n \cos c(\theta)\sin m(\theta)\right] = (np+1)\left(\Sigma^{(0)}(\theta)\right)^n \cos c(\theta)\cos m(\theta). \end{aligned} \tag{4.9}$$

subject to the symmetry and boundary conditions :

$$m(0) = 0, \qquad \cos m(\pi) = 0. \tag{4.10}$$

The dominant equation of equilibrium of forces (2.10) can be solved for $t_{\theta\theta}^{(0)}(\theta)$ and $t_{\theta r}^{(0)}(\theta)$ (with symmetry with respect to the line $\theta = 0$).

$$\begin{aligned} t_{\theta\theta}^{(0)}(\theta) &= D^{(0)}\cos p\theta + 2(p+1)\cdot \\ &\quad \int_0^\pi \left[s_{rr}^{(0)}(\xi)\sin p(\xi-\theta) - s_{r\theta}^{(0)}(\xi)\cos p(\xi-\theta)\right] d\xi \\ t_{\theta r}^{(0)}(\theta) &= -D^{(0)}\sin p\theta - 2(p+1)\cdot \\ &\quad \int_0^\pi \left[s_{rr}^{(0)}(\xi)\cos p(\xi-\theta) + s_{r\theta}^{(0)}(\xi)\sin p(\xi-\theta)\right] d\xi \end{aligned} \tag{4.11}$$

subject to the traction-free boundary conditions :

$$t_{\theta\theta}^{(0)}(\pi) = 0, \qquad t_{\theta r}^{(0)}(\pi) = 0. \tag{4.12}$$

The finiteness of the $\mathcal{J}$-integral requires $p = -1/(n+1)$. Eliminating $D^{(0)}$ from the expressions of (4.12) obtained by use of (4.11) and (4.6), we get the condition

$$\int_0^{\pi} (\sin c(\xi))^{\frac{n-1}{n}} \cos \xi d\xi = 0. \tag{4.13}$$

Our problem is to solve (4.9) subject to conditions (4.10) and (4.13). Numerical computation shows that no solution exists. Hence we are led to the conclusion that for the material obeying Fleck and Hutchinson's strain gradient theory of plasticity, the Cauchy stress and couple stress do not have the same order of singularity. It can be seen

i. if the Cauchy stress predominates, we have the solution (3.7), (3.11) with the parameter $B^{(0)} \neq 0$ and $A^{(0)} = 0$;

ii. otherwise, if the couple stress predominates, we have the solution (3.9), (3.12) with the parameter $A^{(0)} \neq 0$ and $B^{(0)} = 0$.

Acknowledgments—The authors wish to express their thanks to China Natural Science Foundation and China State Commission of Education for their supports. YH gratefully acknowledges the support from US National Science Foundation (Grant # INT - 94 - 23964).

5. References

Cosserat, E. and F. (1909) *Théorie des Corps Déformables*, A. Hermann et Fils, Paris.

Mindlin, R.D. and Tiersten, H.F. (1962) Effects of couple-stresses in linear elasticity, *Archs Ration. Mech, Analysis* **11**, 415–448.

Sternberg, E. and Muki, R. (1967) The effect of couple-stresses on stress concentration around a crack, *Int. J. Solids Structures* **3**, 69–95.

Fleck, N.A. and Hutchinson, J.W. (1993) A phenomenological theory for strain gradient effects in plasticity, *J. Mech. Phys. Solids* **41**, 1825-1857.

Fleck, N.A., Muller, G.M., Ashby, M.F. and Hutchinson, J.W. (1994) Strain gradient plasticity: theory and experiment, *Acta. metall. mater.* **42**, 475–487.

NON-LINEAR DAMAGE EVOLUTION AND FAILURE IN MATERIALS

W. A. CURTIN, H. SCHER*, and M. PAMEL
Engineering Science & Mechanics
Materials Science & Engineering
Virginia Polytechnic Institute and State University
Blacksburg, VA 24061

**Environmental Sciences and Energy Research*
Weizmann Institute of Science
76100 Rehovoth, Israel

Abstract

Damage evolution and time-to-failure are investigated for a model material in which damage formation is a stochastic event. Specifically, the probability of failure at any site i at time t is $\propto \sigma_i(t)^\eta$ where $\sigma_i(t)$ is the local stress at site i at time t, and differs from the applied stress because of the stress redistribution from prior damage. Numerical simulations in 2-dimensional systems demonstrate interesting and non-linear behavior. Of particular interest is $\eta \geq 3$, for which failure occurs by rapid damage growth after a "nucleation" period during which a large damage cluster develops to the critical size. An analytic model of the damage process predicts this "avalanche" failure, as well as (i) more abrupt failure with increasing η, (ii) failure times scaling inversely with system size, and (iii) broadening of the distribution of failure times, so that the failure becomes less predictable. These features are all observed in the simulations. The model also predicts the onset time for the rapid growth, offering the possibility of early detection of impending failure.

1. Introduction

Failure under load is a major limitation to the application of many materials, especially structural materials at elevated temperatures. Understanding the mechanisms by which damage forms, coalesces, and drives failure, and the time

J. R. Willis (ed.), IUTAM Symposium on Nonlinear Analysis of Fracture, 245–255.

scales for these phenomena, is thus an important area of research. That these phenomena are non-linear in time and stress makes the development of predictive models of damage accumulation particularly challenging.

Here, we focus on understanding the details of damage evolution and failure in systems for which the "damage" is a probabilistic, nucleated event that is driven by local stress. One material which appears to behave in this manner is the composite Si/SiC.[1] Under load at elevated temperatures, cavities form in this material between SiC/SiC grain boundary facets. The cavities extend across the entire grain boundary and the rate of cavitation is strongly dependent on the level of applied stress. Furthermore, the spatial distribution of cavities is not random; clustering occurs in the form of arrays of cavities in planes roughly perpendicular to the tensile load axis. The cavity arrays are not formed by the slow crack growth mechanism operative in many ceramics at elevated temperatures and high loads because the cavities are physically disconnected. The cavities lead to strength degradation with time, and ultimately a sufficiently large cavity forms such that macroscopic failure occurs. The progression of damage in time and the dependence of failure time on microscopic aspects of the damage formation are general features occurring in many materials and motivate the study of models to predict a material's remaining strength and reliability.

2. Model of Damage Evolution

In developing a model which is sufficiently general but relevant to materials such as Si/SiC, we first recognize several key features of the deformation and damage. First, the presence of fully-extended cavities suggests that cavity formation is controlled by a critical nucleation step, and is thus a probabilistic event. Second, the sensitivity of cavitation rate to stress and the observation of cavity clustering both suggest that this cavitation rate is dependent on the *local* stress acting across each grain boundary at any given time. We thus consider an elastic material consisting of an ordered array of connected, cavitatable sites. At each site i, the local (tensile) stress $\sigma_i(t)$ consists of the applied load plus additional loads transferred to site i due to previous cavitation damage at other sites. The cavitation rate $r_i(t)$ (probability of cavitation per unit time) is assumed to have power-law dependence on the local stress:

$$r_i(t) = A\sigma_i(t)^{\eta} \quad . \tag{1}$$

Here, A is a rate prefactor and the exponent η determines the sensitivity of the cavitation rate to applied stress. A power-law rate generates a power-law dependence of strain rate on applied stress, as observed in Si/SiC.[1]

The evolution of damage in an array of sites obeying Eq. 1 can be surprisingly complex. Initially, the material is undamaged and all sites have the same cavitation rate. However, once cavities form the stress distribution in the material is non-uniform and the rates are no longer equal. The stress redistribution around the existing cavities tends to increase the stress at nearby sites in the plane perpendicular to the applied load. The cavities are considered to be blunted cracks or voids so that the stress transfer occurs gradually over a length scale comparable to the grain boundaries themselves. The enhanced stress preferentially drives cavitation at sites near to the existing cavities, but at least initially there are many more sites remote from the existing cavities subject to essentially the applied stress. The location of subsequent cavities must be determined probabilistically, and each site i has a relative rate given by $r_i(t)/\Sigma r_j(t)$. The typical time Δt required to form the next cavity is simply the inverse of the sum of the rates, $\Delta t=1/\Sigma r_j(t)$. As time and cavitation proceed, cavity clusters form and the stress at the tips of larger cavities is generally larger than that at smaller cavities. This stress enhancement continues to drive the formation of larger cavities faster than smaller cavities, but is mitigated against by the greater number of smaller cavities and/or uncavitated sites. Cavitation continues until failure, where one cavity cluster spans the entire length of the system - no explicit condition for the onset of actual crack growth is considered.

Numerical simulations of this failure process provide insight into the richness of the damage evolution versus various parameters. We have utilized a two-dimensional triangular network of elastic springs as a mechanical model [2] for studying the damage evolution. In this model, all springs are linear and cavities are represented by "broken" or removed springs. The stress distribution for the evolving array of "cavities" is determined numerically, and the appearance of each successive cavity is determined by the relative rates as described above. The numerical simulation can monitor elastic strain under constant applied load, failure time, number of cavities vs time, and other measures of the damage.

Results for the elastic strain versus dimensionless time $tA\sigma_{app}{}^{\eta}$ are shown in Figure 1 for a fixed lattice size of 20x20 nodes (760 springs under tensile stress initially) and various values of η. For small η, the damage evolution and accumulated strain are very gradual. For larger η, the accumulated strain increases slowly at early times, but after some "incubation" time begins to increase very rapidly in what we term an "avalanche" failure, wherein one single large cavity cluster propagates rapidly across the entire material in a fairly short time. Figure 1 also indicates the rough range of the observed (stochastic) failure time. Notice that although the failure time becomes shorter with increasing η, it also becomes much more abrupt but with a broad spread in failure times.

Materials with large η thus exhibit very little non-linear deformation until just prior to failure and then rapidly fail, with the distribution of failure times being broad relative to the average failure time. Figure 2 shows the approximate

Figure 1. Elastic strain versus dimensionless time for fixed system size of 20x20 nodes (760 springs under initial tension), at various η. Several statistical realizations are shown at each η.

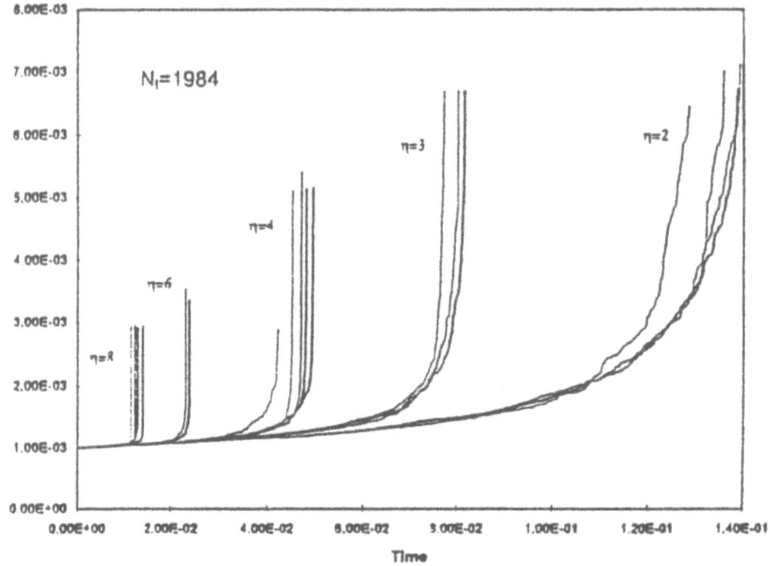

Figure 2. Elastic strain versus dimensionless time at fixed η=8 and various system sizes for realizations failing at about the mean time for each size.

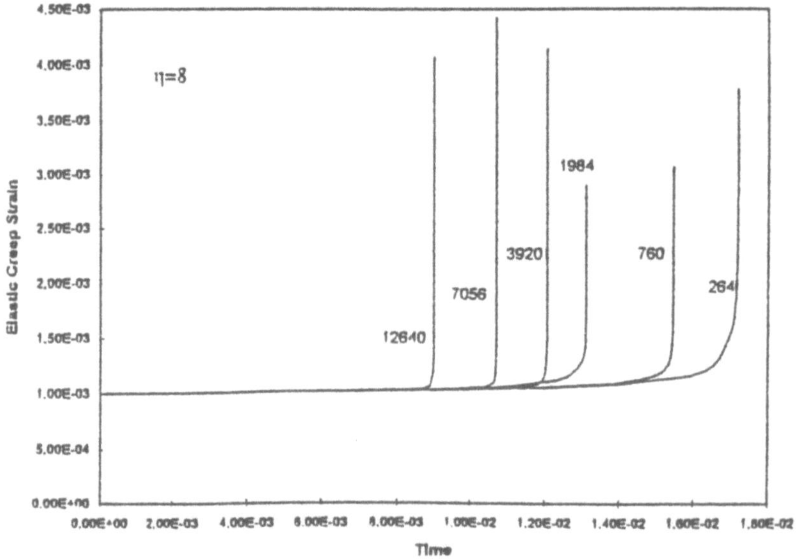

failure time for a fixed η but as a function of system size N_t, from which it is also clear that increasing the system size leads to decreasing failure times.

Hansen et al. studied the accumulated damage at failure in simulation studies of electrical fuse networks [3]. For $\eta \leq 2$, Hansen et al. found a linear dependence of total damage on system size, $N_f \propto N_t$. However, for $\eta > 2$ they found a sublinear dependence, $N_f \propto N_t^{\gamma}$, with $\gamma < 1$ and decreasing with increasing η. These results suggest that these two regimes of η correspond to intrinsically different modes of failure: percolation-like and avalanche-like.

3. Analytic Model for Damage and Failure

To make analytic progress on understanding the subtle aspects uncovered by numerical simulations, and experimental results on materials such as Si/SiC, requires an approximate analysis which retains the key physics. We recognize that the damage rate is highest at the tips (perimeter) of existing damage. We therefore focus on these tip sites by assuming that all sites in the system are either (i) tip sites, at the tip of a cluster of size c units and under an enhanced stress denoted σ_c, (ii) damaged sites in the cavity clusters themselves, which carry no stress, and (iii) all remaining non-tip,non-damaged sites, under only the applied stress σ_{app}. This division of sites into three categories eliminates long-range interactions between existing damage. We also neglect the linking together of two existing clusters. Then, the quantityto determine is the cluster size distribution N(c,t), the number of size c clusters at time t. Within the assumptions above, size c clusters can only form via the growth of size c-1 clusters and can only be lost by growth to size c+1 clusters. Hence, the evolution of N(c,t) follows a Master Equation,

$$\frac{dN(c,t)}{dt} = zA\sigma_{c-1}^{\eta} N(c-1,t) - zA\sigma_c^{\eta} N(c,t) \quad , \qquad (2)$$

where the growth rate of a size c cluster is $zA\sigma_c^{\eta}$ with z being the number of possible growth sites at the tip of the cluster (e.g. z=2 for a 1-d line, z=4 for the triangular lattice). For c=1, we note that $\alpha_0 = 1$ and N(0,t), the number of "size 0" cavities, is precisely the number of non-damaged, non-tip sites:

$$N(0,t) = N_t - \sum (c+z) N(c,t) \quad . \qquad (3)$$

Defining a dimensionless time $\tau = tA\sigma_{app}^{\eta}$ and recognizing that the stress enhancement is generally proportional to the applied field, $\sigma_c = K_c \sigma_{app}$ where K_c is the stress concentration factor at each of the z tip sites a size-c cluster, allows us to rewrite Eq. 2 in the non-dimensional form

$$\frac{dN(c,\tau)}{d\tau} = \alpha_{c-1} N(c-1,\tau) - \alpha_c N(c,\tau) \quad , \qquad (4)$$

where $\alpha_c = zK_c^\eta$ is the dimensionless growth rate. With N(c,t) in hand from Eq. 4, many quantities can be determined. The elastic modulus vs time is given within the mean-field approximation as

$$E(t) = E_0\left(1 - \frac{\beta}{N_t}\sum c^2 N(c,t)\right) \quad , \tag{5}$$

where β is a constant ($(1-\nu)^2\pi/4$ in a continuum elastic model) [4]. The elastic strain at fixed applied load is then simply $\varepsilon(t) = \sigma_{app} / E(t)$. The typical largest cluster of damage at any time, denoted $c^*(t)$, satisfies

$$\sum_{c=c^*} N(c,t) = 1 \quad , \tag{6}$$

i.e. there exists one cluster at least as large as c* at time t. The fracture strength as a function of time depends on the stress concentration factor and the largest crack as $S(t) \propto 1/K_{c^*}$. Complete material failure occurs when the largest crack is equal to the linear length of the system, c*=L. From a probabilistic standpoint, the probability of failure is obtained by simple weak-link considerations to be

$$P_f(t) = 1 - \exp(-N(L,t)) \quad . \tag{7}$$

The major properties of interest thus stem from knowledge of N(c,t).

The solutions to the coupled set of equations for N(c,t) can be obtained analytically by using Laplace transforms and by approximating $N(0,t)=N_t$ [5]. Of particular interest are two limiting cases, corresponding to short and long times for any crack size c. In the limit $\alpha_c\tau/c \ll 1$ the solution is a simple product,

$$N(c,\tau) = \Big(\prod_{c'=1}^{c-1} \frac{\alpha_{c'}}{c'}\Big)\tau^c \tag{8}$$

and so $N(c,\tau)$ is rapidly decreasing for cluster sizes c and times τ in this regime. In the long-time limit $\alpha_c\tau/c \gg 1$ the solution becomes

$$N(c,\tau) \propto \alpha_c^{-1} \quad . \tag{9}$$

The largest cluster size c* can be rewritten from Eq. 6 using Eqs. 4 as

$$\int_0^\tau d\tau' \alpha_{c*-1} N(c*-1,\tau) = 1 \quad . \tag{10}$$

Hence, as the largest cluster enters the long-time regime of Eq. 9, it is clear that to satisfy Eq. 10 the c* must increase rapidly with time. This rapid growth is the avalanche observed in the simulations and always occurs if "long times" can be reached such that Eq. 9 pertains.

We apply the above model to understand the simulation results performed on spring/fuse networks. In these cases, the stress concentration factors scale essentially as the square root of cluster size, and we take the form $K_c=(1+1/z)c^{1/2}$. This form distributes the stress of a size 1 cavity equally to the z neighbors at its tips, and for larger clusters the desired square-root dependence obtains.

We have previously shown that Eqs. 4 explain the fundamentally different failure behaviors observed for $\eta \leq 2$ and $\eta > 2$, and that our predictions agree quantitatively with the simulation results of Hansen et al.[6]. In short, for $\eta \leq 2$, the short-time limit $\alpha_c \tau / c < 1$ applies to *all* cluster sizes up to the failure time. Failure occurs by a gradual proliferation of small damage which ultimately does link together to form a large crack, estimated as the time at which $E(\tau_f)=0$. For $\eta > 2$, the long-time limit can be reached prior to failure, and one large cluster then grows rapidly across the material to cause failure. For larger system sizes a sufficently large cluster, and hence failure, occurs at earlier times.

The evolution of modulus, strain, and failure time distribution can be calculated from Eqs. 4 and subsequent equations, but these do not provide any physical insight into those factors and events which control the time-to-failure and its distribution versus N_t and η for $\eta > 2$. In addition, one would like to identify any signature of incipient failure so as to anticipate failure in any one sample, and one would like such a signature to have the same size and η dependence as the failure time itself. Since in the avalanche regime ($\eta > 2$), one single large "crack" ultimately controls the failure, we focus on this cluster. Now once a single crack of sufficient size (as yet unknown) emerges from the smaller crack distribution, it tends to propagate without any "supply" from smaller crack sizes. To follow the evolution of this one large crack, we take the definition of $c^*(t)$ from Eq. 6 and differentiate with respect to time and then utilize the Master equation itself, Eq. 4, to obtain the growth rate of the largest crack as

$$\frac{dc*}{d\tau} = \alpha_{c*} \left(\frac{\alpha_{c*-1} N(c*-1,\tau)}{\alpha_{c*} N(c*,\tau)} \right) \approx \alpha_{c*} \tag{11}$$

where the final approximation comes from applying Eq. 9 at long times. Eq. 11 can then be integrated directly, given the boundary condition $c^*(\tau_f)=L$ and the K_c from above, to yield

$$c*(\tau) = L \left(1 + \frac{z(1+\frac{1}{z})^{\eta}(\frac{\eta}{2}-1)(\tau_f - \tau)}{L^{1-\frac{\eta}{2}}} \right)^{\frac{1}{1-\frac{\eta}{2}}} . \tag{12}$$

However, such a solution is at best accurate only at times for which $\alpha_{c*}\tau / c* > 1$ because no single large crack appears in the short-time limit. We postulate that Eq. 12 is accurate back to an onset time τ^*,

$$\frac{\alpha_{c*}}{c*}\tau^* = 1 \quad . \tag{13}$$

This is equivalent to assuming that as soon as the largest crack leaves the short-time regime, it then grows forward to failure in a time $\tau_f-\tau^*$. The accuracy of Eq. 12 down to the time determined by Eq. 13 is demonstrated in Figure 3 by comparison to the full results. Solving Eqs. 12 and 13 simultaneously for τ^* leads to the remarkable result

$$\tau^* = (1-\frac{2}{\eta})\tau_f \quad , \tag{14}$$

neglecting factors which are vanishingly small for large system sizes. τ^* marks the onset of the "avalanche" leading to failure, and Eq. 14 implies that the size scaling of the real failure time τ_f is then identical to that of the onset time τ^*.

To obtain the scaling of τ^* with system size, we return to the short-time solutions for $N(c^*,\tau^*)$. Setting $N(c^*,\tau^*)=1$ using Eq. 8 and solving this simultaneously with Eq. 13 then yields

$$\tau^* \propto \left[\ln(N_t)\right]^{1-\frac{\eta}{2}} \quad , \tag{15}$$

and by Eq. 14 the failure time τ_f scales similarly. The critical cluster size $\hat{c}$ which begins the avalanche at time τ^* also is obtained as

$$\hat{c} = \frac{\ln(N_t)}{\frac{\eta}{2}-1} \quad , \tag{16}$$

neglecting corrections which are negligible for large system sizes. From Eqs. 15 and 16 we see that the time-to-failure is predicted to decrease with increasing system size, as observed in the simulations. Also, the critical crack size $\hat{c}$ which initiates the avalanche grows only very slowly with system size, and decreases with increasing η. Neglect of cluster-cluster interactions and limiting cluster shapes to a quasi-linear form are reinforced by the predicted small values of $\hat{c}$ [7].

A final important prediction regards the failure time distribution. Since the mean failure time τ_f is determined by the formation of the $\hat{c}$ cluster at τ^*, we postulate that the failure time distribution is controlled by the probability of appearance of the size $\hat{c}$ crack vs. time. The $\hat{c}$ cluster may appear earlier than, or later than, the typical time τ^*, but once formed grows to failure in the fixed time $\tau_f-\tau^*$. The cumulative probability of forming a size $\hat{c}$ cluster versus time is

$$P(\hat{c},\tau) = 1-\exp(-N(\hat{c},\tau)) \quad , \tag{17}$$

and at time τ^* the probability is 0.632. The corresponding probability of complete failure is then obtained by a *rigid shift* of this cumulative probability forward in time by the amount $\tau_f-\tau^*$:

$$P_f(\tau) = P(\hat{c},\tau-\frac{2}{\eta}\tau_f) \quad , \tag{18}$$

Figure 3. Largest crack c* versus time for $\eta=4$ at sizes $N_t=10^4$ and 10^6: full result from Eq. 4 (-------) ; approximation of Eq. 12 (- - - - -). Also shown as the decreasing solid line is the condition $\alpha_c \tau / c = 1$; the onset of the avalanche occurs at the intersection of this curve with $c^*(\tau)$ and graphically illustrates the onset time τ^* and critical cluster size $\hat{c}$.

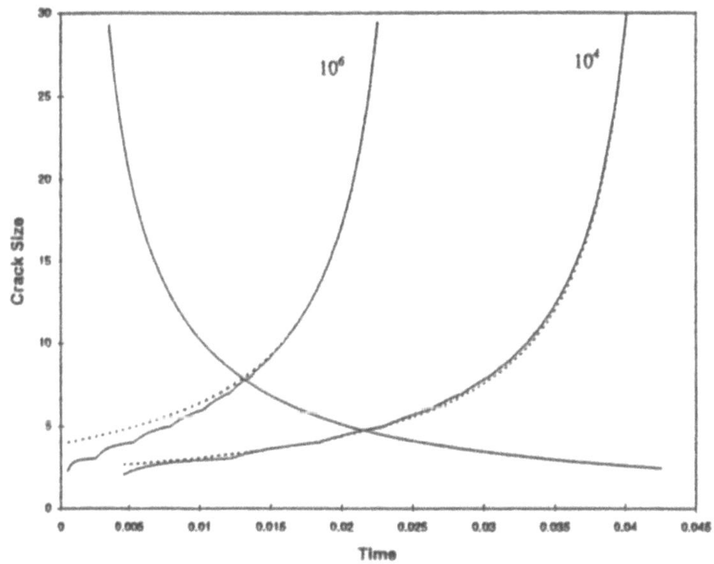

Figure 4. Probability of failure vs time from Eqs. 4 and 7 (-------) for $\eta=4$, $N_t=10^4$; probability of obtaining the critical cluster $\hat{c}$ vs. time (- - - - -); and predicted probability of failure by rigid shift of critical cluster result by τ_f-τ^*.

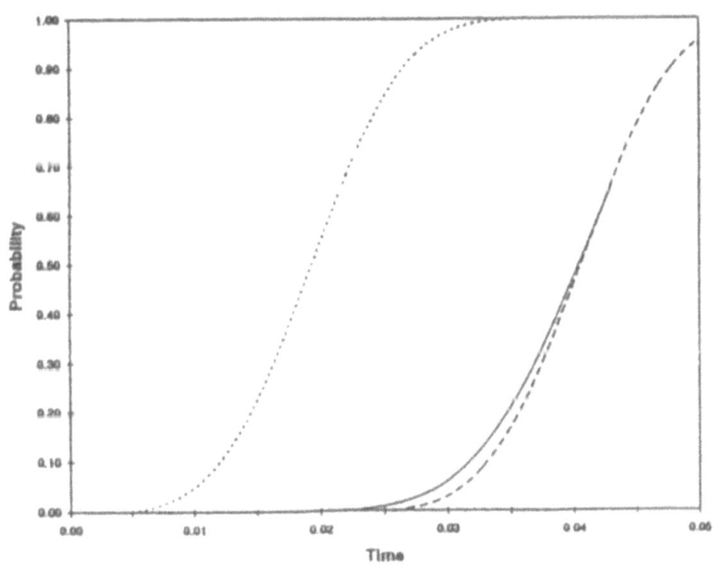

where τ_f refers to the typical failure time so that $P_f(\tau_f)=0.632$. The failure time distribution determined from the full numerical results of Eqs. 4 is shown in Figure 4 along with the values obtained by a rigid shift of Eq. 17, and the agreement is extremely good. Over the middle range of the probability distribution, then, the failure is controlled by the appearance of the $\hat{c}$ cluster. At lower probabilities, there is some deviation and the predicted probability of failure is not conservative, but we are developing an even more accurate description of the full failure time distribution which will rectify this deviation. Using the short-time solution $N(\hat{c},\tau) \propto \tau^{\hat{c}}$, the failure probability is approximately Weibull in form with the critical size $\hat{c}$ as the Weibull modulus.

The predictions of Eqs. 15-18 have the following further implications for the failure time distribution. First, at fixed system size N_t, increasing η decreases the failure time and also makes failure more abrupt (the time span between τ^* and τ_f decreases). This is consistent with the results of Figure 1. Second, at fixed η we find $\hat{c}$ increases with increasing system size such that the probability of failure becomes narrower in absolute terms; however, the variance becomes larger since the distribution width scales like $\hat{c}$ but the failure time scales like τ^*; hence the uncertainty in failure times is generally an increasing function of system size. This behavior is consistent with the results in Figure 2.

4. Discussion

The essence of the complex time-dependent failure by nucleated damage is captured by the analytical model. We have found subtle relationships between the macroscopically measured failure time, failure time distribution, critical cluster size, and the underlying non-linear driving factor η and system size. The details of the predicted scalings have not yet been fully compared to numerical simulations, but the trends in behavior are very consistent with the simulation results to date.

A similar model of damage evolution was previously considered by Phoenix and Tierney (P&T) and various asymptotic results were derived for, essentially, large η [8]. The P&T results are equivalent to the short-time limit only, and therefore do not capture the full dynamics of the avalanche failure. P&T postulated that global failure occurs at the time τ^*, where the short-time solutions breakdown completely. Interestingly, in the present work (which can be formulated as the exact solution to the problem posed by P&T, including cluster linking [5]) we find that the behavior around τ^* *is* the controlling behavior, since τ^* and τ_f scale similarly. In using the short-time solutions for τ^* and $\hat{c}$ to obtain the scaling behavior, we then obtain predictions similar to those of P&T. However, the key Eq. 14 shows that there exists a time margin between reaching the onset time τ^* and the failure time τ_f, a margin that can be critical in *anticipating* failure. The macroscopic conditions (strain, strain rate, etc.) prevailing around τ^* will be the subject of future investigation to determine if the onset of avalanche failure can be identified macroscopically. Such an identification may be difficult, because at lower η the transition is not abrupt, but is critical for applying the theory to assess remaining life in real materials.

Acknowledgment: The authors thank the National Science Foundation for support of this work through grant DMR-9420831.

References

[1] S.M. Wiederhorn, B.J. Hockey, and T-J. Chuang, *Creep and Creep Rupture of Structural Ceramics*, in S.P. Shah (ed) Toughening Mechansims in Quasi-Brittle Materials, Kluwer Academic Publishers, Dordrecht, p. 555 (1991); D.F. Carroll and R.E. Tressler, *Accumulation of Creep Damage in a Siliconized Silicon Carbide*, J. Am. Ceramic Soc. 71, 472 (1988).
[2] Y. Termonia, P. Meakin, and P. Smith, Macromolecules 18, 2246 (1985); P.M. Duxbury, P.D. Beale and P.L. Leath, *Size Effects of Electrical Breakdown in Quenched Random Media*, Phys. Rev. Lett. 57, 1052 (1986); W.A. Curtin and H. Scher, *Brittle Fracture in Disordered Materials: a Spring Network Model*, J. Matls. Res. 5, 535 (1990).
[3] A. Hansen, S. Roux, and E.L. Hinrichsen, *Annealed Model for Breakdown Processes*, Europhysics Lett. 13, 517 (1990).
[4] J.R. Brockenbrough, S. Suresh and J. Duffy, *An Analysis of Dynamic Fracture in Microcracking Brittle Solids*, Philosophical Magazine A 58, 619 (1988).
[5] W.A. Curtin and H. Scher, *Algebraic Scaling of Material Strength*, Phys. Rev. B 45, 2620 (1992).
[6] W.A. Curtin and H. Scher, *Analytic Model for Scaling of Breakdown*, Phys. Rev. Lett. 67, 1991 (1991).
[7] S. Roux, A. Hansen, and E.L. Hinrichsen, *Comment on Analytic Model for Scaling of Breakdown*, Phys. Rev. Lett. 70, 100 (1992); W.A. Curtin and H. Scher, *Curtin and Scher Reply*, Phys. Rev. Lett. 70, 101 (1992).
[8] S. L. Phoenix and L-J. Tierney, *A Statistical Model for the Time Dependent Failure of Unidirectional Composite Materials under Local Elastic Load-Sharing Among Fibers*, Engineering Fracture Mech. 18, 193 (1983).

STRUCTURE-SENSITIVE PROPERTIES OF MATERIALS

P.M. DUXBURY AND E. RZEPNIEWSKI
Department of Physics and Astronomy and CFMR
Michigan State University, East Lansing, MI 48824

AND

C. MOUKARZEL
Höchstleitungsrechenzentrum
Forschungzentrum, D-52425 Jülich, Germany

1. Introduction

The division between "structure-sensitive properties", such as brittle fracture and "structure-insensitive properties", such as the effective elastic moduli of a weakly disordered materials is a rather murky one. There are many properties which lie on the border between these limits and some properties which are of one sort on one length scale while being of the other sort on longer length scales. In this paper, we make the definition of "structure-sensitive" to mean that the property in question depends on the length scale in a "non-linear" or "non-Euclidian" manner. We take as three examples: the tensile fracture strength of inhomogeneous materials; the elastic moduli and topology of stress-bearing paths near the rigidity threshold; and the topology of minimal surfaces in disordered systems. These properties and others which are "structure sensitive" are not amenable to treatment by usual "homogenization" methods and so require a new set of tools and ideas in their analysis. As well as large scale computation, the ideas of scaling, self-similarity, fractals, and extreme statistics provide a sound basis for the analysis of these problems. In this paper, we discuss a simple analytic method for finding the scaling behavior of the tensile fracture stress of disordered networks (section II), and then outline some innovative numerical methods for finding the scaling behavior of: the rigid backbone in central force systems (section II), and the topology of minimal

J. R. Willis (ed.), IUTAM Symposium on Nonlinear Analysis of Fracture, 257–264.

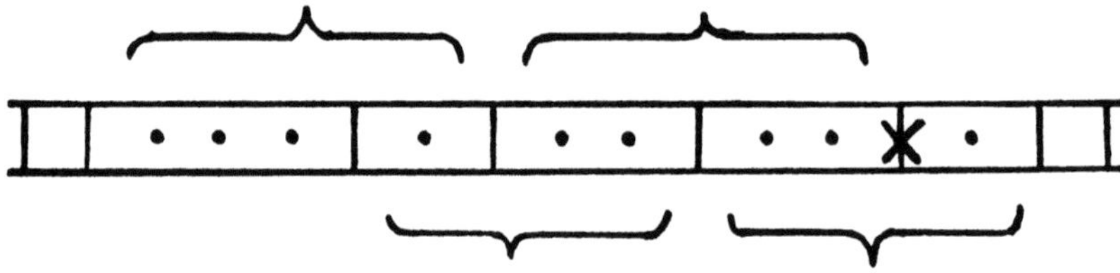

Figure 1. The parallel bar model

surfaces relevant to the plasticity of random materials and the topology of fracture surfaces (section III). The latter sections profit from results in graph theory and combinatorial optimisation.

2. Brittle Fracture of Inhomogeneous Materials

The effect of length scale on tensile fracture strength[1] has been studied since the time of Leonardo da Vinci, and is reflected in the large variability in fracture strength observed in experiment. Rigorous analytic results are only available for very special models[2–4], but precise numerical results are now available for many systems. Unfortunately the size effect is usually logarithmic, so many orders of magnitude are needed to properly test this effect. Most of the simulations in the literature are unreliable in this regard. Only the quasi-one-dimensional "fiber-bundle" models can be studied with sufficient precision to find asymptotic size effects. The results of these precise calculations are consistent with the following simple arguments based of a statistical extension of fracture mechanics.

2.1. THE MODEL

Consider a parallel bar (fiber-bundle) model[1] as illustrated in Fig. 1. Each bar has a strength drawn from a cumulative distribution $F(\sigma_b)$, which. Initially all of the bars expreience the applied stress σ. However, if a bar exceeds its fracture stress, its load is distributed to its neighbours. If a surviving bar is surrounded by k failed bars, it experiences a magnified stress given by,

$$\sigma_k = \sigma(1 + k/2) \tag{1}$$

For example, the bond marked (x) in Fig. 1 has $k = 3$. We want to find the external stress at which a system with L bars experiences complete failure. Initially, weak bars fail randomly, but eventually a dominant crack evolves and fracture occurs. If we assume that fracture occurs once the stress intensity at the tip of the largest crack exceeds the average strength of a bar, σ_{mean}, then it is possible to get a simple estimate of the fracture stress as follows.

First find the fraction f of bonds that break at applied stress σ *assuming no stress enhancement effects*,

$$f = F(\sigma) \tag{2}$$

The typical size of the largest crack, c, at this value of f is given by[5],

$$f^c L \sim 1 \tag{3}$$

or,

$$c \sim -\ln L / \ln f \tag{4}$$

The stress intensity at the tip of this crack is (from (1)),

$$\sigma_{tip} = \sigma(1 - \ln L / \ln f) \tag{5}$$

Fracture occurs when $\sigma_{tip} = \sigma_{mean}$, which leads to the implicit equation,

$$\sigma_f(1 - \ln L / \ln F(\sigma_f)) = \sigma_{mean} \tag{6}$$

For example with a Weibull distribution

$$F(\sigma) = 1 - \exp(-(\sigma/\sigma_0)^m), \tag{7}$$

we have, $\sigma_{mean} = \Gamma(1 + 1/m)\sigma_0$, so that writing $x = \sigma_0/\sigma$, Eq. () becomes,

$$1 + \frac{\ln L}{m \ln x} = \Gamma(1 + 1/m)x \tag{8}$$

At sufficiently large L, the solution to this equation is approximately

$$x \sim \frac{\ln L}{m\Gamma(1 + 1/m)}, \tag{9}$$

which implies that for any m, the size effect is logarithmic. Direct solutions to the parallel bar models are presented in Fig. 2. Fig. 2a shows that the distribution of failure strengths is not Weibull, while Fig. 2b illustrates that asymptotically the size effect is logarithmic, in agreement with Eq. (9). We refer to the original papers for the details of how these calculations were performed and more complete results[6].

3. Fractal Geometry of the Stressed Backbone

An important feature of granular materials is the topology of the stress-bearing paths which occur in these materials. Similarly, if a structure is

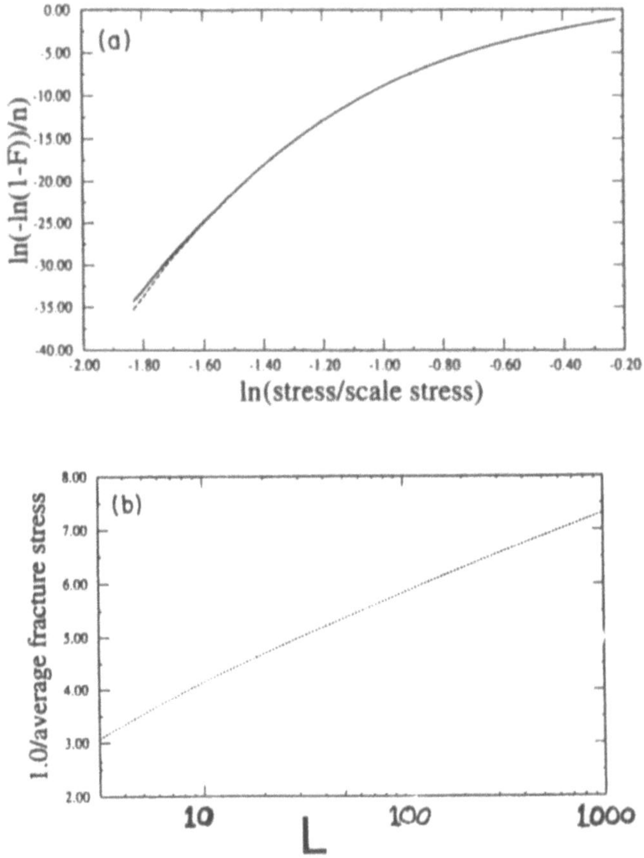

Figure 2. **Numerical results for the parallel bar model whose bars have strengths drawn from a Weibull distribution (with $m = 5$) a) The distribution of breaking strengths (with $L = 1000$); b) A test of logarithmic scaling of the average strength (Eq.(9)) for an $L \times L$ series-parallel model.**

supported primarily by central forces, the geometry of the most weakly connected structures which support stress are non-trivial. Unlike the case of systems with angle forces, where simple connectivity ensures that stress can be transmitted, in systems supported by only central forces higher order connectivity is required in order to transmit stress. There has been a recent breakthrough in the analysis of stress-bearing backbones in two dimensions using ideas based on graph theory[7-9]. The physical idea is as follows. Consider a triangular network in two dimensions. Each node of the network has 2 degrees of freedom. Each bond provides one constraint. Provided none of the constraints is redundant, for the network to be rigid,

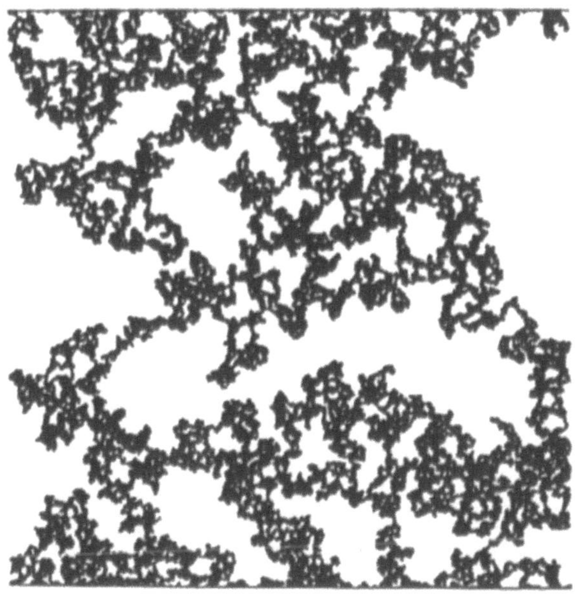

Figure 3. The fractal stress-bearing backbone of a site-diluted random triangular lattice of central-force springs. The lattice has $N \sim 870,000$ nodes

we need,

$$2V - B = 3 \tag{10}$$

This "constraint counting" argument was used by Maxwell and more recently by others to predict what sort of structures are rigid. However only recently has the "constraint-counting" method been used as a precise numerical procedure in 2-d. The key problem is that one must ensure there are no "redundant" bonds in the network. That is, bonds which are not necessary to ensure rigidity of a structure. In two dimension checking Eq.(10) on all length scales is sufficient to ensure this, while in three dimensions the simple extension of this idea is not exact. Even in two dimensions, checking Eq.(10) on all length scales appears prohibitive and at best an exponential algorithm in the number of nodes. In fact, by using an analogy with the bipartite matching algorithm of graph theory, it is possible to find the stress bearing backbone in $O(N^{1.2})$, where N is the number of nodes in the lattice. In Fig. 3, we present the stress bearing backbone of a site-diluted triangular lattice found using this method. The fractal dimension of the rigid backbone is found from the data of Fig. 4, from which we find $D_f = 1.78 \pm 0.02$. Also shown in Fig. 4 is the scaling behavior of the number of critical bonds in the rigid backbone. If a critical bond is removed, stress ceases to be transmitted across the sample. The scaling exponent of critical bonds is found to be 0.85 ± 0.05. Many other new results are being

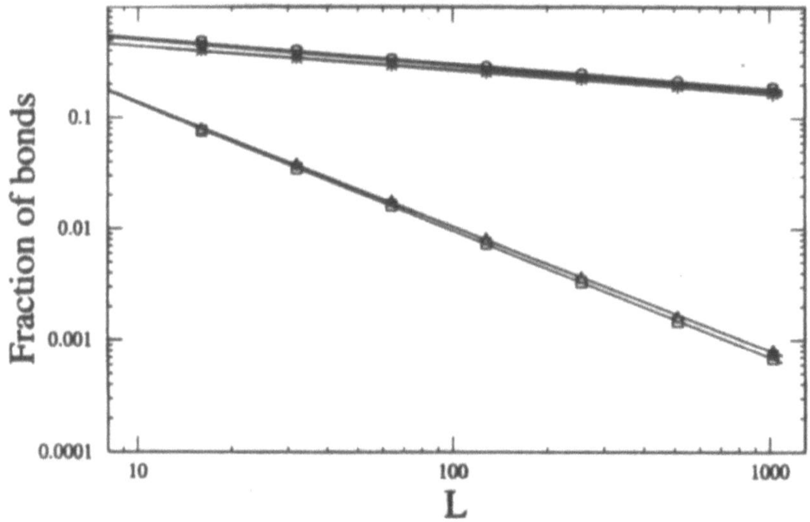

Figure 4. **The scaling behavior of the number of bonds on the backbone (upper data), and the number of critical bonds on the backbone (lower figure). Each point is an average of a number of configurations ranging from 100 for the largest lattices to 10000 for the smallest lattice**

found with this and other new algorithms for finding the geometry of rigid clusters.

4. The Topology of Minimal Surfaces

As a first approximation to finding the topology of slip bands in perfectly plastic materials, and to the topology of fracture surfaces, we construct minimal surfaces. In two dimensions this problem is "dual" to finding the minimum gap in dielectric breakdown, but in three dimensions, the minimum gap is a path problem, while in fracture, we are interested in a minimum surface. Nevertheless the procedures outlined below can be adapted to all of these problems. This work is in a preliminary stage, and has not yet been published elsewhere. In the following we ouline the method and give some preliminary results.

Consider a network, for concreteness a square lattice, whose bonds are perfectly plastic (see Fig. 5).

We outline the procedure for the electrical case, so the discussion is in terms of flow. We want to find the connected path, transverse to the direction of net flow, along which the flow first reaches the plastic limit. We call this path the *minimum cut*, and the external flow at which it occurs the *yield flow*. This problem has been studied intensively in the engineering of optimal flow in a variety of transportation, data systems and integrated circuit contexts. The connection to perfect plasticity has not been made before, and instead prior work on this problem has resorted

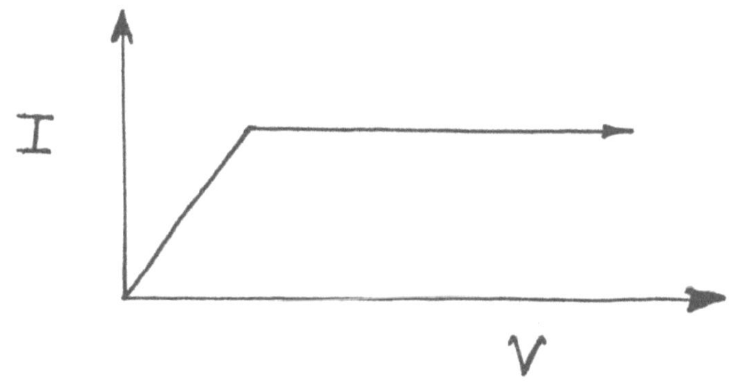

Figure 5. **A perfectly plastic I-V characteristic**

to direct solution of the electrical current flow in the network[11,12]. The latter papers draw the important connections with the topology of many other random surfaces problems, such as those occuring in epitaxial growth. In fact it has been claimed[14] that 2-d fracture surfaces (e.g. paper) have the same roughness exponent (2/3) as the KPZ equation, a contention that is still a matter of some dispute.

The identification of the minimum flow in a network with capacities can be found using the Ford-Fulkinson algorithm[10]. This algorithm works by finding paths between a source and a sink, and then updating the flow on that path. All bonds on a path are updated by the smallest remaining flow capacities on the path. The capacities on the path are then updated to take into account the updated flow. We find the candidate paths by a breadth first search. If we assign integer capacities to the bonds of the network, the entire calculation is in integer arithmetic.

If we randomly assign two integer capacities c_1 and c_2 to the lattice, it is like a composite problem. Clearly the yield flow $f_y \geq c_1$. However, the topology of the minimal cut is independent of c_1, or c_2 and is universal. This is easily seen if one first injects flow c_1 into the system. Since the system is linear at this point, that flow and hence the associated capacity can be subtracted from the total. What remains is a composite with capacities 0 and $c_2 - c_1$. This is a percolation problem. The "interface energy", is in this case the same as the maximal flow, and hence

$$f_y \sim c_1 + (c_2 - c_1)(p - p_c)^{(d-1)\nu} \tag{11}$$

The topology of the minimal cut is also readily found from work on percolation theory, so that at the percolation point(50/50 mixture for the 2-d square lattice), the interface width w is known to be

$$w \sim L^{\zeta} \tag{12}$$

where $\zeta = 1$ in two dimensions at the percolation point. It is interesting to note that the interface roughness is *independent* of c_1. If the possible capacities are varied continuously from the 2 possibilities considered above to an infinite number of integer possibilities, then the model reduces to the model considered by Roux and Hansen[11], and which is related to the KPZ equation. In that case, we have the result in 2-d

$$w \sim L^{2/3} \tag{13}$$

A more thorough study of these systems is in progress.

Acknowledgements
We acknowledge support by the DOE under contract number DE-FG02-90ER45418, by the PRF and by the CMSC at MSU.

References

[1] P.M. Duxbury, S.G. Kim and P.L. Leath, Mat. Sci. and Eng. **A176**, 25 (1994) and references therein.
[2] D.G. Harlow and S.L. Phoenix, J. Mech. Phys. Solids. **39**, 173 (1991).
[3] P.M. Duxbury and P.L. Leath, Phys. Rev. **B49**, 12676 (1994) and references therein.
[4] W.A. Curtin, J. Amer. Ceram. Soc. **74**, 2837 (1991) and references therein.
[5] P.M. Duxbury and P.L. Leath, Phys. Rev. **B36**, 367 (1987).
[6] P.M. Duxbury and P.L. Leath, Phys. Rev. Lett. **72**, 2805 (1994); P.L. Leath and P.M. Duxbury, Phys. Rev. **B49**, 14905 (1994).
[7] B. Hendrickson, Siam J. Comput. **21**, 65 (1992).
[8] D. Jacobs and M.F. Thorpe, Phys. Rev. Lett. **75**, 4051 (1995).
[9] C. Moukarzel and P.M. Duxbury, Phys. Rev. Lett. **75**, 4055 (1995).
[10] see e.g. M. Gondran and M. Minoux "Graphs and Algorithms", Wiley (1984).
[11] A. Hansen, E.L. Hensicksen and S. Roux, Phys. Rev. Lett. **23**, 2476 (1991).
[12] S. Roux and A. Hansen, J. de Physique **2**, 1007 (1992).
[13] E. Medina, T. Hwa, M. Kardar and Y.C. Zhang, Phys. Rev. **A39**, 3053 (1989).
[14] K.J. Maloy, A. Hansen, H.J. Herrmann and S. Roux, Phys. Rev. Lett. **68**, 213 (1992).

THE STATISTICS OF FAILURE PROBABILITY IN HETEROGENEOUS MATERIALS WITH TOUGH–TO–BRITTLE CROSSOVER

P. L. LEATH AND NIU–NIU CHEN
Department of Physics and Astronomy
Rutgers University
Piscataway, NJ 08855

We study and review the statistics of fracture due to randomly disordered and heterogeneous microstructures. The samples to be considered will have a uniform tensile stress applied across them. But, unlike most of the samples considered at this conference, these samples will have no macroscopic notch cut in them but only the cracks and defects due to the disordered microstructure. We shall review very briefly the work of our group on brittle systems[1], dating from almost a decade ago, along with our more recent recursion relation method for heterogeneous samples[2], and finally present our latest work on surface effects and the analytic form of the failure probabilities[3].

Fig.1 a) A random lattice configuration with $p = 0.90$ *after rigidity failure has occurred. b) The configuration after complete fracture has occurred. [Fig.1 from Ref. 4]*

An example of a brittle system is shown in Fig. 1, from the work of Beale and Srolovitz[4]. Here a uniform, uniaxial tensile stress is applied from top to bottom along the vertical axis. The sample is a triangular lattice occupied randomly by a concentration $p = 0.90$ of identical breakable Hookean springs. The disorder is provided by the ten percent missing bonds. This system is brittle in the sense that there is a weakest link (a randomly occurring worst or largest cluster of missing bonds which breaks first due to the stress enhancement at its edges. Once broken this weak area is even weaker thereby nucleating an unstable crack which cleaves the sample. The failure probability distribution $F_L(\sigma)$ for such a randomly diluted lattice is the modified–

J. R. Willis (ed.), IUTAM Symposium on Nonlinear Analysis of Fracture, 265–274.

Gumbel or double–exponential distribution[1] given by

$$1 - F_L(\sigma) = \exp[-AL\exp(-B/\sigma^{1/\alpha})], \tag{1}$$

where L is the sample size, σ is the applied, external stress, A and B are constants (or weak functions of σ) and α is an exponent of order one which describes the stress enhancement distribution around a crack. A failure distribution of this double–exponential form occurs whenever the worst defect of size n (in the above case, this is a large cluster of n vacant bonds) occurs with an exponentially small probability [i.e.– proportional to $\exp(-an)$]. This is the case for all brittle systems where the critical crack is nucleated by the coincidence of randomly located defects. This double–exponential distribution, Eq.(1), implies a mean fracture stress $\langle\sigma\rangle$ which scales logarithmically to zero with sample size as $(\ln L)^\alpha$ in the thermodynamic limit $L \to \infty$.

The statistics of extremes[5] tells us that there are basically only two general analytic forms for the failure distribution function if it is determined by the weakest link. The first is the double exponential form of Eq.(1) which prevails when the weakest link occurs with exponentially small probability. The second is the Weibull distribution, which is of the form

$$1 - F_L(\sigma) = \exp(-AL\sigma^\gamma). \tag{2}$$

This form prevails when the weakest link occurs with an algebraically small probability [i.e.– proportional to $n^{-\gamma}$]. In this case the mean fracture stress $\langle\sigma\rangle$ scales algebraically to zero as $L^{-1/\gamma}$ in the thermodynamic limit $L \to \infty$.

There have been many numerical and experimental confirmations of the modified–Gumbel distribution, Eq.(1), for brittle systems[1,4,6,7]. For example, the system studied by Sahimi and Arbabi[6] corresponds to a “superelastic” triangular lattice with 90% of the bonds occupied by identical breakable Hookean springs and the remaining 10% of the bonds occupied by *unbreakable* springs. The result is that a nucleated crack generally is stopped when it hits a randomly located unbreakable spring. So the system fills with many small cracks which only traverse the sample by a kind of percolation process. This percolation process apparently leads to the power law correlations and a Weibull failure distribution. A somewhat similar behavior occurs for the case of a heterogeneous system with a continuous distribution of local breaking strengths, where for sufficiently small samples, at low applied stress, the system is tough with almost all the cracks (which are caused by the failure of the weaker bonds) being stopped or pinned by the stronger bonds. But at higher applied stress even the strongest bonds break (assuming an upper cut off in the local bond–breaking strength) and the sample becomes brittle. A similar transition from tough to brittle behavior occurs even at low applied stress, as the sample size increases, when one reaches a sample size such that there is the likelihood of occurrence of a cluster of weak bonds so large that the large crack the cluster forms has a sufficiently large stress enhancement at the crack's edges to break even the strongest bonds.

Fig.2 A two-dimensional sample on a square lattice with only some of the vertical bonds broken so that all cracks are one-dimensional. This also represents the "chain of bundles" model for fiber composites.

Specifically our model for heterogeneous systems is to assume that each bond has a breaking strength s chosen randomly from a probability distribution $P(s)$. For the purpose of this article we shall choose the uniform distribution $P(s) = 1/w$, for $0 \leq s \leq w$, and otherwise $P(s) = 0$. This is a rather extreme case (a rather better behaved case is the local Weibull distribution[2]) but any normalized distribution can be used.

We also assume that all cracks are linear and perpendicular to the applied uniaxial stress. This means that only vertical bonds (as illustrated in Fig. 2) break. And the model becomes essentially the "chain of bundles" model of Harlow and Phoenix[8]. In particular, the failure distribution $F_{L\times L}(\sigma)$ of an $L \times L$ sample is given by

$$F_{L\times L}(\sigma) = 1 - (1 - F_L(\sigma))^L, \tag{3}$$

where $F_L(\sigma)$ is the failure probability for a single row of length L. The problem is then reduced to the one–dimensional one for $F_L(\sigma)$.

Finally, following Harlow and Phoenix[8], we assume strictly local–load sharing, where the local stress applied to an intact bond, when an external stress σ has been applied, is given by

$$\sigma_m = (1 + m/2)\,\sigma, \tag{4}$$

where m is the total number of broken bonds on either side of the bond. For example, in Fig. 3, the $\times$ marks a bond with $m = 3$ (Since this bond has broken bonds on both sides, it is one we shall call a lone bond).

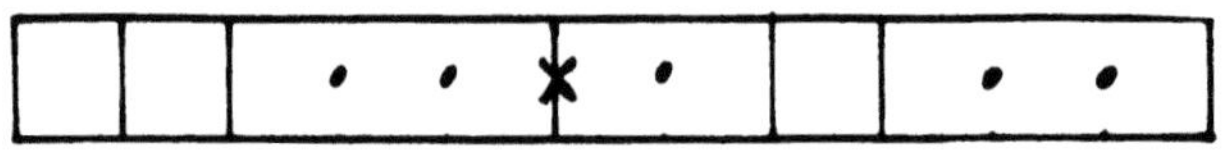

Fig.3 A row of the sample after some bonds have broken. The lone fiber, marked with an $\times$, has a load corresponding to $m = 3$.

But $F_L(\sigma)$ for a single row can be enumerated exactly since $F_L(\sigma) = 1 - S_L(\sigma)$, where $S_L(\sigma)$ is the survival probability of the row, and since $S_L(\sigma)$ is the sum over the $2^L - 1$ possible survival configuration probabilities for L bonds. For example, in Table I we show the elements of $S_L(\sigma)$ for $L = 1$, 2, and 3, where $W_m(\sigma)$ is the probability that a bond survives under local stress σ_m given by Eq.(4) above.

Configuration (degeneracy)	*Survival Probability*
1	W_0
11,10(2)	$W_0^2, 2W_1F_1$
111,110(2),101,100(2).010	$W_0^3, 2W_0W_1F_1, W_1F_1W_1, 2W_2F_2, F_1W_2F_1$

Table I. Survival configurations and their survival probabilities.

In reference 2, we derived an exact recursion relation for $F_L(\sigma)$ of the general form

$$F_L(\sigma) = \mathbf{F}[F_1(\sigma), F_2(\sigma), \ldots, F_{L-1}(\sigma)]. \tag{5}$$

This recursion relation is sufficiently simple to easily allow our exact calculation of $F_L(\sigma)$ up to $L = 5000$. With the enumeration of the exact recursion relation[2], we find a deep minimum in $F_L(\sigma)$, at low σ, at an optimum sample size, as is shown in Fig. 4. To the left of the minimum the sample is tough, and is filled with cracks which formed at the weaker bonds broke, but which stopped upon meeting stronger bonds. With local–load sharing, the probability that an m-bond survives is given by $W_m(\sigma) = 1 - (1 + m/2)\sigma/w$, and thus goes to zero at $m = 2w/\sigma - 2$. Thus, whenever a crack of this critical size, or greater, appears no bond can stop it and the sample cleaves. For samples of size $L < 2w/\sigma - 2$ no such cluster can appear, while for larger samples there is an increasing probability of the occurrence of such large cracks. So the optimum sample size is at $L \sim 2w/\sigma$ and above this size there is a crossover to brittle behavior. Of course, for large applied stress σ, all samples are brittle.

From the differential equation, we have shown that $F_L(\sigma)$ is of the general form

$$F_L(\sigma) \sim \left(\frac{\sigma L}{aw}\right)^{bL}, \tag{6}$$

where a and b are, approximately, of the form $a \sim 5 - 11\sigma/w$ and $b \sim 0.5+0.4\exp[-25\sigma/w]$. The minimum fracture probability occours at

$$L_{\min} \sim \left(\frac{aw}{e\sigma}\right), \tag{7}$$

with a value

$$F_{L_{\min}}(\sigma) \sim \exp(-bL_{\min}). \tag{8}$$

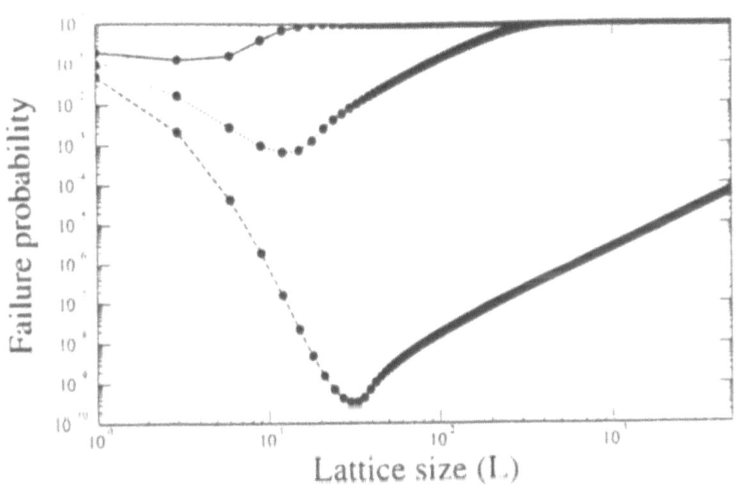

Fig.4 A log-log plot of $F_L(\sigma)$ vs. L, for $\sigma/w = 0.2$ (top curve), 0.1 (middle), and 0.05 (bottom), evaluated numerically from the exact recursion relation, Eq.5. [Fig.2a from Ref.1a]

For samples smaller in size than $L_{\min}$ the sample fills with small cracks and failure occurs only in those samples when these cracks merge together and percolate across the sample. Survival of a tough sample then depends upon the existence of strong bonds or pins to stop the growing cracks. For samples larger than $L_{\min}$, failure occurs when a crack can grow to a size $L_{\min}$ which then is necessarily unstable and cleaves the sample. Failure of these brittle samples depends upon the weakest link which nucleates a critical crack.

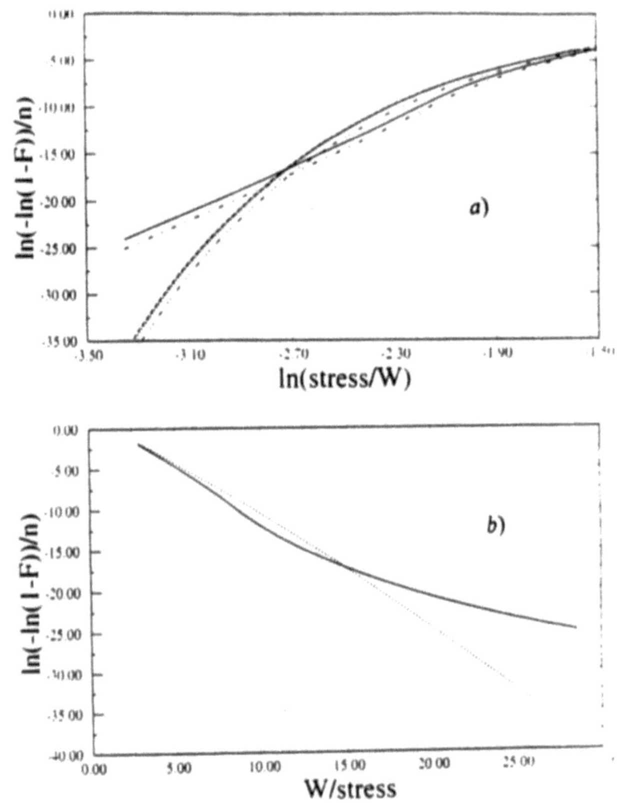

Fig.5 a) Weibull plots of $\ln[-\ln(1-F_L(\sigma))]/L$ *vs.* $\ln(\sigma/w)$. *The curves (two slightly different approximations) for* $L=15$ *are straight (Weibull-like) at low* σ/w, *on the left follow a universal curve over the* σ/w *range shown on this graph. b) Modified–Gumbel (double-exponential) plots of* $\ln[-\ln(1-F_L(\sigma))]/L$ *vs.* w/σ. *The dotted curve, for* $L=150$, *is nearly straight (double-exponential-like), and the solid line, for* $L=15$, *is curved but approaches the same universal straight line at large stress (small* w/σ *). [Fig.3 from Ref.1b]*

We have studied the statistics of failure in each region by Weibull plots (Fig. 5a) and modified-Gumbel plots (Fig. 5b) of $F_L(\sigma)$ versus stress. In Fig. 5a it is shown that, at low applied stress and for small samples below $L_{\min}(\sigma)$, there is a straight line Weibull plot with a slope that is proportional to sample size (as would be expected from our analytic formula). And for larger samples and/or higher applied stress, there is a roughly straight line on the modified-Gumbel plot (Fig. 5a). This Weibull behavior in the tough region is consistent with the results of Sahimi and Arbabi[6] for the breakable/superelastic sample. And the modified-Gumbel behavior in the brittle region is consistent with the weak-link behavior in other brittle systems. Nevertheless, the mean fracture stress goes to zero logarithmically as $L \to \infty$ (see Fig. 6) since the mean fracture stress is almost always sampling the brittle region. This is because the tough region is always at applied stresses well below the mean breaking stress.

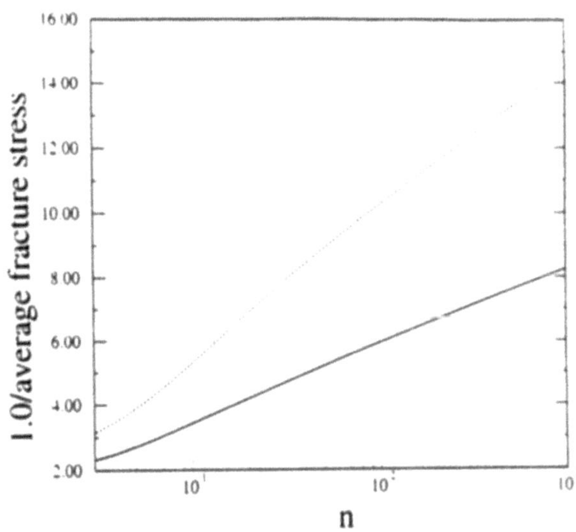

Fig.6 The inverse average fracture stress $\langle\sigma\rangle^{-1}$ *vs.* L, *as a test of logarithmic size scaling. [Fig.6b from Ref.1b]*

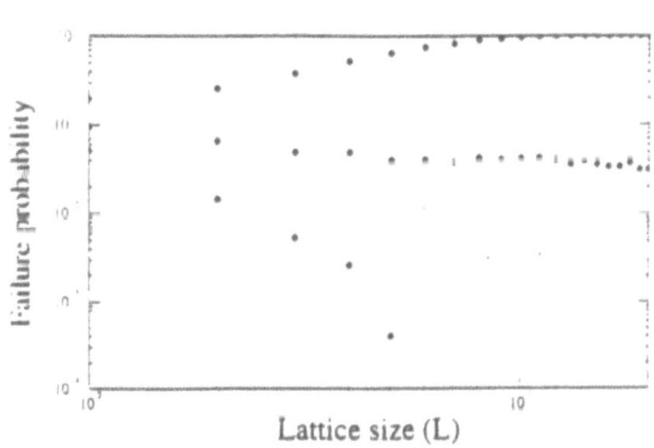

Fig.7 The failure probability $F_L(i)$ *of* $L \times L$ *fuse networks on a square lattice, with a uniform distribution of fuse strength, from numerical simulations with* 5000 *realizations at each value of* L, *and for* $i/w =$ 0.2 *(top curve),* 0.1 *(middle), and* 0.05 *(bottom). [Fig.2b from Ref.1a]*

We have also done numerical simulations on two–dimensional heterogeneous fuse networks. The results are shown in Fig. 7, and are generally consistent with the chain–of–bundles model above, except that we did not succeed numerically in traversing the minimum in $F_L(i)$ at low applied current i. First, the statistics are very difficult to handle numerically when $F_L < 10^{-4}$ since thousands of configurations must be studied to find one that fails. Secondly, the minimum seems shallow compared with the results in Fig. 4. It turns out that this latter effect is due to surface cracks on the sample, which we shall now discuss.

A particularly convenient method of studying the effects surface disorder on fracture properties in the linear crack approximation is the stochastic transfer matrix (STM) method which we have developed by generalizing the approach of Duxbury and Leath[9] to the present case. This approach will be described in detail in a separate publication, and we shall only present the numerical results here.

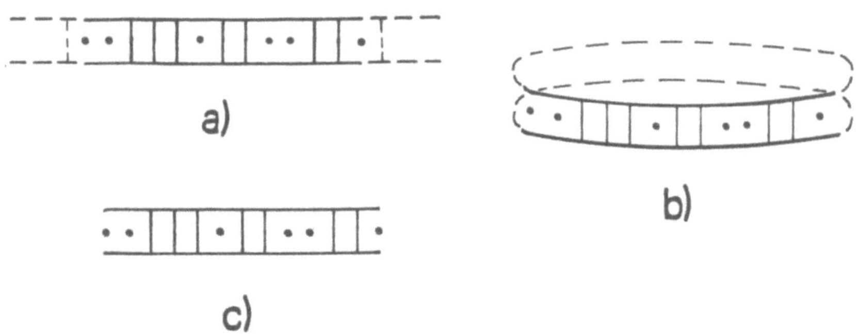

Fig.8 Illustrations of a partially broken row in the interior (a) of a larger sample, with periodic boundary conditions (b), and with open boundary conditions (c). The vertical lines represent intact bonds, and the dots represent broken bonds.

We consider three types of boundary conditions for a row or bundle, in the chain–of–bundles model with local load sharing as are illustrated in Fig. 8. First, interior boundary conditions (Fig. 8a) assume that the load of the cracks, if any, on the end of the sample or segment are shared with an intact bond on each end. This is true if the segment is part of the interior of a larger sample and corresponds to the $F_n(\sigma)$ of the recursion relation of Eq. (5). Second, we consider periodic boundary conditions (Fig. 8b) where the cracks on each end are parts of the same crack. And, finally, open boundary conditions (Fig. 8c) are where the cracks on each end are open and the entire load on the crack must be borne by the intact bounds closest to the surface.

The results of our calculations, when configurations with lone bonds are ignored, are shown in Fig. 9, where $F_L(\sigma)$ versus L is given in log-log plots for each form of boundary condition at the ends of the sample, for $\sigma/w = 0.07$. Clearly, the strongest samples and the ones with the deepest minima in $F_L(\sigma)$ at $L_{\min}$ are those with interior boundary conditions (since there is an intact bond on each end to make the end cracks just like interior cracks). And the weakest samples, with the shallowest minima are those that are open on each end. Also, the reason for the shallow minimum in our numerical simulations for real two-dimensional samples is now clear, since that was done with open boundary conditions.

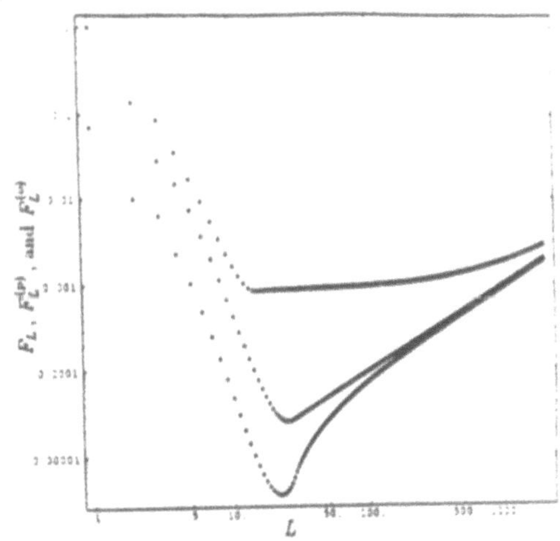

Fig.9 The failure probabilities $F_L(\sigma)$ with interior boundary condition (lower curve), periodic boundary condition (middle curve), and open boundary condition (upper curve), for $\sigma/w = 0.07$.

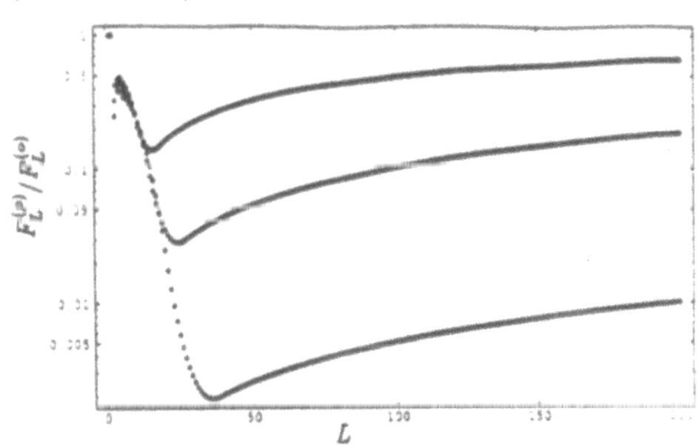

Fig.10 The ratio $F_L^{(p)}(\sigma)/F_L^{(o)}(\sigma)$ vs. L, which is a measure of the fraction of critical fractures which are nucleated on the interior of a sample with open boundaries.

In order to quantify the effect of surfaces we compare $F_L^{(p)}(\sigma)$ for periodic boundary conditions, which has no surfaces, with $F_L^{(o)}(\sigma)$ for the open boundary conditions. In Fig. 10, we plot $F_L^{(p)}(\sigma)/F_L^{(o)}(\sigma)$ which is a measure of the fraction of failures in an open system which occur on the interior (since $F_L^{(p)}(\sigma)$ has only interior cracks). At the minima $F_{L_{\min}}^{(p)}(\sigma/w = 0.05)$, only about two failures in 1000 originate at defects in the interior. The message is clear, the surface cannot be ignored in calculations of failure probability, since so many of the fatal cracks start at surface defects.

In conclusion, we have found that a rather simple failure model of linear cracks in a two–dimensional sample with local load sharing is sufficient to produce many of the features which are seen in real heterogeneous systems. The tough–to–brittle transition is characterized by an optimum sample size, which varies inversely with the applied stress. The tough region is characterized by Weibull statistics and a kind of percolation of cracks in those samples that fail, while the brittle region is a weak–link system characterized by modified–Gumbel statistics and a critical crack which cleaves the system. The effects of open surfaces are profound, even without modification of the surface bonds, and over a wide range of sample sizes, the failure probability minimum is much more shallow for the case of open boundary conditions and the preponderance of fatal cracks originate on the surfaces.

Acknowledgments

The authors would like to thank Phil Duxbury for illuminating discussions.

References

1. P.M. Duxbury, P.D. Beale, and P.L. Leath, Phys. Rev. Lett. **57**, 1052 (1986); P.M. Duxbury, P.L. Leath, and P.D. Beale, Phys. Rev. B**36**, 367 (1987).
2. P.M. Duxbury and P.L. Leath, Phys. Rev. Lett. **27**, 2805 (1994); P.L. Leath and P.M. Duxbury, Phys. Rev. B**49**, 14905 (1994).
3. Niu-Niu Chen and P.L. Leath, 1995, to be published.
4. P.D. Beale and D.J. Srolovitz, Phys. Rev. B**37**, 5500 (1988).
5. E.J. Gumbel, *Statistics of Extremes*, (columbia University Press, New York, 1958); E. Castillo, *Extreme Value Theory in Engineering*, (Academic Press, San Diego, 1988).
6. M. Sahimi and S. Arbabi, Phys. Rev. B**47**, 713 (1993).
7. I.C. Van den Born, A. Santen, H.D. Hoekstra, and J. Th. M. De Hosson, Phys. Rev. B**43**, 3794 (1991).
8. D.G. Harlow and S.L. Phoenix, Int. J. Fracture **17**, 601 (1981).
9. P.M. Duxbury and P.L. Leath, Phys. Rev. B**49**, 12676 (1994).

COMPUTER MODELLING OF DISORDERED PLASTIC AND VISCOELASTIC SYSTEMS

K. KASKI, M.J. KORTEOJA, A. LUKKARINEN
AND T.T. RAUTIAINEN
Institute of Advanced Computing,
Tampere University of Technology,
P.O. Box 692, FIN-33101 Tampere, Finland

1. Introduction

Fracture phenomena in *disordered* materials have attracted a lot of interest due to their intrinsic technological importance [1]. Examples of such materials are certain polymers, fiber reinforced materials, and fibrous compounds like paper. Typically, their mechanical response shows such behaviour as *plasticity* and a time-dependent phenomenon like *viscoelasticity*. From the physics point of view fracture of disordered materials involves a variety of processes occurring on a wide range of length and time scales. We have studied these processes with simplified models, that are capable of showing some salient features of materials under external loading. In these models we have included plasticity and viscoelasticity, which often have been omitted.

In order to describe elastic-plastic behaviour and fracture in disordered materials under tensile elongation we have devised a quasistatic model that captures the essence of the continuum approach. A finite size analysis of the maximum strain in the stress-strain curves shows significant residue values even for large system sizes and for large disorder. We also show how our model is capable of describing the behaviour of strain localization in case of a disordered fibrous compound, e.g. paper.

In contrast to quasistatic models, we have also devised a network model for describing viscoelastic behaviour and dynamic fracture in the presence of disorder and under constant strain rate loading. In this model elastic interactions are described with the Born Hamiltonian, and the local relaxation of force field is introduced as Maxwell type viscoelasticity. Thus we are able to study fracture dynamics and its dependence on disorder and relative time scales of relaxation and loading. Brittle behaviour is encoun-

J. R. Willis (ed.), IUTAM Symposium on Nonlinear Analysis of Fracture, 275–284.

tered in the adiabatic limit of slow straining. However, for finite strain rates and with increasing relaxation the development of damage shows ductile behaviour.

Figure 1. a) Quasistatic model and b) dynamic model.

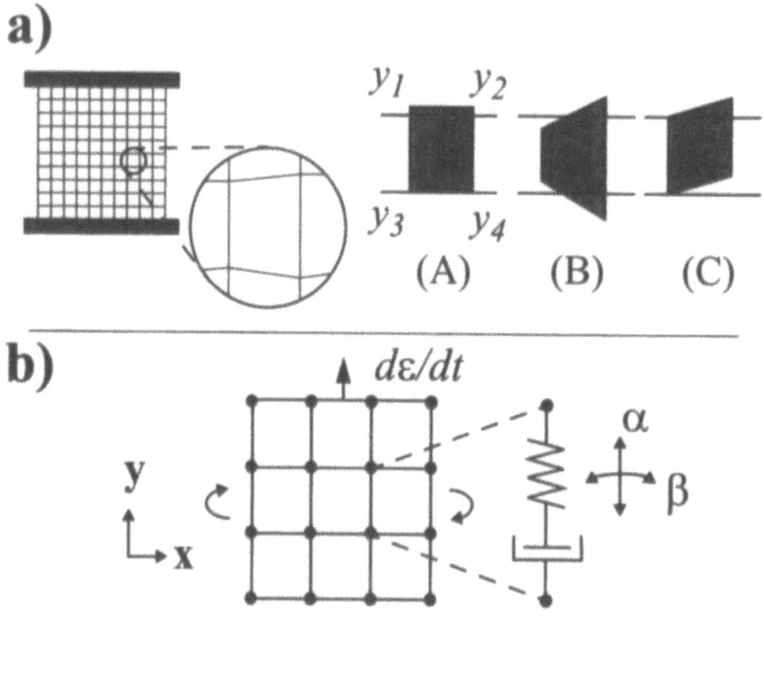

2. Quasistatic Model

In this section we present a quasistatic approach to model elastic-plastic behaviour of disordered planar materials under tensile elongation. The model describes mesoscopic disorder by a 2-dimensional lattice of adjoined initially square cells whose microstructure is averaged out locally. Disorder occurs as coarse-grained density variation, which modulates the local elastic moduli. The cell corners (nodes of the lattice) are allowed to move only parallel to the direction of external strain. Uniaxial displacements of cell corners lead to three basic deformations (see Fig. 1(a)); all other deformations are linear combinations of these three. The deformation of each cell, in response to a given external elongation, is determined by minimizing the total elastic energy of the system:

$$\begin{aligned} H = \sum \alpha_{ij}\rho_{ij} \{ \quad & A \quad (+y_1 + y_2 - y_3 - y_4 - \pi_{A_{ij}})^2 \\ + \quad & B \quad (-y_1 + y_2 + y_3 - y_4 + \pi_{B1_{ij}} - \pi_{B2_{ij}})^2 \\ + \quad & C \quad (+y_1 - y_2 + y_3 - y_4 - \pi_{C1_{ij}} + \pi_{C2_{ij}})^2 \} \end{aligned} \tag{1}$$

Here the variables y_1, y_2, y_3 and y_4 are the longitudinal displacements of nodes defining the cell ij (cf. Fig. 1(a) for the labeling). The quantity ρ_{ij} is the local density and α_{ij} denotes the *integrity* of the cell ij: $\alpha_{ij} = 1$ for an intact cell and $\alpha_{ij} = 0$ for a completely failed cell. The prefactors A, B and C depend on the Young's modulus E and shear modulus G as $A = \frac{1}{8}E$, $B = \frac{1}{24}(E+G)$ and $C = \frac{1}{8}G$, respectively. Since the nodes are free to move only in the direction of loading, longitudinal and shear deformations are allowed but transverse contraction is not.

If one of the deformation components exceeds a set limit, then the cell either fails or becomes plastic — depending on the choice of parameters. The possible plastic elongations for each cell are determined as follows:

$$\begin{aligned} \pi^k_{A_{ij}} &= max(\pi^{k-1}_{A_{ij}}, (y_1 + y_2 - y_3 - y_4 - 2\theta_A)) \\ \pi^k_{B1_{ij}} &= max(\pi^{k-1}_{B1_{ij}}, (y_1 - y_3 - \theta_B)) \\ \pi^k_{B2_{ij}} &= max(\pi^{k-1}_{B2_{ij}}, (y_2 - y_4 - \theta_B)) \\ \pi^k_{C1_{ij}} &= max(\pi^{k-1}_{C1_{ij}}, (y_1 - y_4 - \theta_C)) \\ \pi^k_{C2_{ij}} &= max(\pi^{k-1}_{C2_{ij}}, (y_2 - y_3 - \theta_C)) \end{aligned} \tag{2}$$

where θ_A, θ_B and θ_C are the yield thresholds for the three deformation modes. In this ideally elastic-plastic model it is simple to introduce the failure mechanism by defining the failure limit θ_{fr} of a cell as a certain amount of plastic deformation. If the failure threshold is set equal to the yield limit, the model describes failure in an elastic material.

Thus this model is defined by the following four factors: 1) specimen geometry (in units of the number of cells), 2) the statistical distribution of local moduli, 3) the ratio G/E, between the shear modulus G and the Young's modulus E, and 4) the ratio θ_{fr}/θ_{pl} between the critical strain for rupture and the plastic yielding. The values specified for Young's modulus and yield strain merely set the scale of stress and strain. The ratios G/E and θ_{fr}/θ_{pl} are material parameters of the model, that are taken to be constants.

2.1. FINITE SIZE SCALING OF ELASTIC FRACTURE

As a first application of this quasistatic model we study the scaling behavior of elastic fracture in a disordered system [2]. For this purpose we varied the linear system size L over two orders of magnitude up to $L = 400$. In the transverse direction we used either free or periodic boundary conditions, but saw no significant effect on the finite size scaling. From the stress-strain curves we found that the decay of maximum or fracture stress σ_f as a function of L seems to slow down significantly and even stop for large values of L. This tempted us to try a non-linear least-square fit of the form $\sigma_f \sim a + bL^c$. Of the fitting parameters (a, b and c) a could be interpreted as the remaining fracture stress in the limit of infinite system size (should it be non-zero) and c as the effective scaling exponent.

Figure 2. Finite size scaling of fracture stress.

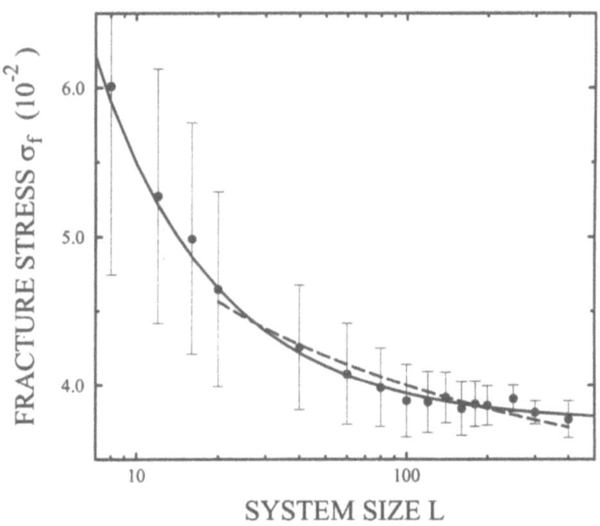

As seen in Fig. 2 the fitting, with the exponent $c \approx -0.94$, is accurate over two orders of magnitude in L. The non-zero limiting value ($\sigma_f \to 0.037$ when $L \to \infty$ and $G/E = 0.25$, $E = 1$, $\theta_{fr} = 0.2$ and $d = 2$) indicated by this power-law-fit is consistent with our previous work [3], in which $\sigma_f \to 0$ only when the "shear" coupling between cells was zero. However, for percolation systems it has been suggested [4, 5], that the fracture stress σ_f should vanish logarithmically as $\sigma_f \sim 1/(a + b(lnL)^c)$, with $1/2 \leq c \leq 1$ in two dimensions. An attempt to do this kind of fit by demanding $c \geq 0$ and omitting data points for $L \leq 20$, is shown in Fig. 2 by the dashed line. Although this fit falls within the error-bars, it is not as good as the power-law fit. It is noted, however, that our present model differs from the percolation model in that the probability of a cell to have zero density is extremely small (in fact of zero measure) while in the percolation case it is not. In addition, due to its continuum mechanical nature, our model has greater tendency to distribute stress to the wider neighbourhood of a cell than bond models do. Whether these differences are sufficient to cause different scaling behaviour is beyond the studied system sizes. Nevertheless, since our model of mesoscopic or macroscopic cells corresponds to realistic system sizes the large residue failure stress is a novel finding.

2.2. CASE STUDY: LOCAL STRAINS IN PAPER

As another application of our model we study the strain localization in disordered fibrous compounds. Paper is a good example of such a material, since its fibrous nature can be controlled during the manufacturing process. Typically this process begins with colloidal suspension of fibers. In the suspension phase, fibers may cling together to form larger fiber clusters or 'flocs'. As the fiber suspension is later deposited on a surface and the fluid is drained out, the spatial distribution of local density is seen to be 'flocculated'.

We have modelled the effect of flocculation on the local strain distribution [6, 7]. Pronounced flocculation in the transverse direction against external load was found to be most critical, i.e. it led to large variation in local strains for a given variation in the local elastic moduli. We have also shown that the strain distribution remains essentially the same throughout the elastic regime. As a natural continuation, one might ask what happens to the strain distribution when the external elongation is increased beyond the yield point and the system starts to undergo large plastic deformations. It seems obvious that as external elongation increases the cells with the highest strain will yield and eventually fail. In the case of highly yielding material, the coefficient of variation of local strains is first seen to be constant, but starts to grow as the specimen is loaded beyond the yield point.

The reason for this can be seen both in simulations and experiments, in which local strains concentrate on narrow plastic bands or strain lines (cf. Fig. 3). When a small area of paper starts to yield plastically, it becomes very difficult to further increase the local stress of the neigbouring cells in the longitudinal direction. This means that plastic yielding in paper takes place only in a small fraction of the specimen area. In both simulations and experiments the site of eventual failure can often be visually located long before the failure actually occurs, i.e. the plastic strain lines can be viewed as precursors for fracture.

Figure 3. Distribution of plastic strain in a simulation (a), and in an experiment (b). The light areas in (a) indicate plastic strains, and in (b) the location of plastic debonding of fibers [6].

a) b)

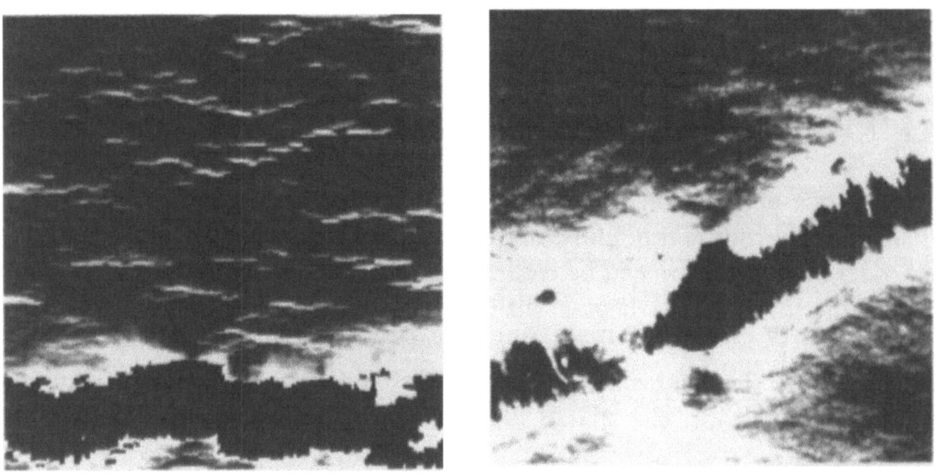

3. Dynamical Model

In this section we present a 2-dimensional dynamical model of fracture and analyze the interplay between disorder and force relaxation. Each site in an $L \times L$ lattice (initially a square lattice) is assigned a unit mass, which is connected to nearest neighbour mass sites by bonds, as in quasistatic lattice models for fracture [8]. The bottom-most row of masses is always kept fixed while the top-most row of masses is moved in unison with a constant rate to introduce loading as tensile elongation in the y-direction. In the x-direction we apply periodic boundary conditions (see Fig. 1(b)).

The elastic interaction strength of a bond ij is described by the Born Hamiltonian [9, 10]

$$H_{ij} = \frac{\alpha}{2}((\vec{u}_i - \vec{u}_j) \cdot \vec{d}_{||})^2 + \frac{\beta}{2}((\vec{u}_i - \vec{u}_j) \cdot \vec{d}_{\perp})^2, \tag{3}$$

in which α and β are elastic tensile and bending coefficients of a bond, respectively, $\vec{u}_j$ is the displacement vector of site j, and $\vec{d}_{||,\perp}$ is the unit vector either parallel or perpendicular to the vector connecting neighbouring sites in the undeformed lattice. For the purpose of describing the local adjustment of the material structure to the stress we introduce a Maxwell type relaxation or dissipation mechanism to the forces acting on mass sites [10]. Now the constitutive equation for the forces reads as follows

$$\frac{\partial f_{ij}}{\partial t} = \frac{\partial f_H}{\partial t} - \frac{1}{\tau} f_{ij}, \tag{4}$$

in which f_{ij} is the dynamic interaction force between nearest neighbour sites i and j, f_H is the elastic force derived from the Born Hamiltonian and τ is a phenomenological force relaxation constant. Then τ and the elastic interactions (α and β) set the internal time scales of the system. Our choice of local dynamics is equivalent to a Born spring with tensile and bending stiffness in series with a viscous dashpot, as depicted in Fig. 1(b). In this model disorder is introduced by randomly removing a fraction $q = (1 - p)$ of the bonds, i.e. quenched disorder (see Ref. [8]).

Figure 4. Time-development of the number of cut bonds N_c averaged over several simulation runs and for strain rate $d\epsilon/dt = 2.5\times10^{-4}$. The end-points of the curves indicate the final fracture.

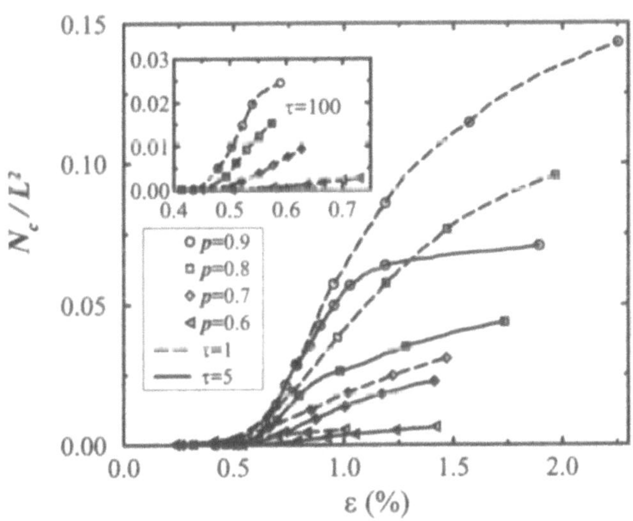

3.1. DISORDERED SYSTEM

In disordered systems the fracture behaviour is expected to depend on the amount of disorder and on the interplay between the strain rate ($d\epsilon/dt$) and internal time scales, especially the force relaxation constant (τ). At a finite strain rate, different regimes are observed in the fracture process, which can be illustrated by the increase in damage during tensile elongation, see Fig. 4. This figure shows the time-development of the number of cut bonds, when the force relaxation is varied. Fracture in a system with negligible relaxation (large τ) is clearly affected by the loading procedure. An abrupt application of constant strain rate at $t = 0$ introduces a propagating shock wave into a system, which decays slowly in a system with weak relaxation ($\tau = 100$). When travelling back and forth, the propagating shock wave may break individual bonds here and there in an uncorrelated manner. Due to the lack of relaxation, and hence, to the increased stress enhancement, these microruptures, interacting with the pre-existing disorder, soon start to correlate. As soon as one of the cracks begins to dominate, the system suffers rapidly a macroscopic breakdown. At finite strain rates with very small relaxation, there may exist competing microcracks, which have nucleated from several defects — not only from the most critical one [4], as in models of brittle fracture. The microcracks exhibit correlated growth and can eventually coalesce.

When $\tau = 1$ the strong relaxation damps effectively all the disturbances in the system, and the initial loading-induced shock wave has no effect on the fracture behaviour. The cracks start to propagate very slowly from initial defects, if at all, because the stress enhancement at the crack tip is strongly reduced compared with systems of weak or moderate relaxation. Since in this case there is practically no potential energy stored in the bonds, an individual rupture does not necessarily cause the next bond-rupture to take place close to the current one. The interactions between neighboring mass sites diminish and the local dynamics becomes increasingly independent of the dynamics of the surroundings. The lack of strain-dependent and reversible energy makes the sample very ductile. As a result, a system with strong relaxation evolves towards the macroscopic breakdown via numerous microruptures in an uncorrelated fashion. The system is capable of withstanding a considerable amount of external strain, unlike systems with weak relaxation. Eventually, this drives the system to a drastically deformed final configuration. In real materials this behaviour can be interpreted as permanent plastic deformation, which in our model is captured by the non-recoverable response of the dashpot elements.

When the force relaxation is weakened towards a moderate value of $\tau \simeq 5 - 10$, features of systems with both weak and strong relaxation are captured. As Fig. 4 shows, in weakly disordered systems of $p = 0.8-0.9$, the proportion of ruptured bonds grows rapidly in the beginning of the fracture history. Then the accumulation of damage becomes slower, and the system is able to accommodate additional straining. However, in strongly disordered systems this tendency is lost. Judging from the width of the fracture regime in strain or strain to final fracture (in Fig. 4), the system shows an increased ductility for strong relaxation. In addition, for weak relaxation a strongly disordered system is able to accommodate more external strain than a weakly disordered one (see the insert of Fig. 4). For increasing relaxation, this is reversed: a weakly disordered system withstands straining most.

4. Summary

Here we have demonstrated how computer modelling gives interesting qualitative information on fracture in disordered materials. First we developed a mesoscopic elastic-plastic model to study two-dimensional systems under adiabatic loading conditions. In the case of an elastic system the finite size analysis of the maximum strain showed a significant residual value for large system sizes and large disorder. Another demonstration of the versatility of this quasistatic model is its capability to show similar strain localization as seen experimentally in paper.

We also developed a dynamical model to study the interplay between force relaxation and disorder in systems under constant strain rate. In disordered systems, the strong relaxation was reflected in the slow crack nucleation and uncorrelated microruptures, leading to very ductile characteristics. A more detailed description of the results from this dynamical model will be published elsewhere [11].

Acknowledgements

This study was partly supported by the Technology Development Centre of Finland and the Academy of Finland. We would also like to thank Dr. M. Alava and Dr. K. Niskanen for very fruitful collaboration in some of the studies reported here.

References

1. Lawn B. (1993) *Fracture of Brittle Solids*, 2nd Ed., Cambridge University Press, Cambridge.
2. Korteoja, M.J., Lukkarinen, A., Kaski, K. and Niskanen, K.J. (1995) Model for Plastic Deformation and Fracture in Planar Disordered Materials, *Phys. Rev. E* **51**, pp. 1055–1058.

3. Karttunen, M.E.J., Niskanen, K.J. and Kaski, K. (1994) Fracture in Mesoscopic Disordered Systems, *Phys. Rev. B* **49**, p. 9453.
4. Duxbury, P.M.,Beale,B.D. and Leath, P.L. (1986) Size Effects of Electrical Breakdown in Quenched Random Media, *Phys. Rev. Lett.* **57**, pp. 1052–1055.
5. Duxbury, P. M., Leath, P. L., Beale, P. D., (1987) Breakdown properties of quenched random systems: The random-fuse network, *Phys. Rev. B*, **36**, pp. 367–380.
6. Korteoja, M.J., Lukkarinen, A.,Kaski, K., Gunderson, D.E., Dahlke, J.L. and Niskanen, (1996) Local Strain Fields in Paper, *TAPPI Journal*, **79**, pp. 217–223.
7. Korteoja, M.J., Lukkarinen, A., Kaski, K. and Niskanen, K.J. (1995) Computational Study of Formation Effects on Paper Strength, *submitted for publication.*
8. Herrmann, H.J. and Roux, S., (1990) *Statistical Models for The Fracture of Disordered Media*, North-Holland, Amsterdam.
9. Born, M. and Huang, K. (1954) *Dynamical Theory of Crystal Lattices*, Oxford University Press, London.
10. Rautiainen, T.T., Alava, M.J. and Kaski, K. (1995) Dissipative dynamic fracture of disordered systems, *Phys. Rev. E* **51**, pp. R2727–R2730.
11. Rautiainen, T.T., Alava, M.J. and Kaski, K. (1996) Dynamics of Fracture in Dissipative Systems, *submitted for publication.*

KINETIC TRANSITIONS IN ENSEMBLES OF DEFECTS (MICROCRACKS) AND SOME NONLINEAR ASPECTS OF FRACTURE

O. B. NAIMARK
Institute of Continuous Media Mechanics
of the Russian Academy of Sciences
1 Acad. Korolev street, 614061 Perm, Russia

1. Introduction

During the last decades experimental studies have revealed some specific nonlinear regularities in the behaviour of microcracks due to the transition from the dispersive accumulation of microcracks to the formation of a main crack, in the interaction of defects with macrocracks, and in statistical self-similarity of defect distribution. These effects are developed as the consequence of collective behaviour in the defect system and are caused both by typical features of the system "solid with defects" and by some universal laws of nonlinear system behaviour. In the last case the structure evolution is accompanied by various forms of structural (kinetic) transitions that lead to the generation of localized modes of deformation and failure, and sharp changes of the symmetry properties (topological transitions).

2. Statistical Model and Constitutive Equations

2.1. STRUCTURAL PARAMETERS IN SOLIDS WITH MICROCRACKS

It has been established that deformation and failure of solids are accompanied by multiple nucleation and growth of microcracks. The existence of different types of microcracks and diversity of mechanisms of their generation and development requires the adequate choice of parameters characterizing the microcracks. It should be noted that microcracks, both in ductile and brittle materials, are orientated by the stress field and are characterized by the anisotropy of their microgeometry, which is higher for brittle

J. R. Willis (ed.), IUTAM Symposium on Nonlinear Analysis of Fracture, 285–298.

solids ($\sim 1:10$) and lower for ductile ones ($\sim$1:2) [1]. The material density change ($\Delta\rho/\rho \approx 10^{-4} - 10^{-3}$), which accompanies plastic deformation and precedes the formation of a macroscopic crack is another significant characteristic of the microcrack generation. The parameter, determining the volume concentration and the preferential orientation of the microcrack may be represented by the symmetric tensor $p_{ik} = n\ \langle s_{ik} \rangle$, where n is the number of microcracks in unit volume. The "microscopic quantity"

$$s_{ik} = s\ \nu_i \nu_k$$

corresponds to the density tensor of dislocations modelling the considered single defect. Tensor s_{ik} characterizes the volume and orientation of the disk–shaped microcracks with the base $\vec{S_0} = S_0 \vec{\nu}$ and vector of the displacement jump (normal to the base) in transition from one edge to another $\vec{b} = b \vec{\nu}$. The volume of the microcrack is $s = Sp\ s_{ik}$.

2.2. STATISTICAL MODEL

Evolution of defects is caused by the statistical distribution of microcrack nuclei and interaction between defects. The evolution equation for the microscopic parameter s_{ik} is given by the Langevin equation [2]

$$\dot{\mathbf{s}} = \mathbf{K}(\mathbf{s}) - \mathbf{F}(t),$$

where $\mathbf{K}(\mathbf{s})$ and $\mathbf{F}(t)$ are deterministic and fluctuating parts of the interaction forces. Fluctuating forces are characterized as usual by the following properties

$$\langle \mathbf{F}(t) \rangle = 0, \quad \langle \mathbf{F}(t)\mathbf{F}(t') \rangle = Q\delta(t - t'),$$

where the averaging is taken over the stochastic process; Q is the correlator of fluctuating forces. In general, and it is typical in the considered case of defect accumulation, fluctuating forces depend on the variable $\mathbf{s}$. But for our applications it is more convenient to study statistical properties using the Fokker-Planck equation for the distribution function $W(\mathbf{s}, t)$. The Fokker-Planck equation corresponding to the Langevin equation has the form

$$\frac{\partial}{\partial t} W = -\frac{\partial}{\partial \mathbf{s}} (\mathbf{K}(\mathbf{s}) W) + \frac{1}{2} Q \frac{\partial^2}{\partial \mathbf{s}^2} W, \tag{1}$$

where the first term is the so-called drifting term and the second is the diffusion term. The distribution function $W(s_{ik}, t)$ describes the probability of the state in phase space having coordinates $(s_{ik}, \dot{s}_{ik})$.

In [3] the process of damage accumulation was described assuming the self-similarity of the microcrack distribution at the various damage stages. The analysis of experimental data showed that self-similarity of defect accumulation may be observed in essentially distinguishing conditions of material deformation. Following Barenblatt we will assume statistical self-similarity of the microcrack distribution. The latter corresponds to the spectrum of stationary solutions of the Fokker-Planck equation (1). It is quite easy to get the stationary solution for the distribution function W ($\dot{W} = 0$) if the condition $W(\mathbf{s}) \to 0$ is satisfied for $\mathbf{s} \to \pm\infty$. The stationary solution is determined by the expression

$$W = Z^{-1} \exp\left(-\frac{\int_0^{\mathbf{s}} 2\mathbf{K}(\mathbf{s}')\, d\mathbf{s}'}{Q} \right),$$

where Z^{-1} is the normalizing parameter given by the integral $\int W(\mathbf{s})\, d\mathbf{s} = 1$.

The self-similarity hypothesis introduces the spatial distributions of defect ensembles into consideration when the ratio of the energy $E = \int_0^{\mathbf{s}} 2\mathbf{K}(\mathbf{s}')\, d\mathbf{s}'$ to the correlator Q is constant. The parameter Q characterizes the energy relief of the initial structure (grain boundary energy, energy of dislocation pile-ups) representing the microcrack nuclei. This case is valid for quasi-brittle damage which is accompanied by change of the mean size of microcracks but not concentration. For the damage accumulation under the ductile failure the self-similarity assumption may also be extended if we introduce the spectrum of Q_i for ensembles of defects of each mesoscopic level. Experimental data presented in [3] support this assumption.

A dislocational microcrack is the possible form of the hollow nucleus of a superdislocation with Burgers vector $\mathbf{B} = n\mathbf{b}$ ($n \sim 20 - 30$ is the number of unit dislocations with the Burgers vector $\mathbf{b}$). The microcrack energy may be represented as

$$E = \frac{\mu B^2 L}{4\pi} \ln \frac{R}{h} + E_{sf}. \tag{2}$$

Here L is the length of dislocations modelling the microcrack, h is the width of the microcrack, $E_{sf} = 2\gamma_{sf} hL$, ($\gamma \approx \mu b/4\pi$) is the microcrack surface energy, R is the size of the overstress zone (so-called the "head" zone where the local overstresses decrease as $\sigma_l \sim r^{-\kappa}, \kappa \geq 1/2$).

The microcrack growth is continued until the "head" energy does not relax. The equilibrium size of microcrack $h_e = 1/2n^2 b$ is estimated from (2) and for $n = 20$, $h \approx 0.1\mu m$. A close range of sizes is observed experimentally [4]. In studying the microcrack growth in the field of external stresses we drop in (2) the surface energy part and rewrite the energy expression in the form

$$E = \frac{\mu}{4\pi}\left(\frac{B}{hL}\right)^2 (hL)^2 L \ln\frac{R}{h}. \tag{3}$$

As it is obvious the expression B/hL is the scalar magnitude of the dislocation density.

Thus, the self-similarity allows the average dimensions of defects and their energy characteristics and their correspondence to the applied loading to be introduced into consideration.. The energy of microcracks may be calculated in accordance with the energy expression (3). We will consider the microcrack growth under the external loading and separate the contributions from the energy of the dislocation set modelling microcracks (tensor s_{ik} plays the part of the dislocation density) and that of interaction between the microscopic gauge field s_{ik}, adjacent defects and external stress field [5]

$$E = E_0 - H_{ik}\, s_{ik} + \alpha\, s_{ik}^2,$$

where E_0 is the term depending on p_{ik}; $H_{ik} = \gamma\,\sigma_{ik} + \lambda\, p_{ik}$ is the effective (mean) force field, acting on the microcrack, σ_{ik} is the macroscopic stress tensor; α, λ and γ are the material parameters.

Parameter α characterizes the susceptibility of the material to the defect generation. Low values of α allow the development of larger microdefects. The form of H_{ik} reflects the fact that reconstructions of the material structure are determined by the local stresses, which may differ considerably from the macroscopic ones. The term $\lambda\, p_{ik}$ describes the force action, which causes defect growth in the fields of overstresses from the adjacent defects.

It is to be noted that the solid with defects is a non–linear system which is far from equilibrium and p_{ik} plays the part of the order parameter. Some analogies in the behaviour of such systems are discussed in [6] and the distribution function for the order parameter was taken as $W = Z^{-1}\exp(-E\,/\,Q)$, where Q is the fluctuating force intensity determined above which characterizes the potential relief of the initial defect structure. Averaging s_{ik} with the distribution function W, we obtain the self-consistency equation for p_{ik}

$$p_{ik} = n \int s_{ik}\, Z^{-1} \exp(-E\,/\,Q)\, d\,s\, d^3\vec{v}\,, \tag{4}$$

where $Z = \int \exp(-E/Q) ds\, d^3\vec{v}$. Equation (4) is a constitutive equation for the medium with microcracks for the case of the quasi–brittle behaviour of solids, when there occurs preferential growth of the mean size of defects at constant number of the latter n [7]. The analysis of (4) for the case of uni-axial extension has shown that depending on the value of the dimensionless parameter $\delta = 2\alpha/\lambda n$ there are three responses of material to the defect growth (Fig.1): monotonic ($\delta > \delta_*$), metastable ($\delta_c < \delta < \delta_*$) and unstable ($\delta < \delta_c$); δ_c and δ_* being the bifurcation points corresponding to the change of asymptotes. The value of δ is determined by the natural scale of structural heterogeneity (grain or block sizes) and the correlation radius of the overstress field from microcracks. In the case of a monotonic reaction $\delta > \delta_*$, which is characteristic of fine-grained materials, the applied stress corresponds to the constant value of the microcrack concentration. These values of the p_{ik} provide the minimum of free energy of the material with characteristic initial structure under applied stresses.

In the case of metastable reaction ($\delta_c < \delta < \delta_*$) the change of p_{ik} is jump-like in the metastable region and is accompanied by the ordering of the defect system (the orientational transition). The unstable reaction ($\delta < \delta_c$) is characteristic of coarse-grained materials, which initially contain large nuclei of microcracks. The stress scale in this case is divided into two regions: the metastability region ($\sigma < \sigma_f$) with infinite value of change of p_{ik} at the existence of structural disturbances and the region of absolute instability ($\sigma > \sigma_f$) which allows the existence of arbitrarily large defects. The energy of material with defects may be represented taking into consideration a qualitative analogy between microcrack development in the metastable range and phase transition of the first kind. Free energy F can be expressed in terms of fourth order expansion in p_{ik} by analogy with Landau expression

$$F = 1/2\, A p_{ik}^2 + 1/3\, B p_{ik}^3 + 1/4\, C p_{ik}^4 - D \sigma_{ik} p_{ik} \,. \tag{5}$$

3. Some Non–Linear Aspects of Transition to Fracture

3.1 KINETICS OF DAMAGE ACCUMULATION

Kinetics of the damage accumulation was studied in [8, 7] and based on the assumption that free energy Ψ of materials with the considered type of defects is determined by the statistical model and depends on the parameter p_{ik} and elasticity tensor u_{ik}. When analyzing nonlinear regularities of the microcrack accumulation, especially, damage localization, it is very important to take into consideration the spatial non-homogeneity of the defect distribution. The non-locality effect appears due to the high gradient of internal stresses caused by the non-homogeneous defect distribution at the mesoscopic level [9]. These gradients are determined by the scales from $1\mu m$ to $1cm$ and more.

Non-local potential is written in the form $\Psi^* = \Psi + (1/2)\chi(\partial p_{ik}/\partial x_l)^2$, where the quadratic gradient term describes the non-local effect in the so-called long-wave approximation, χ is the nonlocality parameter. To follow the Ginsburg-Landau approach [2] we obtain as the consequence of the evolution inequality $\delta\Psi/\delta t = (\delta\Psi/\delta p_{ik})\,(\partial p_{ik}/\partial t) \le 0$ ($\delta\Psi/\delta p_{ik}$ is the variational derivative) the kinetic equation for the p_{ik} tensor

$$\frac{\partial p_{ik}}{\partial t} = -\frac{1}{\tau_p}\frac{\partial\Psi}{\partial p_{ik}} + \frac{\partial}{\partial x_l}(\chi\frac{\partial p_{ik}}{\partial x_l}), \tag{6}$$

where τ_p is the relaxation time for p_{ik}. This equation coupled with the constitutive equation of an elastic medium with microcracks [5], initial and boundary conditions closes the system of equations defining the continuum problem of the quasi-brittle failure. Since the transition from the dispersive microcrack accumulation to the damage localization is described by the set of curves with $\delta \le \delta_c$ we will examine in more detail the nonlinear behaviour of system in this regime. Let us consider specific features of the non-linear behaviour of the system due to the transition through the instability threshold $\mathbf{p_f}$. In some neighbourhood of this point ($\mathbf{p} > \mathbf{p_f}$) the kinetics is determined mainly by the higher terms of the expansion (5) and the kinetic equation is presented in the form:

$$\frac{\partial p}{\partial t} = \frac{1}{\tau_p}Sp^{\alpha} + \frac{\partial}{\partial x}(\chi_0 p^{\beta}\frac{\partial p}{\partial x}). \tag{7}$$

Here for simplicity the scalar parameter $p = Sp\, p_{ik}$ is introduced, for analyzing the cross section of a uni-axially loaded specimen. S, α, β and χ_0 are the parameters of expansions assuming the dependence of the non-locality parameter ϱ on p. Two main properties of the solution of (7) were clarified in [10]. At the developed kinetic stage the p-growth occurs in the so-called "peak-regime". It means that p grows infinitely as $t \to \tau_c$ and is localized in some area. The spatial distribution of p has self-similar profiles at the developed stage of **p**-kinetics and does not depend on the initial disturbances. Maintenance of the profiles allows us to assume the existence of the self-similar solution in the form

$$\mathbf{p}(\mathbf{x},t) = \mathbf{g}(t)\,\mathbf{f}(\zeta), \quad \zeta = \mathbf{x}\,/\,\phi(t), \tag{8}$$

where $\mathbf{g}(t)$ governs the growth law of parameter **p** and $\phi(t)$ defines the variations of the half-width of the localization region. For the function $\mathbf{f}(\zeta)$ determining the localized profile (dissipative structure) we have a nonlinear boundary problem. It follows from (8) that the time dependence of **p** remains self-similar: this is simply an extension along the **x** and **p**-axes. Substitution of (8) into the equation (7) allows clarification of the form of the functions $\mathbf{g}(t)$ and $\phi(t)$

$$\mathbf{g}(t) = G_1(1 - \frac{t}{\tau_c})^{-1/(\alpha-1)}, \quad \phi(t) = G_2(1 - \frac{t}{t_c})^{0.5(\alpha-\beta-1)/(\alpha-1)},$$

where τ_c is "peak time" ($p \to \infty$ at $t \to \tau_c$ [10]); $G_1 > 0$, $G_2 > 0$ are the parameters of nonlinearity, which characterize the rate of the free energy release $\partial\Psi/\partial\mathbf{p}$ with the increase of the microcrack density in the region $\mathbf{p} > \mathbf{p}_f$ ($\delta < \delta_c$) and the law of the half-width change. The similarity transformation allows to find the profiles $\mathbf{f}(\zeta)$ for other τ_c. The spectrum of the eigen-functions determines inherent properties of non-linear media. Its solution gives the spectrum of eigenforms $\mathbf{f}_i(\zeta)$ "living" during $\tau^i{}_c$ time in discrete ranges of eigenvalues ζ_i, specifying the damage localization scales. The solution (8) refers to the class of nonlinear singular solutions, describing an infinite growth of $\mathbf{p}(t)$ over localization scale ζ_i (fundamental lengths [10]) as $t \to \tau_c{}^i$.

3.2. TOPOLOGICAL ASPECTS OF TRANSITION TO FRACTURE IN QUASI-BRITTLE MATERIALS

The nonlinear character of the microcrack interaction, followed by the instability and spatial localization leads to stochasticity which is typical for the behaviour of an essentially nonequilibrium system. The instability involving localization is known to be fractal in nature [11]. Attempts at crack modelling based on fractal sets are proposed in [12, 13, 14]. The common features of these approaches are the usage of geometrical images of fractal sets, for example the Koch curves. But here we will consider the fractal aspect of failure by studying the nonlinear property of ensembles of defects. We connect the fractal set with the "limit set" of the localized scales of the self-similar solution in some area of the phase space of the system behaviour (the attractor area [11]). Stochastic behaviour of the microcrack system is observed after the instability threshold p_f. Small variation of the initial data near this value leads to qualitative changes of topological properties and various scenaria of damage localization follow due to the localized instability on the set of spatial scales ζ_i with the different peak times τ_i^c. Such qualitative regeneration of the dynamic system "solid with microcracks" due to the localized growth of the microcrack density on the limit set of the spatial scales ζ_i is accompanied by the qualitative change of topological properties of this system. Topological properties may be characterized by the fractal dimension

$$d_H = \lim \frac{\ln N(r)}{\ln(1/r)},$$

where r is characteristic dimension of cubes, filling the phase volume, $N(r)$ is the number of cubes, enclosing at least one element of the system. The dimension d_H introduced in this way allows differentiation of the extent of complexity and intricacy of trajectories of the system in phase space and represents the scale invariance of the considered system.

The topological regularities of transition from damage to fracture were examined in [15, 16] for strength estimation of carbon-carbon composites. Fracture of a carbon–carbon composite reveals two main aspects of the quasi–brittle fracture problem: the presence of the dispersive microcrack accumulation and the well–defined statistical character of fracture. The developed approach was used to simulate the failure in carbon-carbon plane specimens with initial macroscopic defect located in the center (the macrocrack edges are normal to the extension direction) with chararacteristic size N_a (N is number of the finite elements in the bulk of the composite plate). Simulation of damage has demonstrated that under loading the initial stage is accompanied by preferential failure of elements located in the vicinity of the macroscopic defect. The percolation cluster across the specimen results from coalescence of the cluster originated from the initial macrodefect and clusters in its immediate neighbourhood. The cluster appears to be fractal in character and with increase of linear dimension L of the damaged array its mass M (the number of failed elements) increases on the average as

$$M(L) = A\, L^D,$$

where D is the fractal dimension and A is the effective amplitude. The mean value of A is obtained by averaging over the manifold of realizations of the percolation cluster. Numerical predictions of topological properties of the cluster being generated in the vicinity of the initial macrodefect are presented in Fig. 2 as the logarithmic dependence of the cluster mass M on the side length of the initial "cell" L enclosing the cluster.

The dependence $M(L)$ is seen to consist of two linear portions, the slope of which is determined by the fractal dimension D. The cluster growth characterized by the fractal dimension $D = 1$ occurs predominantly normal to the loading direction. The growth of the cluster not associated with the initial macrodefect leads to the increased value of the fractal dimension D, which may be as high as $1.4 - 1.7$. This is indicative of qualitative changes in the topology of the damage accumulation process and the fracture mechanism replacement.

The measure of fractal dimension $D = 1$ supports the validity of the approaches of linear fracture mechanics only at the initial stage of crack evolution, when the crack propagation is defined by disturbance of the stress fields in the vicinity of the crack. However, this holds true for sufficiently large cracks when the extension of the disturbed stress region d is essentially smaller than crack length l ($d/l << 1$). In the general case when the initial macrocrack size is below some critical value $l < l_t$ the process of fracture evolution lacks "the intermediate–asymptotic mode" of crack propagation [17]. This is important for understanding of the fact that linear fracture mechanics generalizations used as the basis for describing the fracture stage can not be valid for each case. The evidence for this is the increase of the fractal dimension up to value $D > 1$, in the presence of the initial macrocrack with $l > l_t$ and $l < l_t$.

Attention must be paid to the essential difference of fractals characterizing spatial ordering from the sets arising due to the space-time ordering caused by the existence of the attractors considered above (strange attractors [11]). In the first case the fractals are determined in a space of low dimension and the fractal dimension can be determined from relatively small computations. The presented results were obtained from numerous realizations (the number attained 120 realizations) since the numerical analogue of the set of invariant-group solutions (6) is inserted in the space of large dimension.

3.3. RESONANCE CONDITIONS OF DAMAGE LOCALIZATION

In this part we will consider some aspects of damage evolution under dynamic (impact) loadings which in our opinion corresponds to the resonance excitation of damage localization. One of the most interesting failure phenomena is spall fracture produced by impact loadings [18]. The spall failure is characterized by the small times ($10^{-7} - 10^{-6}$ *sec.*) and large amplitude of tensile stresses, exceeding by several times the quasi–static limit of strength. In spite of this peculiarity, spall failure is similar in its structural features to the process of quasi–static failure of brittle materials.

Fracture due to impact wave loading occurs in an essentially different situation relative to the quasi–static loading. Since the physical reason for the macrocrack nucleation lies in generating an appropriate profile of microcrack concentration, the intensity of impact wave may produce different resonance–type excitations of the damage localization region. This can be observed in the regularities of transition from a single to a multiple spall at the increased amplitude of the impact wave. In spalling the failure usually spreads over a considerable portion of the material. However, the peak regime can develop only within typical (fundamental) lengths with a minimum value of fracture time τ_c. As the load impulse increases, several structures localized on fundamental lengths are generated. In this case the appearance of the localized damage zones (dissipative structures) corresponds to the experimentally observed transition from single to multiple spalling [8, 19, 20, 21].

In reference to the equation (6) the self–similarity of its solution, corresponding to the peak regime asymptotes, may occur for a weak de-

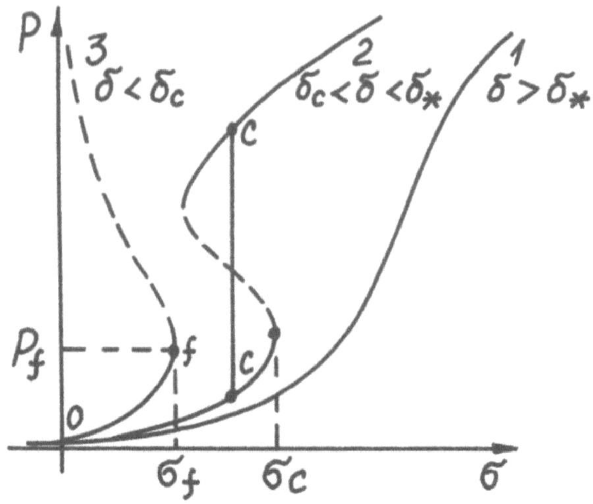

Figure 1. Characteristic responses of solid on the microcrack grouth.

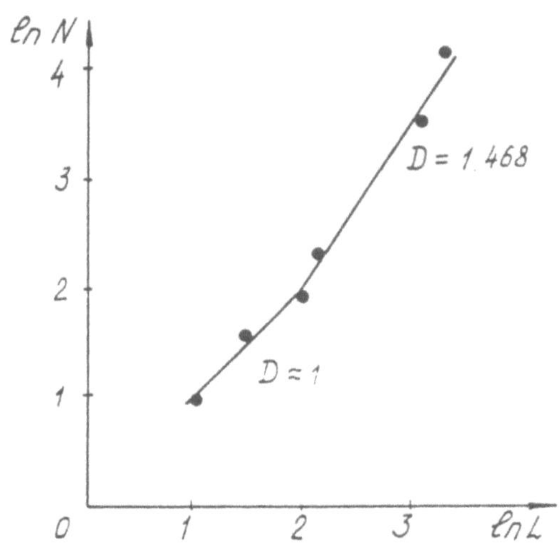

Figure 2. Topological characteristics (fractal dimension) of the damage cluster growth.

pendence of the time of failure on the impulse amplitude – the effect of "dynamic branch" under spalling conditions [22, 19].

The regularities of the formation of localized dissipative structures are specially vivid under shock wave loading with the duration of about 1 μsec. Experiments were carried out on rods ($10-12\,mm$ in diameter and $100-200\,mm$ long) of PMMA and ultraporcelain. A compression pulse was

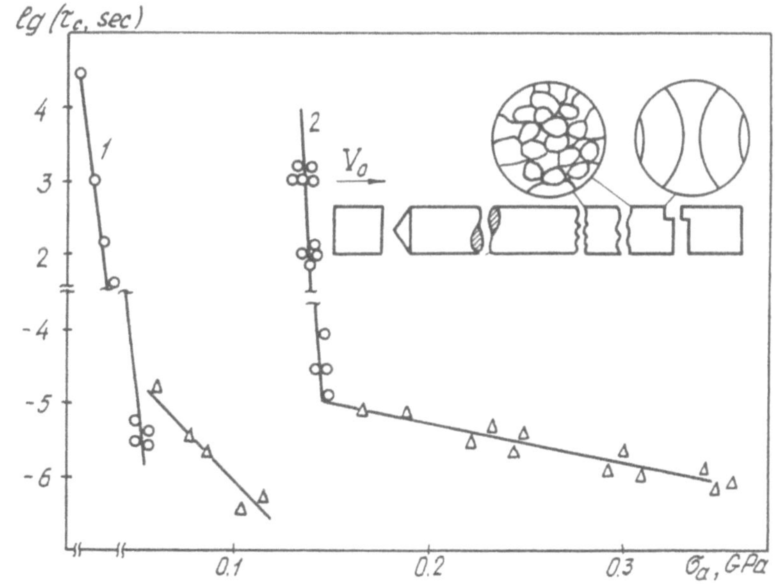

Figure 3. Strength time dependence of(1) polymethyl methacrylate (PMMA) and(2) ultraporcelain (○ and △ correspond to quasi-static and dynamic branches respectively).The inset shows a schematic diagram of the fractured sample [19]

initiated in the rods by impact from a light–gas cannon. The parameters of the compression pulse were measured with a laser differential interferometer. From the results of experimental studies of the spall fracture of rods, we plotted the logarithm of the fracture time τ_c versus the amplitude of the tensile stress σ_a (Fig. 3). Also shown in this figure are the results of quasi–static experiments. At values $\tau_c \sim 10^{-4}s$, according to a fractographic analysis of the surfaces of the spall sections, the development of fracture occurs in the same way as during quasi–static extension. The dependence $\log \tau_c(\sigma_a)$ agrees with the time dependence of the strength of the materials which were studied during quasi–static loading. An increase in the level of the acting stress leads to a deviation of the $\log \tau_c(\sigma)$ curves toward a longer fracture time (Fig. 3). At the same time we observed a transition from a single–center fracture starting from the surface to a characteristic multicenter fracture (multiple mirror zones) in the spall sections. It may be suggested that the time dependence found for the brittle solids in the region of pulsed extension is related to a transition to a multicenter internal fracture and is determined by the nonlinear behaviour in the ensemble of microcracks. Fractographic pictures of fracture are of great interest in different sections of the spalling (Fig. 4). In the first section (Fig. 4a) where the amplitude of loading impulse is maximal many mirror zones are seen

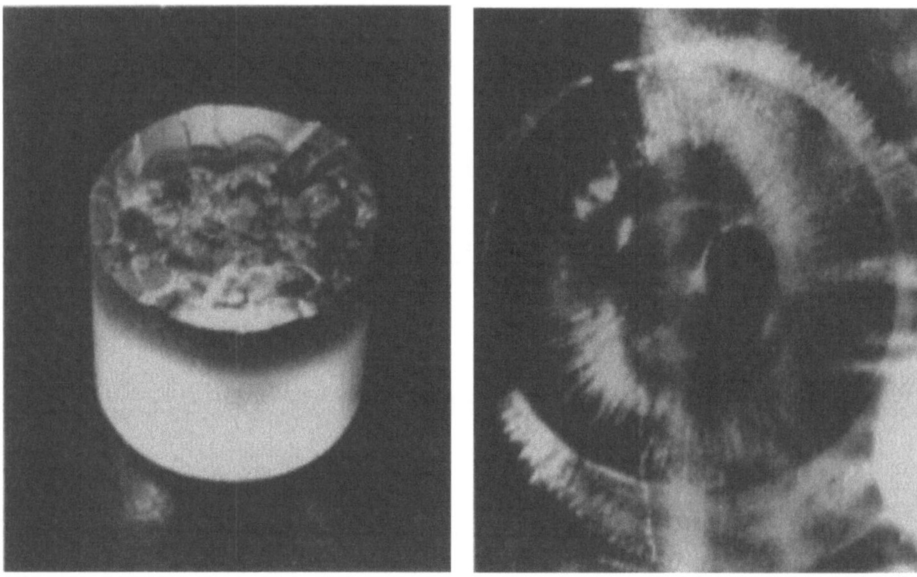

Figure 4. Fractographic pictures of the fracture surfaces in the spall sections of PMMA (experiment was kindly given by E.N.Bellendir).

on the spalling surface. Mirror zones appear to be zones of localized damage. In the section of spalling the picture is similar but the scale of mirror zones increases. In the last spall section only one or two mirror zones are formed. The transition from single–center fracture to multicenter fracture is observed in the interval of loading times $10^{-4} - 10^{-6} sec$. Resonance excitation of damage localization is the main factor of the weak dependence of the fracture time on the amplitude of the tensile stress ("dynamic branch", Fig. 3) under spalling. This result explains the "overloading" effect under dynamic fracture.

Under quasi–static loading the random distribution of initial defects produces only simple structures, e.g. there are mirror zones near surfaces of specimen. Initiation of the complex structures may be realized by the resonance excitation. The example of the same structure was observed in some spall section (Fig. 4b), where the characteristic slope of the stress wave provided for the resonance regime of the complex structure excitation.

ACKNOWLEDGEMENTS

The research described in the publication was made possible in part by Grant $JSK100$ from the International Science Foundation and the Russian Foundation for Basic Research (project 94-01-00481-a).

References

1. BETECHTIN V.I., NAIMARK O.B., SILBERSHMIDT V.V. 1989, The fracture of solids with microcracks: experiment, statistical thermodynamics and constitutive equations, *Proc. 7th Int. Conf. Fract. (ICF-7)*, **6**, 38–45.

2. LANDAU L.D., LIFSHITZ E.M., *Course of Theoretical Physics*, Vol.5: Statistical Physics, Pergamon Press, Oxford (1980).

3. BARENBLATT G.I., BOTVINA L.R., 1983, Self–similarity of fatigue fracture. Defect accumulation, *Izv. AN SSSR. Mech. Tv. Tela*, **4**, 161–165 (in Russian).

4. BETECHTIN V.I., VLADIMIROV V.I., 1979, Kinetics of microfracture of crystalline bodies, *Problems of Strength and Plasticity of Solids*, 142–154 (in Russian).

5. NAIMARK O.B., 1995, *Structural Transitions in Solids and Mechanism of Plasticity and Failure*, Preprint of the Institute of Continuous Media Mechanics of the Russian Academy of Sciences, Perm.

6. HAKEN H., 1983, *Advanced Synergetics: Instability Hierarchies of Self-Organizing Systems and Devices*, Springer-Verlag, Berlin.

7. NAIMARK O.B., SILBERSHMIDT V.V., 1991, On the fracture of solids with microcracks, *Eur. J. Mech., A/Solids*, **10**, 607–619.

8. NAIMARK O.B., BELJAEV V.V., 1989, Kinetics of accumulation of microcracks and the staged nature of the fracture process under shock loading, *Phys. Combustion and Explosion,* **25**, 115–123.

9. KRÖNER E., 1960, Allgemeine Kontinuumstheorie der Versetzungen und Eigenspannungen, *Arch. Rat. Mech.Anal.*, **4**, pp.273-280.

10. KURDYUMOV S.P., 1988, Dissipative structures and chaos in non–linear space, *Plasma theory, non–linear and turbulent processes in physics* , **1**, 431–459.

11. ZASLAVSKY G.M., SAGDEEV R.Z., 1988, *Introduction to non-linear physics*, Nauka, Moscow (in Russian).

12. MANDELBROT B.B., PASSOJA D.E., PAULLAY A.J.,1984, Fractal character of fracture surfaces, *Nature*, **308**, 5961, 721-722.

13. CARPINTERI A., 1994, Scaling laws and renormalization groups for strength and toughness of disordered materials, *Int. J. Solids Structures*, **21**, 3, 291-302.

14. BAZANT Z.P. Scaling of quasi-brittle fracture. *Int.J. of Fracture*, in press.

15. NAIMARK O.B., DAVYDOVA M.M., 1993, Fracture of the carbon–carbon composites as non–equilibrium kinetic transition in ensemble of microcracks, *IX Int. Conf. Comp. Mat.*, **2**, 576–583, University of Zaragoza, Woodhead Publishing Limited,

16. NAIMARK O.B., DAVYDOVA M.M., 1994, Topological (fractal) analysis of defect accumulation in evaluation of the carbon–composite strength, *Mech. Comp. Mat.*,**1**, 97–113.

17. BARENBLATT G.I., 1979, *Similarity, Self-Similarity, Intermediate Asymptotics*, Gidrometeoizdat, Leningrad, (in Russian).

18. ZEL'DOVICH Ya.B., RAISER Yu.P., 1966, *Physics of Shock Waves and High-Temperature Hydrodynamic Phenomena*, Nauka, Moscow (in Russian).

19. BELLENDIR E.N., BELJAEV V.V., NAIMARK O.B., 1989, Kinetics of multicenter fracture under spalling conditions, *Sov. Tech. Phys. Lett.*, **15**, 13, 90–93.

20. BELJAEV V.V., NAIMARK O.B., 1985, Kinetic transitions in media with microcracks and metal fracture in stress waves. *J. Appl. Mech. Tech. Phys.*,**1**, 163–171.

21. BELJAEV V.V., NAIMARK O.B., 1990, Kinetics of multicenter fracture under shock wave loadings, *Sov. Phys. Dokl.*, **312**, 289–293.

22. ZLATIN N.A., PUGACHOV G.S., BELLENDIR E.N., 1981, On the question of the overstresses of solids under microsecond duration loadings, *Sov. Tech. Phys. Lett.*, **7**, 65–69.

ON CONTINUED VOID GROWTH IN DUCTILE METALS SUBJECTED TO CYCLIC LOADINGS

J. DEVAUX [(1)], M. GOLOGANU[(2)],
J.B. LEBLOND[(2)], G. PERRIN[(3)]

[(1)] *FRAMASOFT+CSI*
10, rue Juliette Récamier, 69006 Lyon, France
[(2)] *Laboratoire de Modélisation en Mécanique*
Université Paris VI, Tour 66, 4, place Jussieu,
75005 Paris, France
[(3)] *Bureau de Contrôle des Chaudières Nucléaires*
15, rue Jean Bertin, 21000 Dijon, France

1. Introduction

Experiments conducted on both compact tension specimens (by Kobayashi *et al.*, 1991) and cracked pipes (by Schmidt *et al.*, 1991) made of ductile materials have demonstrated a considerably reduced fracture strength under cyclic loadings; that is, fracture occurs for a lower overall deformation if that deformation is reached under cyclic conditions than if it is reached monotonically. The experiments of Schmidt *et al.* (1991) also seem to indicate that the effect is connected in some way to strain hardening: indeed it is more marked for stainless steels, which exhibit considerable hardening, than for low-alloy steels, which have lower hardening slopes.

It was quickly concluded by Gilles *et al.* (1992), after due elimination of other possible explanations, that the phenomenon should arise from enhanced void growth under cyclic conditions. This interpretation was supported by the results of some preliminary finite element calculations performed by these authors. These calculations simulated the cyclic behaviour of elementary representative porous volumes subjected to cyclically and proportionally varying overall stresses. They demonstrated a ratchet effect of

J. R. Willis (ed.), IUTAM Symposium on Nonlinear Analysis of Fracture, 299–310.

the porosity (progressive increase of its average value during one cycle as successive cycles were performed).

The aim of the present paper is to confirm and complete these numerical results and see what theoretical models can say about the effect observed. It should be stated at once that the hardening behaviour of the matrix will be described in a simplistic way as purely isotropic. We are well aware of the fact that in reality, the matrix behaviour under cyclic conditions is certainly more complex and that this is bound to be important in actual experiments such as those of Kobayashi *et al.* (1991) and Schmidt *et al.* (1991). However, we do not claim to fully explain these experiments; our purpose here is solely to concentrate on a single feature, namely the influence of the porosity upon the cyclic behaviour of ductile metals, which is why strain hardening is described in the simplest envisageable way.

2. The Absence of a Ratchet Effect for the Porosity in Gurson's Model

It is instructive, prior to performing any numerical simulation, to see what Gurson's (1977) famous model says about the evolution of the porosity under cyclic conditions (although it was admittedly not designed for such loadings).

The essential feature of the predictions of Gurson's model is encapsulated in the following statement:

Gurson's model predicts that the porosity f of an elementary volume subjected to cyclically, proportionally varying overall stresses, depends only on the initial porosity f_0, the (fixed) absolute value $|T|$ of the triaxiality $T \equiv \Sigma_m/\Sigma_{eq}$ (Σ_m : overall mean stress, Σ_{eq} : overall Von Mises equivalent stress) *and the* algebraic equivalent plastic strain $\tilde{E}^p_{eq}$ *defined by*

$$\tilde{E}^p_{eq} \equiv \int_0^t \mathrm{sgn}(\Sigma_m(\tau)) D^p_{eq}(\tau) \mathrm{d}\tau$$

(D^p_{eq} : overall equivalent plastic strain rate).

This means that during successive cycles, the points $(\tilde{E}^p_{eq}, f)$ always remain on the same, single curve; in other words, the porosity is instantaneously stabilized (no ratchet effect).

This statement can be proved by adapting a simple argument due to Perrin (1992) to the case of cyclic loadings. In fact, it holds not only for Gurson's model but more generally for any model involving a yield function depending only on Σ_{eq}, Σ_m, f and a *single* hardening parameter $\bar{\sigma}$, of the form $\Phi(\Sigma_{eq}/\bar{\sigma}, \Sigma_m/\bar{\sigma}, f)$. Indeed, let us first note that the function Φ must be even with respect to its second argument, since the yield locus must be invariant under change of sign of the stresses, as usual for metal plasticity. Thus the yield condition may be written in the form $\Phi(X, TX, f) = \Phi(X, |T|X, f) = 0$ where $X \equiv \Sigma_{eq}/\bar{\sigma}$; this shows that X depends only on f and $|T|$ (which is fixed since the overall stress Σ is assumed to vary proportionally). Moreover combination of the classical evolution equation for f resulting from plastic incompressibility of the matrix and the associated plastic flow rule yields (denoting by D^p_m the overall mean plastic strain rate and accounting for the fact that the functions $\partial\Phi/\partial\Sigma_m$ and $\partial\Phi/\partial\Sigma_{eq}$ are odd and even, respectively, with respect to Σ_m):

$$\begin{aligned} \dot{f} = 3(1-f)D^p_m &= (1-f)\frac{\partial\Phi/\partial\Sigma_m}{\partial\Phi/\partial\Sigma_{eq}}(X, TX, f)D^p_{eq} \\ &= (1-f)\mathrm{sgn}(T)\frac{\partial\Phi/\partial\Sigma_m}{\partial\Phi/\partial\Sigma_{eq}}(X, |T|X, f)D^p_{eq} \\ \Rightarrow \quad \frac{\mathrm{d}f}{\mathrm{d}\tilde{E}^p_{eq}} = \frac{\dot{f}}{\mathrm{sgn}(\Sigma_m)D^p_{eq}} &= \frac{\dot{f}}{\mathrm{sgn}(T)D^p_{eq}} = (1-f)\frac{\partial\Phi/\partial\Sigma_m}{\partial\Phi/\partial\Sigma_{eq}}(X, |T|X, f)\,. \end{aligned}$$

This equation implies that $\mathrm{d}f/\mathrm{d}\tilde{E}^p_{eq}$ depends only on f, $|T|$ and X, and hence only on f and $|T|$ by what precedes. It follows upon integration that f depends solely on f_0, $|T|$ and $\tilde{E}^p_{eq}$, as announced[1].

Remarks. 1. Observe that the property established holds *whatever the evolution law for the hardening parameter* $\bar{\sigma}$, provided that *only one such parameter appears in the criterion.*

2. Provided that strain hardening intervenes only through the value of $\bar{\sigma}$, *i.e.* that the function Φ is independent of the hardening law, the $f - \tilde{E}^p_{eq}$ curve is also independent of it. On the other hand, if Φ depends upon the hardening law (for instance through the introduction of some Tvergaard (1981) parameter q_1 depending upon the hardening exponent, as suggested by Koplik and Needleman (1988)), the same is of course true of the $f - \tilde{E}^p_{eq}$ curve. But this does not change the main point which is that that curve is identical for all cycles.

3. Numerical Simulations of the Cyclic Behaviour of an Elementary Porous Volume Element

The property of Gurson's model just established strongly suggests to precisely consider conditions of *proportional overall stressing* in the numerical simulations envisaged. Thus the triaxiality will remain constant throughout except for a change of sign at the beginning of each semi-cycle. In fact,

[1] In fact, this argument only considers the plastic phases of the mechanical history (since it uses the yield condition); but this is sufficient since both $\dot{\tilde{E}}^p_{eq}$ and $\dot{f}$ vanish during the elastic phases.

this was exactly the choice made by Gilles *et al.* (1991), for that precise reason. Their simulations however suffered from the fact that they involved fluctuations of the triaxiality of the order of a percent (in relative value). Since void growth is known to be extremely sensitive to the triaxiality, such fluctuations might be sufficient to explain at least part of the ratchet effect of the porosity observed by Gilles *et al.* Thus, new simulations involving smaller fluctuations of the triaxiality are desirable. Better constancy of the triaxiality is achieved here by means of an algorithm described in (Gologanu *et al.*, 1994); the (absolute) difference between the target triaxiality and that actually imposed is smaller than 10^{-4}.

It may be noted incidentally that there is no reason why the triaxiality should be considered as constant in the experiments of Kobayashi *et al.* (1981) and Schmidt *et al.* (1991). Thus at least part of the reduced fracture strength they observed in cyclic loadings may have arisen from a larger triaxiality (in absolute value) during the tensile phases of the cycles than during the compressive phases. Again, we do not claim to fully explain these experiments, but concentrate instead on a single feature, namely the progressive increase of the average porosity during one cycle with the number of cycles, under admittedly special cyclic loading conditions (triaxiality fixed in absolute value).

It is also useful to make the following remark: *for an elementary porous volume element with matrix made of some rigid-ideal plastic material* (*i.e.*, deprived of elasticity and strain hardening) *and subjected to some cyclic loading under conditions of proportional overall stressing, the porosity will stabilize instantaneously* (*i.e.*, no ratchet effect will occur). (Unlike the statement made in the preceding section, this remark is independent of any homogenized model for the overall behaviour of such an element, and only requires the local behaviour of the sound matrix to obey the classical rules of metal plasticity). To show that this is indeed true, it suffices to note that for such a material, starting from one possible mechanical evolution, one can obtain another possible evolution by changing the signs of both the (local) velocities and stresses. Thus the tensile and compressive phases of each cycle will be exactly symmetric (even if one takes geometry changes into account). The argument does not hold in the presence of elasticity, because there is a change of regime from a plastic loading phase to an elastic unloading one at the end of each semi-cycle which destroys the symmetry. It is also wrong if (isotropic) hardening takes place, because the hardening parameter continuously increases, which introduces some irreversibility.

Put in different terms, this remark means that the ratchet effect of the porosity under proportional cyclic overall stressing is fundamentally tied to

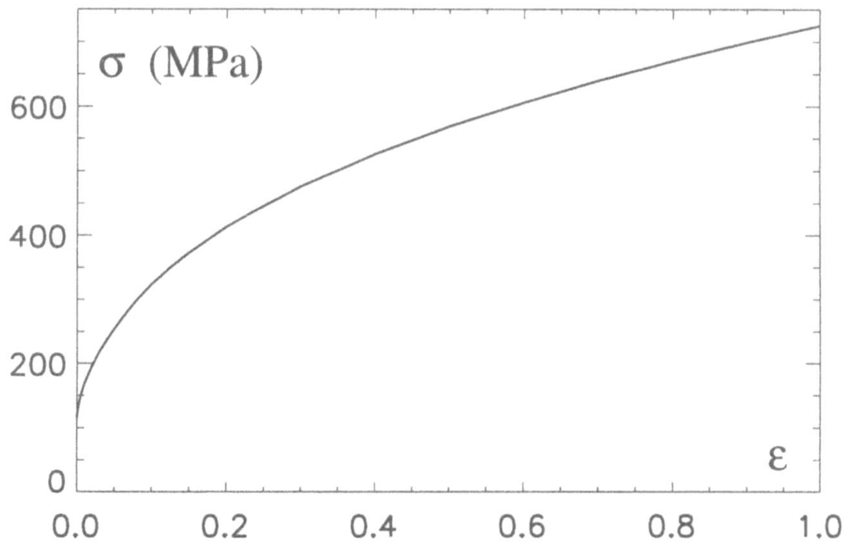

Figure 1: Stress-strain curve of the sound matrix used in the calculations.

two features of the material behaviour: *strain hardening* and *elasticity*. (The influence of the second factor is anticipated to be much smaller than that of the first one, however, since elasticity effects are generally quite small in ductile rupture). There is a strong connection here with the observation of Schmidt *et al.* (1991) that the steeper the hardening curve, the stronger the reduction in fracture strength in cyclic experiments.

Because of that remark, it was decided to consider a strongly hardenable material in the present study, corresponding to the (Cauchy) stress - (logarithmic) strain curve shown in *Figure 1*. The beginning of the curve (first 34% of deformation) corresponds to the actual behaviour of the SA 376 TP 304 stainless steel at 290°C (this was the steel and temperature of the experiments of Schmidt *et al.* (1991)). The rest of it is obtained by extrapolation using a (reasonable) hardening exponent of 0.35. Also, the influence of elasticity is studied by considering two values for Young's modulus: a normal (for a steel) value of 200,000 MPa, and an artificially enhanced one of 2,000,000 MPa (nearly rigid material).

Except for the fact that the loadings considered are cyclic, the simulations are very similar to those of Koplik and Needleman (1988). The geometry considered is a cylinder with equal initial diameter and height containing an initially spherical void, the initial porosity being $f_0 = 10^{-3}$. A constant $|T|$ of 2.7 (typical value near the tip of a crack) is imposed by means of uniform horizontal and vertical displacements prescribed on the lateral surface and

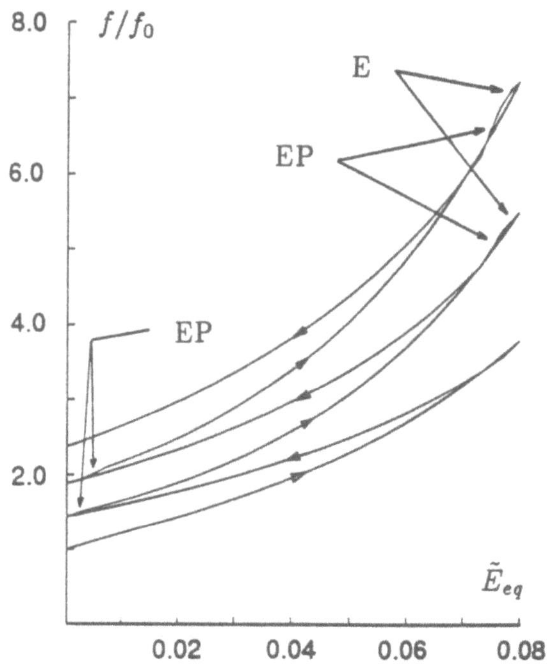

Figure 2: $\frac{f}{f_0}$ as a function of $\tilde{E}_{eq}$; $E = 2 \times 10^5$ MPa, tensile first semi-cycle.

the top cap, respectively, all surfaces being free of shear forces. The axial overall stress is larger than the lateral one. Three cycles are simulated, starting either with a tensile phase or a compressive one. It may seem that these two cases are identical except for a shift of one semi-cycle, but they are not in fact because the first semi-cycle plays a special role owing to the fact that it is the only one that starts with a pristine material. The value of the amplitude (in overall strain) of all cycles is $\Delta\tilde{E}_{eq} = 0.08$. $\tilde{E}_{eq} \equiv \frac{2}{3}\ell n \frac{h}{r}$ (h : current half-height of the cylinder; r : current radius) here is the algebraic equivalent *total* overall strain, which differs from the algebraic equivalent *plastic* overall strain $\tilde{E}^p_{eq}$ defined in the preceding section only because of elasticity effects. Indeed in the absence of elasticity, $\dot{\tilde{E}}^p_{eq} = \text{sgn}(\Sigma_m).\left|\frac{2}{3}\left(\frac{\dot{h}}{h} - \frac{\dot{r}}{r}\right)\right| = \frac{2}{3}\left(\frac{\dot{h}}{h} - \frac{\dot{r}}{r}\right)$ because the sign of Σ_m is always the same as that of $\frac{2}{3}\left(\frac{\dot{h}}{h} - \frac{\dot{r}}{r}\right)$.

All computations use the *Large Strain Plasticity* option of the SYSWELD FE code developed by the FRAMASOFT+CSI Company (Leblond, 1989).

Figure 2 shows the relative porosity f/f_0 as a function of $\tilde{E}_{eq}$ for all three cycles in the case where $E = 200,000$ MPa and the first semi-cycle is tensile. One observes a typical Christmas tree shape, exactly as in the work of Gilles *et al.* (1991). This eliminates all possible remaining doubts about the validity of the conclusions of that work due to the fluctuations

	f_{max}/f_0 cycle 1	f_{max}/f_0 cycle 2	f_{max}/f_0 cycle 3	f_{min}/f_0 cycle 1	f_{min}/f_0 cycle 2	f_{min}/f_0 cycle 3
FE result	3.81	5.51	7.21	1.44	1.88	2.38
DLP model	5.19	6.99	7.81	0.77	0.60	0.51
"improved" model	5.19	5.51	5.58	1.33	1.38	1.41

Table 1: $\frac{f_{max}}{f_0}$, $\frac{f_{min}}{f_0}$ values; $E = 2 \times 10^5$ MPa, tensile first semi-cycle.

	f_{max}/f_0 cycle 1	f_{max}/f_0 cycle 2	f_{max}/f_0 cycle 3	f_{min}/f_0 cycle 1	f_{min}/f_0 cycle 2	f_{min}/f_0 cycle 3
FE result	3.74	4.79	5.36	1.38	1.59	1.75
DLP model	5.26	7.35	8.46	0.75	0.59	0.50
"improved" model	5.30	5.80	5.99	1.32	1.38	1.42

Table 2: $\frac{f_{max}}{f_0}$, $\frac{f_{min}}{f_0}$ values; $E = 2 \times 10^6$ MPa, tensile first semi-cycle.

of the triaxiality. In order to save space, the results corresponding to the other cases envisaged are shown only in the form of tables indicating the sole maximum and minimum values of f/f_0 for each cycle (*Tables 1 to 4*). Two essential observations can be made. First, both f_{max}/f_0 and f_{min}/f_0 increase with the number of cycles in all cases. Second, the effect is more prominent when E takes on its lower (normal) value; in fact, the influence of elasticity is much larger than one would expect *a priori*.

4. Comparison with the Predictions of the Gurson and DLP Models

In the case of a tensile first semi-cycle, and with a (reasonable) value of 1.35 for the Tvergaard (1981) parameter q_1, the predictions of the Gurson model are as follows (elasticity being disregarded for simplicity, the influence of the elastic moduli on the values found being quite weak): $f_{max}/f_0 = 21.4$, $f_{min}/f_0 = 1$ for all cycles. Comparison with the values given in *Tables 1 and 2* shows that these predictions are utterly erroneous[2]. However they are at least conservative, since the maximum porosities predicted, which govern the onset of coalescence of cavities and final failure, are greater than those found numerically. The situation is worse in the case of a compressive first

[2]One may be surprised by the unusually large extent of the gap; such an extent is to be related to the steepness of the hardening curve considered (which is realistic, however, for stainless steels).

	f_{max}/f_0 cycle 1	f_{max}/f_0 cycle 2	f_{max}/f_0 cycle 3	f_{min}/f_0 cycle 1	f_{min}/f_0 cycle 2	f_{min}/f_0 cycle 3
FE result	1.61	2.51	3.77	0.35	0.55	0.78
DLP model	1.68	2.04	2.24	0.09	0.09	0.08
"improved" model	1.16	1.23	1.26	0.21	0.24	0.26

Table 3: $\frac{f_{max}}{f_0}$, $\frac{f_{min}}{f_0}$ values; $E = 2 \times 10^5$ MPa, compressive first semi-cycle.

	f_{max}/f_0 cycle 1	f_{max}/f_0 cycle 2	f_{max}/f_0 cycle 3	f_{min}/f_0 cycle 1	f_{min}/f_0 cycle 2	f_{min}/f_0 cycle 3
FE result	1.48	1.92	2.31	0.35	0.49	0.60
DLP model	1.83	2.32	2.64	0.09	0.09	0.09
"improved" model	1.25	1.36	1.42	0.21	0.25	0.27

Table 4: $\frac{f_{max}}{f_0}$, $\frac{f_{min}}{f_0}$ values; $E = 2 \times 10^6$ MPa, compressive first semi-cycle.

semi-cycle. Indeed the values resulting from Gurson's model (for a rigid-plastic material), namely $f_{max}/f_0 = 1$, $f_{min}/f_0 = 0.01$ for all cycles, are not only erroneous but non-conservative, the value of f_{max} being smaller than those found numerically (see *Tables 3 and 4*).

It follows from the statement made in Section 2 that the basic explanation of the failure of Gurson's model is not to be found in the analytic form of the criterion or the evolution equation for the hardening parameter $\bar{\sigma}$: it lies in the fact that this parameter is *unique*. A new criterion preserving the essential features of that of Gurson but involving *two* hardening parameters was proposed by Perrin (1992) and Leblond *et al.* (1995):

$$\frac{\Sigma_{eq}^2}{\Sigma_1^2} + 2q_1 f \cosh\left(\frac{3}{2}\frac{\Sigma_m}{\Sigma_2}\right) - 1 - q_1^2 f^2 = 0 \ .$$

The new model (hereafter referred to as the DLP model) does not predict instantaneous stabilization of the porosity in cyclic experiments as that of Gurson, because the history-dependent ratio Σ_1/Σ_2 now appears in the evolution equation for f. However, the prime motivation for the introduction of Σ_1 and Σ_2 was not in fact that one, but the observation that for a hollow sphere prestrained in a purely hydrostatic way, the overall yield stresses Σ_{eq}^Y, Σ_m^Y under purely deviatoric and purely hydrostatic loadings obey the rigorous inequality

$$\frac{\Sigma_{eq}^Y}{1-f} < \frac{\Sigma_m^Y}{-(2/3)\ell n f} .$$

This is incompatible with Gurson's original model (with $q_1 = 1$ for a hollow sphere) which equates both sides of the inequality to $\bar{\sigma}$, but, provided that $\Sigma_1 < \Sigma_2$, compatible with the new variant proposed which equates the left- and right-hand sides to Σ_1 and Σ_2 respectively. The formulae proposed for these parameters were based on an approximate calculation of Σ_{eq}^Y and Σ_m^Y for a hollow sphere after some axisymmetric prestraining. That prestraining was taken as *monotonic and proportional*[3] in order to calculate Σ_1 and Σ_2 as functions of two parameters only (in addition to the initial porosity f_0), namely

$$E_{eq}^p \equiv \int_0^t D_{eq}^p(\tau)\mathrm{d}\tau \quad \text{and} \quad E_m^p \equiv \int_0^t |D_m^p(\tau)|\,\mathrm{d}\tau .$$

The DLP model was not originally intended to reproduce the behaviour under cyclic loadings, because of the assumption of monotonicity of the pre-straining introduced to simplify the calculation of Σ_1 and Σ_2. It is interesting, however, to see what its predictions are for the cyclic cases considered here. These predictions are shown in *Tables 1 to 4*. It can be seen that the effect of elasticity predicted is unfortunately opposite to that actually observed in the numerical simulations: the lower the value of E, the lower the increase of f_{max} per cycle. Also, the f_{min} values predicted are rather poor: they are too low and tend to decrease, instead of increase, with the number of cycles. However, the major feature of interest is the value of f_{max} (which governs coalescence), not that of f_{min}, in the sole case of a normal E, and it can be seen that the model gives satisfactory predictions in that respect.

5. Can One Improve on the DLP Model?

One seemingly obvious improvement of the DLP model consists in dropping the hypothesis of monotonicity and proportionality of the prestraining in the calculation of Σ_1 and Σ_2. It is then no longer possible to express these quantities in terms of only two parameters, E_{eq}^p and E_m^p. However, it remains possible to calculate them as functions of the whole sequence of successive values of $D_{eq}^p(\tau)$ and $D_m^p(\tau)$, $0 \leq \tau \leq t$. *Tables 1 to 4* show the results obtained in that way. It can be seen that the prediction of f_{min} values is

[3] Of course, this hypothesis was temporary and introduced only to simplify the calculation of Σ_1 and Σ_2; the model was later used for totally arbitrary mechanical histories, the values of those parameters being then assumed to be the same as if the volume element studied had been strained monotonically and proportionally up to the instant considered.

indeed better than before; in particular these values now increase with the number of cycles, as desired. However, the increase of f_{max} per cycle is still wrongly predicted to decrease when E decreases. Also, more importantly and quite deceivingly, it is now substantially underestimated.

With regard to the effect of elasticity, prior to envisaging any improvement of the DLP model, one must answer the following two questions: first, why does it predict (just as Gurson's original model) that the increase of f_{max} per cycle decreases when E decreases? Second, why does precisely the contrary occur in the numerical simulations?

The answer to the first question is simple: both models say that when E takes on lower and lower values, more and more of the amplitude (in total strain) of each semi-cycle is lost in the starting elastic phase involving no variation of the porosity, resulting in a smaller increase of the porosity per cycle. The answer to the second, less obvious question can be found by carefully examining the numerical $f/f_0 - \tilde{E}_{eq}$ curves for the normal value of E. One thus observes on *Figure 2* that at the beginning of all semi-cycles from the 3rd one, there is a short phase (noted EP) where the porosity varies more than before[4]. During that phase, the inner zones of the spherical shell are already plastic but the outer ones are still elastic. Therefore numerical experience, if not theory, reveals that *partial plasticity* of that shell favours variations of f. Since the lower the value of E, the longer the phase of partial plasticity, this explains why the increase of f per cycle is maximum when E is low. Unfortunately, all existing models make the simplifying hypothesis that each porous volume element is either entirely plastic or entirely elastic; therefore they cannot incorporate this effect. A major obstacle is to be encountered in the derivation, based on homogenization, of some model including it. Indeed one is faced with the difficult task of extending Gurson's classical analysis of a hollow, *rigid*-plastic sphere subjected to some axisymmetric loading so as to include the possible existence of an external *elastic* region surrounding the plastic core around the void.

It should be noted that a similar surprisingly strong effect of elasticity was encountered by Tvergaard *et al.* (1992) in the slightly different but related context of cavitation instabilities in plastic solids. The extent of that effect was explained by the fact that large amounts of elastic energy can be stored in the outer elastic region and subsequently transferred to the inner plastic core, thereby promoting cavitation. Although this explanation may seem to be quite different from that expounded here at first sight, there is in fact a

[4] In the 4th and 6th semi-cycles, this occurs after an even shorter initial phase, noted E, where it varies *less* than before; this is because the material is undergoing purely elastic unloading then.

close connection between the two. Indeed in both of them, emphasis is not placed on elasticity effects in the elastoplastic region, but on the existence of a *large purely elastic zone around it.*

In conclusion, the use of the DLP model seems the only possible choice in the case of cyclic loadings, in view of the breakdown of Gurson's original proposal and the difficulties encountered in the search for more refined models. That choice is not absurd, in spite of the deficiencies of some of the predictions of the DLP model, since the parameter of major interest, namely the maximum porosity during each cycle, is predicted with a reasonably good accuracy.

Acknowledgement

The authors wish to express their sincere thanks to Dr. Ph. Gilles and the FRAMATOME Company for both stimulating discussions and financial support.

References

Gilles, Ph., Jullien, B. and Mottet, G. (1992) Analysis of cyclic effects on ductile tearing strength by a local approach of fracture, *Advances in Fracture/Damage Models for the Analysis of Engineering Problems*, publication AMD-Vol. 137 of the ASME, 269-284.

Gologanu, M., Leblond, J.B. and Devaux, J. (1994) Numerical and theoretical study of coalescence of cavities in periodically voided solids, *Computational Material Modeling*, publication AD-Vol. 42/PVP-Vol. 294 of the ASME, 223-244.

Gurson, A.L. (1977) Continuum theory of ductile rupture by void nucleation and growth - Part I: Yield criteria and flow rules for porous ductile media, *ASME Journal of Engineering Materials and Technology* **99**, 2-15.

Kobayashi, H., Kusumoto, T. and Nakazawa, H. (1991) Cyclic $J-R$ curve and upper limit characteristic of fatigue crack growth in 2-1/2 Cr-Mo steel, *11th International Conference on Structural Mechanics in Reactor Technology*, Tokyo, Japan, paper # G27/1.

Koplik, J. and Needleman, A. (1988) Void growth and coalescence in porous plastic solids, *International Journal of Solids and Structures* **24**, 835-853.

Leblond, J.B. (1989) Grandes transformations - Contraintes planes - Déformations planes généralisées, *FRAMASOFT+CSI Internal Report # CSS.L. NT.89/4014.*

Leblond, J.B., Perrin, G. and Devaux, J. (1995) An improved Gurson-type model for hardenable ductile metals, *European Journal of Mechanics A/Solids* **14**, 499-527.

Perrin, G. (1992) *Contribution à l'étude théorique et numérique de la rupture ductile des métaux*, Ph.D. thesis, Ecole Polytechnique, Paris, France.

Schmidt, R.A., Wilkowski, G.M. and Mayfield, M.E. (1991) The International Piping Research Group (IPIRG) program. An overview, *11th International Conference on Structural Mechanics in Reactor Technology*, Tokyo, Japan, paper # G23/1.

Tvergaard, V. (1981) Influence of voids on shear band instabilities under plane strain conditions, *International Journal of Fracture* **17**, 389-407.

Tvergaard, V., Huang, Y. and Hutchinson, J.W. (1992) Cavitation instabilities in a power hardening elastic-plastic solid, *European Journal of Mechanics A/Solids* **11**, 215-231.

PHASE TRANSITIONS IN FRACTURE AND CRYSTALLIZATION PROCESSES

L.R.BOTVINA
Professor, A.A.Baikov Institute of Metallurgy, Russian Academy of Science, 117911, Leninskii prospekt 49, Moscow, Russia

1. Introduction

Comparative analysis of the basic properties of fracture processes and crystallization, in particular their stepwise character, shows that fracture may be considered like crystallization, from the standpoint of phase transition theory [1].

In Fig.1 we present schematically typical fracture kinetic curves $K = f(\tau)$ obtained under a constant value of the parameter P defining the test conditions (temperature, loading rate, etc.; K is a certain strain and/or fracture property; τ is the time). It is clear that these curves remain similar to each other until a certain critical value of the parameter P is reached. For $P = P_C$, the second steady-state stage disappears on the fracture kinetic curves and they become closer to straight lines.

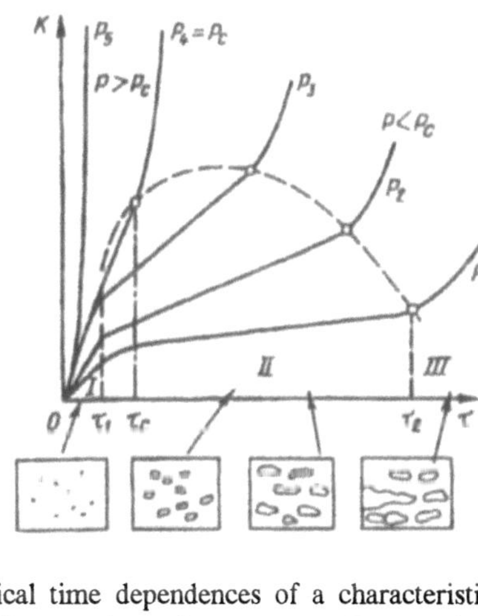

Figure 1. Typical time dependences of a characteristics (K) of deformation and/or fracture for different values of a parameter P governing the loading conditions : *I, II, III* — stages of damage accumulation.

Many dependencies (both kinetic and parametric) of the deformation and fracture characteristics are similar to the curves shown in Fig.1. These include creep curves (for different stresses or temperatures), deformation curves (for

J. R. Willis (ed.), IUTAM Symposium on Nonlinear Analysis of Fracture, 311–320.

different loading rates or temperatures), kinetic curves of fatigue crack growth, wear curves, and the dependencies of internal friction in material, and swelling after an irradiation dose, etc.

The initial stage (*I*) on the fracture kinetic curves may be absent. This is most often due to the difficulty in the recording of this stage because it requires finer measurements than registration of fracture processes at stages *II* and *III*. However, the other trends are preserved. The size of stage *II* decreases as P increases over a rather broad range. Close to P_C, the dependence of the lifetime on P increases, while the scatter of experimental points for lifetime also increases. Similar kinetic curves for fatigue failure are presented in Fig.2a.

The comparison of the fracture kinetic curves presented in Fig.1 with the isotherms for the liquid-gas phase transition plotted in pressure (p) vs. density (ρ) coordinates [2], shows that the fracture curves have a shape similar to the isotherms at different temperatures. On each of the fracture curves, as on the liquid-gas isotherms, we can isolate an initial section (0 - τ_1) within which the first derivative of the characteristic K decreases; the second steady-state stage (τ_2 - τ_1), for which $dK/d\tau$ is almost constant and the third stage where the development of the process is unstable and the derivative $dK/d\tau$ increases.

The effect of the parameter P is characterized by the fact that with an increase in P, the size of the second steady-state stage of the fracture process decreases and the first derivative of the characteristic K increases. For the critical values of P_C and τ_C, the second stage on the curves disappears ($\tau_2 - \tau_1 = 0$) and the third stage follows the first stage. In this case, the time dependence of the characteristic K becomes close to linear, as does the isotherm $p = f(\rho)$ when the critical temperature is achieved under the liquid-gas transition conditions.

By analogy to the liquid-gas phase transition, the closeness to the critical point of the fracture process (in this case, to the point with coordinates P_C, τ_C) is defined by the difference τ_2 - τ_1.

This difference may serve as an order parameter for the fracture process of a solid. In this case, we can probably assume that the process itself undergoes the phase transition at the point with coordinates P_C, τ_C. What can we say about phase transition in this case? The fracture mechanisms are different under different conditions. However, all of them are connected with nucleation and gradual accumulation of different types of damage. At the initial stage of fracture (region *I* in Fig.1), pores are only nucleated and the damage process is not yet developed, a dislocation structure is formed which causes strain-hardening of the material. Accordingly, we can consider that at stage *I* the material does not contain pores or contains significantly less of them in comparison with the number of defects at the other stages. At stage *II*, the accumulation of pores is going on and the material can be considered as a "two-phase" one, i.e.

consisting of damaged and undamaged volumes. At stage *III* the rapid growth of pores, the formation of macrocracks and fracture are going on.

Thus, the dashed curve in Fig.1, similar to the coexistence curve for the liquid-gas phase transition, separates the "two-phase" region in which the specimen consists of pores or microcracks and sections of undamaged material.

Such a representation of the two-phase structure of a solid under loads is close to Frenkel's idea [3] which considers a typical crystal of a monatomic substance as a binary material consisting of proper atoms and vacancies which are "atoms" of the second substance (voids). If the material is a two-phase one at the steady-state stage of fracture, we can assume that upon achievement of critical conditions, a "phase transition" occurs with formation of a new phase: a macrocrack.

In the crystallization process, the role of pores is played by particles of the solid phase already crystallized from the melt; therefore the steady-state stage on the crystallization curves (plotted in viscosity of the melt vs. fraction of solid phase coordinates) is connected with development of a two-phase structure, and the size of this stage (determined by the volume fraction of the solid phase) is the order parameter of the crystallization process.

Let us estimate the order parameters and the critical exponents for fracture processes under fatigue, creep, and impact loading conditions, and also for the crystallization.

2. Fatigue

In Fig.2a, we present the dependencies of the fatigue crack length in notched specimens of aluminium alloy on the number of cycles.

As we approach the critical stress, the sensitivity of the alloy to a stress variation abruptly increases (Fig.2b). This is accompanied with the transition of fatigue fracture mode from the normal rupture under low stress amplitudes to slantwise shear mode under high stresses. When the specimens without notches are tested, at the critical stress amplitude the number of origins on the fracture surfaces from which fatigue cracks develop increases. In contrast to the fracture surfaces at low stresses, on which only a single origin is observed, the fracture surface becomes multiorigin. One more important feature is observed: the fatigue curve and the kinetic diagram for fatigue failure, plotted in the crack rate vs. the range of the stress intensity factor coordinates, have discontinuities corresponding to the stress at which the fracture mechanism changes and a high sensitivity of the lifetime to the stress appears [1].

In estimating the order parameter (N) and its normalized value $(N-N_C)/N_C$, the following power-law relation has been established (Fig.2c):

$$|(N-N_C)/N_C| = A\,|[(\sigma-\sigma_C)/\sigma_C]|^{\beta}$$

where σ and σ_C are the instantaneous and critical stress amplitudes respectively.

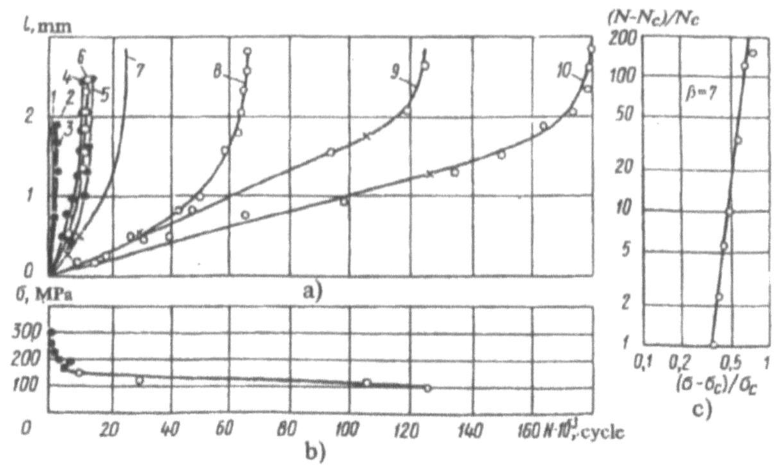

Figure 2. Dependencies of the fatigue crack length (l) on the number of loading cycles (N) for notch specimens of B95T1 aluminium alloy, tested at loading frequency of 120 Hz, cycle asymmetry R=1, and stress amplitudes (MPa):
300 (*1*), 260(*2*), 220(*3*), 200 (*4*), 190 (*5*), 180 (*6*), 160 (*7*), 140 (*8*), 120 (*9*) and 100 (*10*) (a), the relation between the order parameter and the stress (b), and between the normalized values of the order parameter and stress (c): × — points of the curve from whose coordinates we plotted the dependencies in Fig.2b and 2c.

The critical exponent for the given alloy under the investigated loading conditions is β = 7. An increase in the frequency of cyclic loading leads to decreasing the critical exponent.

3. Creep

In Fig.3a, we present the creep curves for steel, plotted for different stresses [4].

An estimate of the order parameter from these curves also leads to a power-law equation (Fig. 3b):

$$|(\tau - \tau_C)/\tau_C| = B|[(\sigma - \sigma_C/\sigma_C])^{\gamma}$$

In this case, the critical exponent γ varies from 1.9 to 4 as a function of the testing temperature : the lower the temperature, the lower the critical exponent.

For the creep close to the critical point, as for the fatigue, the sensitivity of the deformation values to the stress variation increases and the fracture mechanism changes.

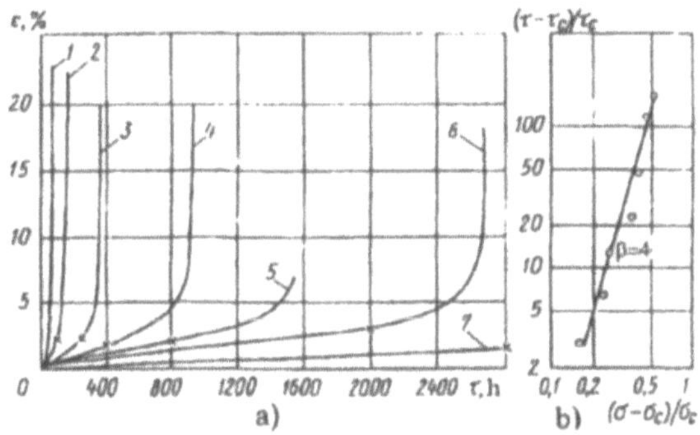

Figure 3. Creep curves for the steel ЭП 44 [4] at 585°C and different stresses (a) and the relation between the normalized values of the order parameter and the stress (b) (MPa): 240 (*1*), 200 (*2*), 180 (*3*), 160 (*4*), 150 (*5*), 140 (*6*) and 130 (*7*).

4. Impact loading

In Fig.4, we present the temperature dependencies of the impact toughness A of carbon steel with different carbon contents [5], and also the dependence of the normalized order parameter $(A\text{-}A_C)/A_C$ on the normalized carbon concentration. The critical exponent in impact loading is somewhat lower than that for fatigue and creep.

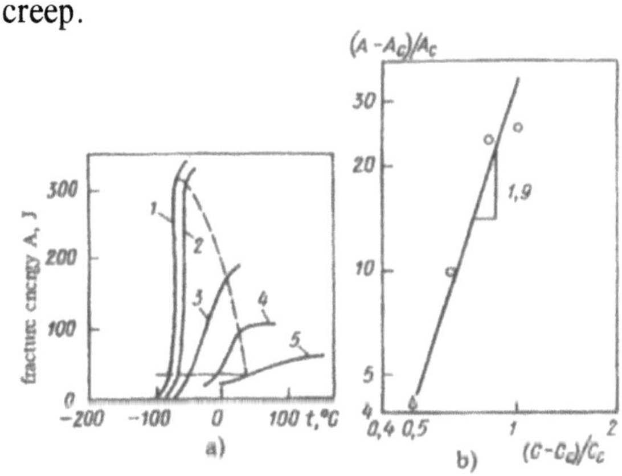

Figure 4. Temperature dependencies of the energy of impact fracture (A) for different carbon steels [6] and relation between normalized values of the order parameter and the carbon concentration (C):
0.01%C (*1*), 0.11%C (*2*), 0.22%C (*3*), 0.31%C (*4*), 0.63 %C (*5*).

Furthermore, the impact toughness curves, contrary to the fatigue and creep curves, are not time-dependent; the two-phase region bounded by the dashed coexistence curves is connected with different fractions of brittle and ductile regions on the fracture surface. With an increase in the carbon content of the steel, the extent of the two-phase region and the slope of the impact toughness curve decreases, which leads to convergence of the horizontal sections corresponding to ductile and brittle fractures (Fig. 4). This is probably due to a transition from a localized ductile-brittle fracture (in which we can visually observe a region of ductile fracture located at the tip of the notch and a region of a "brittle square") to a delocalized or diffuse fracture [6], on the fracture surface of which we observe alternating brittle fracture facets and ductile dimples. Formation of a diffuse fracture may be the result of an increase in the strength of material [5].

Thus the phase transition in the given case is caused by the appearance of a macroscopically brittle but microscopically delocalized diffuse fracture. Accordingly, we can consider that the process of transition from the ductile state to the brittle state had been completed at the macroscopic level, but continues at the level of 2-3 orders of magnitude higher (if the level of the process is characterized by the length of the brittle crack). The above-mentioned power-law relations obtained for fracture processes probably stem from power-law strength characteristics, but they may prove to be useful in estimating the lifetime of a material or structure in the stage preceding generation of the critical damage or the critical macrocrack length.

5. Crystallization

In crystallization, the role of pores is played by particles of the solid phase. Therefore, as the order parameter, in this case we will use the length of the steady-state stage of crystallization, determined by the volume fraction of the solid phase.

Let us consider crystallization of semi-solid materials, studied in detail in [7]. Semi-solid materials include those containing 50-60% already crystallized phase and are used to obtain composite materials or for other final shaping processes. The major factors determining the mechanism of crystallization in such materials are the cooling rate and the rate of shear of the liquid metal. Shearing the liquid metal during the early stages of solidification promotes a breakdown of the dendritic structure, and as a result the viscosity decreases and a spheroidal structure forms. Thus, the phase transition at the critical point is connected with transition (under decreasing the shear rate) from a spheroidal structure to a dendritic one.

As we see from Fig. 5b, the crystallization process of the alloy Al-4.5%Cu--1.5%Mg is characterized by a significantly smaller critical exponent compared with the exponents characterizing the fracture processes but the relation for the order parameter is similar to the preceding relations:

$$|(f_S - f_S^c)/f_S^c| = C\,|[(\dot{\gamma} - \dot{\gamma}_c)/\dot{\gamma}_c]|^{\delta}$$

where f_S is the volume fraction of the solid phase; $\dot{\gamma}$ is the shear rate of the melt. The phase transition during crystallization in which the first derivatives of some physical quantities change in a jumpwise manner and heat is evolved is traditionally classified as the first-order phase transition [2]. As far as the phase transitions during fracture are considered, they have features of both first-order and second-order types of phase transitions. In fact, at the critical point for cyclic loading, we observe an abrupt increase in the temperature of the specimen surface [1], the crack growth rate, and the acceleration of the crack growth, i.e. the first and the second derivatives of the crack length. In other cases of failure, no increase in the temperature of the specimen surface is observed at the critical stress.

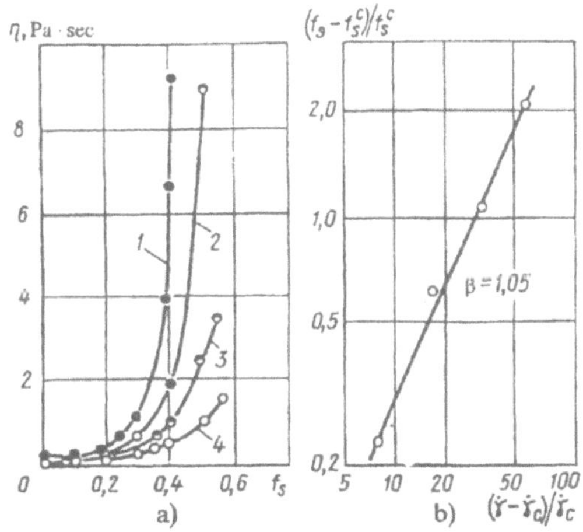

Figure 5. The viscosity of the alloy Al—4.5%Cu—1.5%Mg for constant cooling rate (0.03 C/sec) and different shear rates ($\dot{\gamma}$) of the melt [7] (a) and the relation between the normalized values of the order parameter and the shear rate of the melt (b): $\dot{\gamma}$ = 90 1/sec (*1*), 180 1/sec (*2*), 330 1/sec (*3*), 560 1/sec (*4*).

However, the approach used and the proposed estimate of the critical exponents for fracture and crystallization make it possible to compare the characteristics of phase transitions in different media.

The estimations of the order parameters and the critical exponents characterizing transition from laminar to turbulent flow in fluids and solids confirm the above opportunity [8].

6. Plastic Deformation of Monocrystals

Plastic deformation of monocrystals with face centered cubic structure occurs in several stages.

During the initial stage, slip develops along planes of one slip system because only this system operates. Such a plastic flow has been called by Cottrell "laminar plastic flow". During this stage the stress depends linearly on the strain, and the material is not strengthened.

With increasing plastic deformation, two or more slip systems start to operate, therefore a slip on the intersecting slip system develops. This is accompanied with the interaction of dislocations of two slip systems, a cross-slip, a turn of lattice and local microvolumes of the crystal. Such a plastic flow is "turbulent" according to Cottrell's definition.

At the next stage these processes result in a developed turbulent plastic flow and strengthening of the crystal. The stress-strain dependence becomes a power law one.

Thus during the intermediate stage, the crystal structure may be considered as a two-phase one, i.e. consisting of the laminar and turbulent plastic flow regions. Then the strain value, corresponding to this stage may serve as the order parameter of the plastic deformation process.

With increase in the test temperature this intermediate stage disappears, the order parameter becomes equal to zero and the turbulent plastic flow follows the laminar one.

As it can be seen in Fig.6, in this case the order parameter is also a power law function of the test temperature.

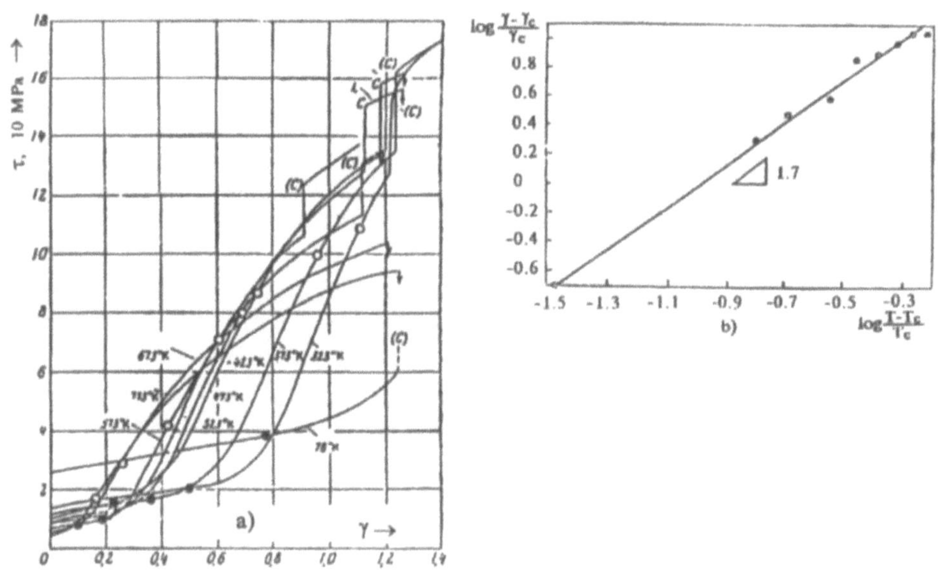

Figure 6. Plastic deformation curves for the Ni-50%Co monocrystals at shear rate 0.003 1/s [9] and the relation between normalized values of the order parameter and temperature (b): (●) - beginning transition stage, (○) - end of transition stage , ⌐↓- neck formation , (*C*)- turning crystal axis

7. Flow of Fluid in Smooth Pipes

By analogy with the laminar-turbulent transition in solids, it is of interest to consider similar transition in fluids. In Fig. 7, dependencies of the pressure variation rate on the fluid flow rate in pipes of different diameters are presented, plotted by Reynolds [10] in physical and logarithmic coordinates.

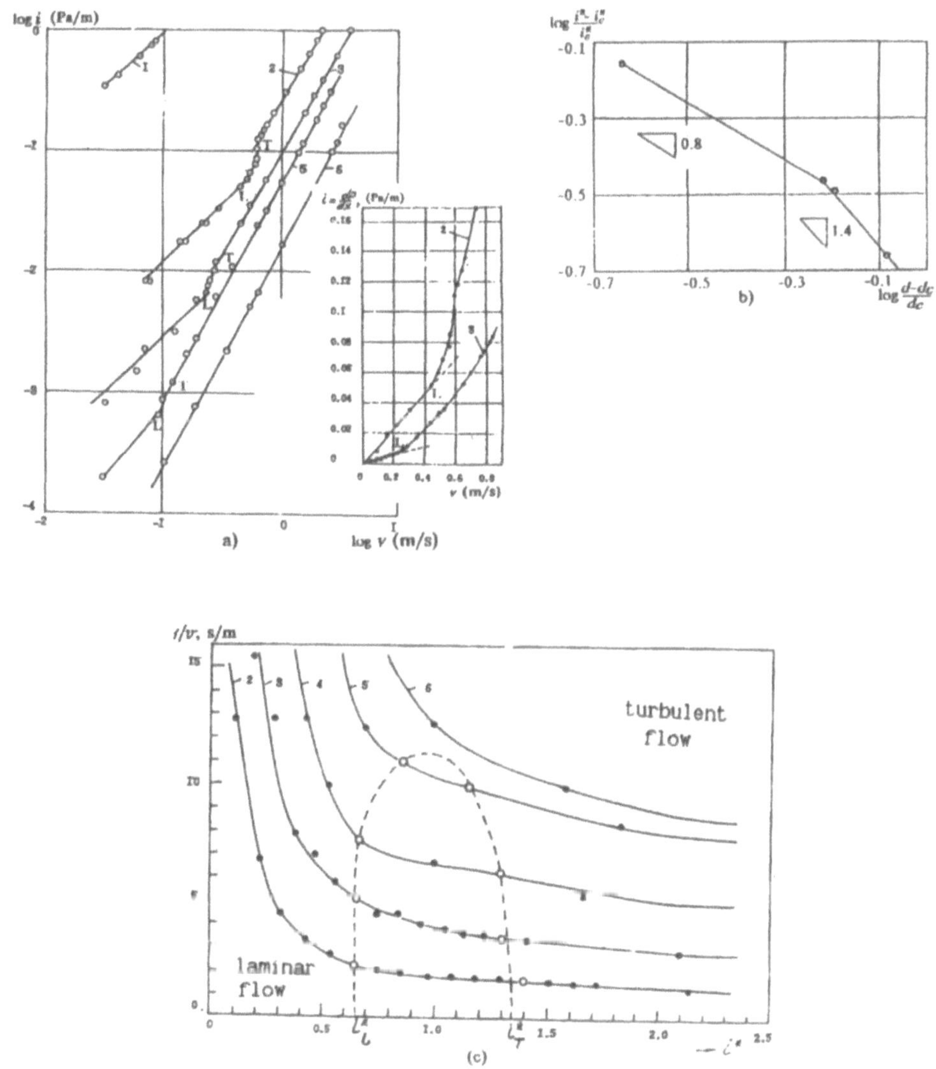

Figure 7. Curves of Reynolds for the fluid flow in pipes of different diameteres (a) [10] , the relation between the normalized values of the order parameters and pipes diameteres (b), and the coexistence curve for the laminar-turbulent phase transition (c):
P - pressure, i = dP/dx, $i^*= 2i/(i_T - i_L)$, v- flow rate, L - end of laminar flow, T - beginning turbulent flow ; d -pipes diameters: 0.65mm (*1*), 6.15 mm (*2*), 12.7 mm (*3*) , 14 mm (*4*) 27 mm (*5*), 81,9 mm (*6*).

The curves in physical coordinates, similar to the deformation curves of monocrystals, consist of a linear region of laminar flow, a laminar-turbulent transition region and a turbulent one.

With increase in the pipe diameter the laminar and laminar- turbulent regions decrease. In Fig. 7.c the two-phase transition region bounded by the dashed coexistence curve is shown. Here i_T^* and i_L^* are the relative values of the pressure variation rate corresponding to the end and the initiation of the transition region. If we use the difference $i_T^* - i_L^*$ as the order parameter, we obtain the curve in Fig. 7 consisting of two linear parts with two critical exponents. In case of such choice of the order parameter, the critical exponents are negative and their absolute values decrease near the critical point.

The noted analogy of processes occuring in different media is probably connected with similar behaviour of the change in entropy characterizing the indicated processes.

This work was carried out with the financial support of the Russian Foundation for Basic Research, project 94-01-00475-a.

REFERENCES

1. Botvina, L.R.: Common characteristics of fracture and crystallization processes, *Metal Science and Heat Treatment*, **36**, (1994), 393—403.
2. Stanley, H.E.: *Introduction to Phase Transitions and Critical Phenomena* [Russian Translation], Mir, Moscow, 1973.
3. Frenkel, J.: *Statistical Physics* [in Russian], Izd. Akad. Nauk SSSR, Moscow-Leningrad, 1948.
4. Lepin, G.F.: *Creep of Metals and Heat Resistance Criteria* [in Russian], Metallurgiya, Moscow, 1976.
5. Gulyaev, A.P.: Metal Science [in Russian], *Metallurgiya*, Moscow, 1986.
6. Goritskii, V.M.: Ductile dimple failure and the width of the ductile-brittle transition in structural steels with a bcc lattice", *Problems of Strength*, **3**, (1985), 78—85.
7. Flemings, M.C. Behavior of metal alloys in the semisolid state, *Metallurgical Transactions*, **22B**, (1993), 269—293.
8. Botvina, L.R. and Zaigraev, M.M.: Phase transitions in processes of plastic deformation and flow of fluid, *Reports of Russian Acad.Sci.*, (in press).
9. Berner, R. and Kronmüller, H.: *Plastische verformung von Einkristallen*, Springer, Berlin, 1965.
10. Reynolds, O. : An experimental investigation of the circumstancies which determine whether the motion of water shall be direct or sinuous, and of the law of resistance in parallel channels, *Philosophical Transactions*, Royal Soc., London, **174** (1883).

NONLINEAR CRACK-BRIDGING PROBLEMS

L.R. FRANCIS ROSE
Aeronautical and Maritime Research Laboratory
Melbourne, Australia

Abstract. The influence of a progressive, time-dependent, decay in crack-bridging efficiency is investigated, with particular attention to two configurations involving a fully bridged semi-infinite crack. First, the evolution of the stress intensity factor K with time is considered, assuming that no crack growth occurs. It is shown that, at any given time t, K can be expressed in terms of the complementary energy for a stress/stretch curve applicable for that instant t. However, that curve is not prescribed *a priori*: it must be derived by first solving for the spring stretch, and then calculating the associated transmitted stress at every location along the bridged crack. The resulting stress/stretch curve at any instant t is generally non-linear, even when the bridging is modelled by linear viscoelastic springs, for example; this nonlinearity results from the load-history dependence of the degradation. Similar difficulties are also encountered for the second case considered, *viz.* steady-state crack growth. A physically plausible approximation, leading to a useful, conservative, estimate for K is proposed for the first case; a comparable estimate for the second case is not yet apparent.

1. Introduction

Crack-bridging models have attracted considerable attention recently because they arise in a variety of contexts, and they provide a computationally efficient approach for characterising the overall material or structural response to applied loads. The essential features of these models have now been well characterised for time-independent bridging laws (Rose, 1987). The present work considers the much more difficult, but also more widely applicable case, where the bridging law is time-dependent, and generally non-linear, reflecting a progressive degradation, or a relaxation, of the bridging mechanism. Efficient numerical methods have been developed for solving the non-linear integral equation governing these problems (Cox and Rose, 1994), but the emphasis is placed here on characterising the possible regimes of behaviour and identifying

J. R. Willis (ed.), IUTAM Symposium on Nonlinear Analysis of Fracture, 321–331.

the relevant non-dimensional combinations of parameters, rather than on detailed numerical calculations for specific cases.

A key result for the case of time-independent bridging is the existence of a finite upper bound K_∞ for the stress intensity factor (under a fixed uniform applied stress and assuming a fully bridged crack). This upper bound is approached asymptotically (from below) with increasing crack length. Once this asymptotic character of a bridged-crack response is recognised, an explicit estimate can be derived for the limiting value K_∞ by energy-balance arguments applied to an incremental extension of a fully-bridged semi-infinite crack. This approach appears to have been first developed in the context of crack repair by bonded reinforcements (Rose, 1982), and independently in the context of fibre-bridging of matrix cracking in brittle matrix composites (Marshall *et al.*, 1985; Budiansky *et al.*, 1986; McCartney, 1987). From a practical viewpoint, this upper bound K_∞ can be used as a convenient, conservative estimate for K for repair design, and indeed this approach has now been applied and thoroughly validated through extensive testing and comparison with finite-element analysis in relation to a safety-critical repair to primary structure for a military aircraft (Rose *et al.*, 1995). An important issue in extending the scope of this analysis and design procedure is to assess the influence of a progressive decay in crack-bridging efficiency. This paper presents some preliminary results obtained to date on this issue, and formulates some definite problems requiring further detailed analysis.

2. Problem Statement

Consider a standard centre-cracked panel, with a fully bridged crack of length 2a, subjected to a uniform remote stress

$$\sigma_{yy} = \sigma_a \;, \quad x^2 + y^2 \to \infty \;. \tag{2.1}$$

The presence of a crack bridging mechanism (*e.g.* bonded reinforcements or fibre bridging) results in a non-zero load being transmitted across the crack. The bridging mechanism is characterised by a traction law which needs to be specified. In most previous studies, this law is prescribed as a functional relation between the stress transmitted and the local crack-opening displacement (or spring stretch) δ, for example of the form

$$\sigma_{yy}\,(a < x < -a\,,\, y = 0\pm) = E'\,\phi\,(\delta), \tag{2.2}$$

$$\delta\,(x) \equiv u_y\,(x, y \to 0+) - u_y\,(x, y \to 0-) \tag{2.3a}$$

$$\equiv 2\,u_y\,(x, y \to 0+) \tag{2.3b}$$

where E' denotes as usual the reduced Young's modulus (Rose, 1987), and u_y the y-component of the elastic displacement, whereas ϕ is a specified, possibly non-linear, function.

In the present work, it is envisaged that the crack bridging efficiency may suffer a progressive decay. The appropriate traction law will depend on the precise mechanism for this decay, which will vary from one context to another. This opens up a wide range of possible responses which can only be fully characterised through extensive numerical computations on a case-by-case basis. Nevertheless, it is possible to characterise some important general features of the response, albeit qualitatively, as will be shown below. Furthermore, a relatively comprehensive characterisation can be attempted for the case where the bridging is modelled by linear viscoelastic springs. This case retains, in the simplest mathematical form, the essential features of a dependence on the previous history of deformation (or of spring stretch) and of a relaxation time for the decay of crack-bridging tractions. The results obtained for that case can therefore be expected to provide useful insights, applicable to the more general case of non-linear time-degrading springs, at least for cases which do not involve a distinct characteristic stress for the onset of degradation (a damage threshold stress). A specific example of the latter type has been studied by Cox and Rose (1994), where the decay is attributable to cyclic fatigue. As noted in that reference, the results obtained for cyclic degradation can be readily re-interpreted as results applicable for monotonic (static) loading, and *vice versa*.

The problem being addressed is that of determining the crack opening displacement, or spring stretch, $\delta(x;t)$, and the associated stress intensity factor $K(t)$, for a specified time-dependent traction law, and a specified load history. For simplicity, it will be assumed that the load is applied at time $t = 0$, as a step-function in time,

$$\sigma_{yy}(x, y; t) = \sigma_a H(t) \quad , \quad x^2 + y^2 \to \infty \ , \tag{2.4}$$

and that the response is quasi-static, so that the stress state (2.4) would prevail at all points x, y in the absence of the centre crack. Thus, the load history introduces no characteristic time. Furthermore, the material surrounding the crack is assumed to be linearly elastic, with crack growth occurring when $K(t)$ exceeds a threshold value K_{th}, according to a prescribed growth-law of the usual form,

$$da/dt = f\,(K(t) - K_{th}). \tag{2.5}$$

Provided that no crack growth occurs, *ie.* for $K(t) < K_{th}$, a time dependence enters solely through the crack-bridging traction law, which has yet to be specified. This traction law can include a failure criterion. However, the possibility of failure for the crack bridging reinforcement will not be considered in the present work, to restrict the analysis to a more manageable scope. Thus, on load application at $t = 0$, two possibilities arise, depending on whether the instantaneous stress intensity factor $K(t \to 0+)$ exceeds or is less than the threshold K_{th}. In the latter case, the problem is to determine the evolution of K(t) with time for a crack of fixed length, and the incubation time t_i required for K(t) to reach the threshold value K_{th}. In both cases, once the crack begins to extend, it is assumed that freshly created bridging springs appear between the crack faces; the problem is again to determine K(t), but now keeping track of the evolving crack length a(t) as well. An important aspect of the latter problem is to establish the conditions or requirements for crack growth to evolve to a steady state (or a suitably defined quasi-steady state), and to fully characterise this eventual steady state growth, if it can occur.

These problems can be formulated as singular integral equations in a variety of distinct (but ultimately equivalent) forms (McCartney, 1987; Nemat-Nasser and Hori, 1987; Rose, 1987; Budiansky *et al.*, 1988; Willis and Nemat-Nasser, 1990; Cox and Marshall, 1991; Cox and Rose, 1994). Given the insights already mentioned above for time-independent bridging, it can be expected that the most significant results from a practical viewpoint can be derived by further restricting attention to problems involving a semi-infinite crack, rather than a finite-length centre crack. The results obtained to date are summarised in the following sections.

3. Fully-bridged semi-infinite crack

Consider the response of a fully bridged semi-infinite crack, along $-\infty < x < 0$, $y = 0$, to the step-function loading (2.4), assuming first that K remains below K_{th}, so that no crack growth occurs. The primary objective here is to quantify the evolution of K with time.

The most convenient approach is to use the Eshelby-Rice J-integral taken around the closed contour shown in Fig. 1, as proposed by Marshall and Cox (1988) for time-independent bridging. Exploiting the path-independence of the J-integral to deform the contour to a limiting configuration for which the integral can be readily evaluated, one obtains

$$K_{tip}^2(t)/E' = J_{tip}(t), \tag{3.1a}$$

$$= \sigma_a\, \delta_\infty(t) - \int_0^{-\infty} \sigma_{yy}(x, y = 0; t)\, \partial\delta/\partial x\, (x\, ; t)\, dx, \tag{3.1b}$$

where

$$\delta_\infty (t) \equiv \delta (x \to -\infty, t) \; . \tag{3.1c}$$

The main complication, relative to the time-independent case, is that the integral in (3.1b) is now much more difficult to evaluate, because at any given time t, the relation between the transmitted stress σ_{yy} (x, y = 0; t) at a given station x along the crack and the spring stretch δ (x; t) at that location, is not prescribed *a priori*; that relation will generally depend on the prior history of spring stretch at x, which in turn can only be determined by solving an integral equation for δ (x; t) at every previous instant of time in the interval (0, t).

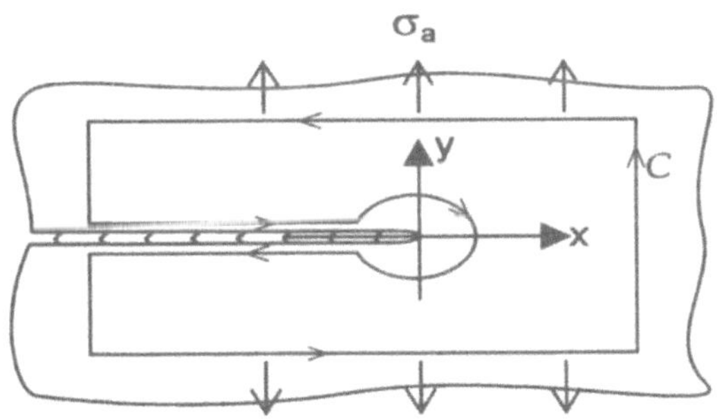

Figure 1. A fully bridged semi-infinite crack subjected to a uniform remote stress σ_a, showing the choice of coordinate axes, and the J-integral contour used for evaluating *K*(*t*).

To proceed further, the constitutive equation prescribing the time-dependent response of the bridging reinforcement must now be specified. It is convenient at this stage to restrict attention to bridging by linear viscoelastic springs, characterised by the following relation between the transmitted stress and the spring stretch,

$$\sigma_{yy} (x, y = 0; t) = \frac{E'}{2} \int_0^t k (t - \tau) \frac{\partial \delta}{\partial t} (x; \tau) \, d\tau \; , \tag{3.2a}$$

$$= \frac{1}{2} E' k * \delta \, , \tag{3.2b}$$

where the asterisk in (3.2b) denotes a Stieltjes convolution (Christensen, 1982), and the time derivative in (3.2a) is to be interpreted as a distributional derivative if the load history involves step discontinuities. Using a circumflex to denote the Laplace transform of a variable with respect to time t, this relation can equivalently be written as follows, for the step-function loading (2.4),

$$\hat{\sigma}_{yy}\,(x, y=0\,;\,s) = \frac{1}{2}\,E'\,s\,\hat{k}(s)\,\hat{\delta}\,(x;\,s)\,, \tag{3.3}$$

where s denotes the transform parameter. A factor ½ is used in (3.2) and (3.3) to retain a close correspondence with the notation previously used for linear elastic springs (Rose, 1987).

Following the procedure outlined by Rose (1987), an integral equation can be derived for the spring stretch at any given time t, in terms of the dislocation density $D\,(x;\,t)$ at time t, where

$$\delta\,(x\,;\,t) = \int_x^0 D\,(\xi\,;\,t)\,d\xi\,, \qquad -\infty < x < 0\,. \tag{3.4}$$

Taking the Laplace transform over time of that integral equation leads to

$$\frac{E'}{4\pi}\int_{-\infty}^{0}\frac{\hat{D}\,(\xi\,;\,s)}{x-\xi}\,d\xi = -\frac{\sigma_a}{s} + \frac{1}{2}E'\,s\,\hat{k}(s)\int_x^0 \hat{D}\,(\xi\,;\,s)\,d\xi \tag{3.5}$$

where the relevant solution is required to have an inverse square-root singularity for $x \to 0$-. This last equation is identical to the one for time-independent linearly elastic springs, provided that $s\hat{k}(s)$ is interpreted as the spring constant and σ_a/s as the applied stress, in keeping with the well-known elastic-viscoelastic correspondence principle (Christensen, 1982). Thus, an explicit analytic solution of (3.5) for $\hat{D}\,(x\,;\,s)$ is available (Rose, 1987; Budiansky *et al.* 1988), albeit in a rather complicated form.

Unfortunately, it would seem to be necessary to obtain the inverse transform of this expression for $\hat{D}(x\,;\,s)$ before the integral in (3.1b) can be evaluated, because the integrand involves a product (rather than a convolution) of functions of time. Thus, it would seem to be quite challenging to characterise accurately the evolution of K(t) even for the relatively simple case of linear viscoelastic springs. However, the general features of the expected response can be illustrated schematically, as shown in Fig. 2.

Both the instantaneous response and the long-term response reduce to that for linear elastic springs, with the spring constant k(0) or k(∞) holding uniformly for all locations x along the crack. For intermediate values of t, however, one must first determine $\delta\,(x\,;\,t)$, and hence $\sigma_{yy}\,(x,\ y\,=\,0;\ t)$ from (3.2), before one can construct the (generally non-linear) σ/δ curve illustrated in

Fig. 2, and hence evaluate (3.1b), which corresponds in effect to the complementary energy for that instantaneous σ/δ relation prevailing at time t. This complication is a result of the history dependence of deformation for viscoelastic behaviour; similar (or worse) complications can therefore be expected for more realistic prescriptions for the degradation in bridging efficiency.

Nevertheless, Fig. 2 suggests a useful approximation: the actual non-linear σ/δ curve can be approximated by a linear relation shown as a dashed line in Fig. 2. This corresponds physically to assuming that the bridging can be modelled by linear springs with a uniform value of spring stiffness k_∞ (t) prevailing for all stations x along the crack, where k_∞ (t) denotes the degraded stiffness which pertains for $x \to -\infty$ at time t, *ie.*

$$k_\infty(t) = 2\sigma_a/\{E'\delta(x \to -\infty\, ;\, t)\} \tag{3.6}$$

This approximation leads to the following estimate for K(t), by adapting the known result for linear springs (Rose, 1987),

$$K(t) = \sigma_a / \sqrt{k_\infty(t)}\ . \tag{3.7}$$

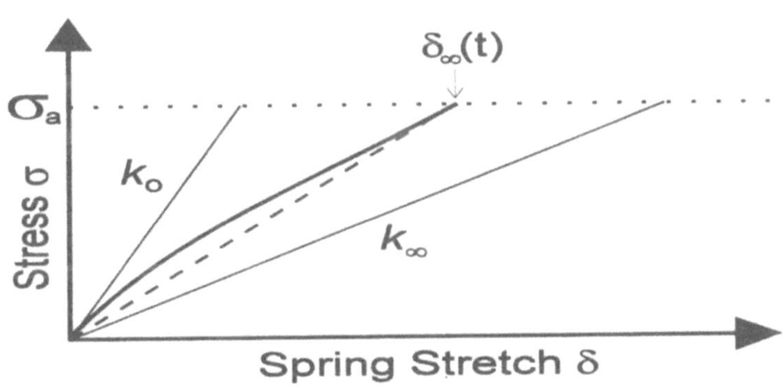

Figure 2. Schematic stress-stretch curve at a given time *t* for a semi-infinite crack reinforced by linear viscoelastic springs. The dashed straight line leads to a useful, conservative, estimate for *K*(*t*).

A physical interpretation of this approximation suggests that the dashed line in Fig. 2 should always lie below the actual σ/δ curve for time t, and therefore that (3.7) should represent an upper bound for K(t). This is particularly useful from a practical viewpoint, because (3.7) would then provide a *conservative estimate* for the actual stress intensity factor for a finite-length fully bridged crack as well. This estimate is similar in spirit to that already proposed by Baker (1993) and Cox and Rose (1994) for fatigue degradation of bonded repairs.

4. Steady-state crack growth

The term "steady state" implies a constant crack-growth rate $\dot{a}$ and translational invariance, so that relative to coordinate axes x, y attached to the moving crack tip, the stress and displacement fields depend only on these spatial coordinates x, y, and not on the time t. Thus, a partial derivative with respect to t in a fixed frame of reference can be replaced by a spatial derivative,

$$\partial/\partial t \rightarrow -\ \dot{a}\ \partial/\partial x \tag{4.1}$$

There would appear to be only two configurations for which a strict steady-state can prevail:

(i) a fully bridged semi-infinite crack, in a uniform remote stress field;

(ii) a semi-infinite crack with near-tip bridging over a fixed, finite-length zone, under the standard conditions for a small-scale bridging approximation.

The latter case has been studied by Pickthall and Rose (1996). In the former case, it is essential that the traction law should exhibit a finite compliance for t $\rightarrow \infty$, to ensure that the crack opening for x $\rightarrow -\infty$ remains finite. This requirement is not satisfied by the traction law proposed by Begley *et al.* (1995a) in the context of fibre bridging, so that a strict steady-state cannot be expected for that case, as noted in the more recent work of Begley *et al.* (1995b).

The stress intensity factor can again be derived by using the J-integral approach of Marshall and Cox (1988), with the integration contour shown in Fig. 1, but x, y now denoting a coordinate system attached to the moving crack tip. It is assumed here that $\dot{a}$ is small compared with the elastic wave speeds, so that a quasi-static analysis is again appropriate. Thus, instead of (3.1) we now obtain

$$\frac{K^2(\dot{a})}{E'} = J_{tip}(\dot{a}) \tag{4.2a}$$

$$= \sigma_a\, \delta_\infty(\dot{a}) - \int_0^{-\infty} \sigma_{yy}(x, y = 0\, ;\, \dot{a})\, \partial\delta/\partial x\, (x\, ;\, \dot{a})\, dx\ . \tag{4.2b}$$

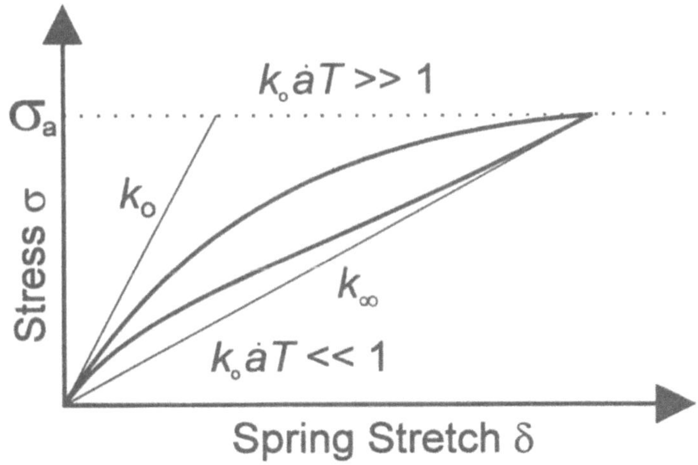

Figure 3. Schematic stress-stretch curves for steady-state growth of a fully bridged semi-infinite crack at various growth rates $\dot{a}$ for bridging by linear viscoelastic springs with a single relaxation time T

Similar difficulties arise in evaluating the integral in (4.2b) to those discussed above in Section 3. The general features of the response can be understood with reference to Fig. 3. With any given growth rate $\dot{a}$ one can associate a definite σ/δ curve, as indicated in Fig. 3, and the right hand side of (4.2b) would correspond to the complementary energy for the hypothetical springs defined by that σ/δ relation. However, that relation is not prescribed: it must be derived by first solving an integral equation for δ (x; $\dot{a}$) and then deriving the corresponding stress (at position x along the crack) from the prescribed, history dependent, constitutive equation.

For the case of linear viscoelastic springs defined by (3.2), the appropriate relation is

$$\sigma_{yy}(\xi, y = 0 ; \dot{a}) = \frac{E'}{2} \int_0^{\xi} k\,[(\xi - \eta)/\dot{a}]\, \frac{\partial \delta}{\partial \xi}(\eta ; \dot{a})\, d\eta\,, \tag{4.3}$$

where $\xi = -x$. The governing integral equation for the dislocation density $D(\xi)$ is

$$\frac{1}{2\pi} \int_0^{\infty} \frac{D(\eta)}{\xi - \eta}\, d\eta + \int_0^{\xi} k\,[(\xi - \eta)/\dot{a}\,]\, D(\eta)\, d\eta = 2\sigma_a / E'\,. \tag{4.4}$$

This equation should be amenable to solution by the Wiener-Hopf technique, but the second integral in (4.4) represents a significant complication relative to the case of linear elastic springs. Work is currently in progress in an attempt to solve this equation for viscoelastic springs with a single relaxation time **T**, specified by

$$k(t) = k_0 - k_1 (1 - e^{-t/T}) = k_\infty + k_1 e^{-t/T}, \tag{4.5}$$

where k_0 and k_∞ denote respectively the instantaneous and the fully relaxed spring stiffness. The solution will then depend only on the non-dimensional parameters k_0 à T and k_0/k_∞ . For the limiting cases k_0 à T « 1 or » 1, the solution can be expected to approach the slow or fast elastic response, respectively, as indicated in Fig. 3, but this asymptotic behaviour may not be uniform in x. It is clear from Fig. 3 that assuming a fully relaxed spring stiffness for all locations x will necessarily lead to an upper bound for K. This provides a conservative estimate for K, but the accuracy could be expected to be much poorer than for the dashed line in Fig. 2.

5. Conclusion

The influence of a progressive decay in crack-bridging efficiency has been discussed, with particular attention to two configurations involving a fully bridged semi-infinite crack. The case of bridging by linear viscoelastic springs illustrates in the simplest mathematical form some of the key features and practical difficulties which can be expected for more realistic (and mathematically more complicated) traction laws. Even for that special case, however, there remain significant challenges in deriving exact analytical results, or efficient numerical procedures, which require further detailed study.

6. Acknowledgement

The author is grateful to Dr C. Pickthall for valuable discussions during the course of the work.

References

Baker, A.A. (1993) Repair efficiency in fatigue-cracked aluminium components reinforced with boron/epoxy patches, *Fatigue Frac. Engng. Mater. Struct.* **16**, 753-765.

Begley, M.R., Cox, B.N. and McMeeking, R.M. (1995b) Time dependent crack growth in ceramic matrix composites with creeping fibers, *Acta Metall. Mater.* (to appear).

Begley, M.R., Evans, A.G. and McMeeking, R.M. (1995a) Creep rupture in ceramic matrix composites with creeping fibers, *J. Mech. Phys. Solids* **43**, 727-740.

Budiansky, B., Amazigo, J.C. and Evans, A.G. (1988) Small-scale crack bridging and the fracture toughness of particulate-reinforced ceramics, *J. Mech. Phys. Solids* **36**, 167-187.

Budiansky, B., Hutchinson, J.W. and Evans, A.G. (1986) *J. Mech. Phys. Solids* **34**, 167-189.

Christensen, R.M. (1982) Theory of Viscoelasticity, 2nd ed., Academic Press, New York.

Cox, B.N. and Rose, L.R.F. (1994) Time or cycle-dependent crack bridging, *Mech. Mater.* **19**, 39-57.

Marshall, D.B. and Cox, B.N. (1988) A J-integral method for calculating steady-state matrix cracking stresses in composites, *Mech. Mater.* **7**, 127-133.

Marshall, D.B., Cox, B.N. and Evans, A.G. (1985) The mechanics of matrix cracking in brittle matrix fiber composites, *Acta Metall.* **33**, 2013-21.

McCartney, L.N. (1987) Mechanics of matrix cracking in brittle-matrix fibre-reinforced composites, *Proc. R. Soc. Lond.* **A409**, 329-350.

Nemat-Nasser, S. and Hori, M. (1987) Toughening by partial or full bridging of cracks in ceramics and fiber reinforced composites, *Mech. Mater.* **6**, 245-269.

Pickthall, C. and Rose, L.R.F. (1996) Viscoelastic relaxation and steady-state crack growth, Proc. 1st Aust. Congress on Applied Mech., R.H. Grzebieta et al. (eds.), Institution of Engineers, Melbourne, Australian.

Rose, L.R.F. (1982) A cracked plate repaired by bonded reinforcements, *Int. J. Fracture* **18**, 135-144.

Rose, L.R.F. (1987) Crack reinforcement by distributed springs, *J. Mech. Phys. Solids* **19**, 39-57.

Rose, L.R.F., Callinan, R.J., Baker, A.A., Sanderson, S. and Wilson, E.S. (1995) Design validation for a bonded composite repair to the F-111 lower wing skin, Proc. Sixth Australian Aeronautical Conference, Institution of Engineers, Melbourne, Australia.

Willis, J.R. and Nemat-Nasser, S. (1990) Singular perturbation of a class of singular integral equations, *Q. App. Math.* **48**, 741-753.

NOTCH STRENGTH OF CERAMIC COMPOSITES: LONG FIBERS, STOCHASTICS, SHORT FIBERS

BERNARD BUDIANSKY
Division of Applied Sciences
Harvard University, Cambridge, MA 02138, USA

JOHN C. AMAZIGO
Department of Mathematics
University of Nigeria, Nsukka, Nigeria

Abstract

The tensile strength of an aligned-fiber ceramic composite containing an initial, through-the-fibers flaw is examined, and comparative studies are made of the effects of long fibers having deterministic tensile strengths, long fibers with stochastic strengths, and short fibers. Long-fiber reinforcement by perfectly uniform fibers having deterministic strengths has already been investigated extensively, and early studies of the effects of statistical variations in fiber strength associated with random fiber flaws (Thouless and Evans, 1988; Sutcu, 1989) indicated that long fibers having stochastic strengths could, paradoxically, provide significantly higher composite notch strengths than comparable fibers having uniform tensile strengths. The essential reason for this is that the randomly flawed fibers can suffer breaks in the interior of the composite, and then continue to carry load as well as enhance energy dissipation as they are pulled out of the matrix. A similar pullout effect can be induced by use of sufficiently short fibers, even when their strength is deterministic (Budiansky and Cui, 1995).

The influences on composite strength of these various fiber types are revisited and reviewed here. Particular attention is given to the question of "notch-sensitivity", which in the present context means the effect of the length of a major crack-like flaw on the overall composite strength.

1. Introduction

The configuration shown in Figure 1 illustrates a 2D model problem for the assessment of notch sensitivity. An infinite aligned-fiber composite containing a through-the-fibers flaw (or "notch") of length $2a_0$ is subjected to uniform tension σ at infinity. The fibers of radius R have volume concentration c_f and are held in the matrix by an interface friction of maximum magnitude τ. We assume further a matrix fracture toughness K_m=0; this simplifying assumption (Suo et al , 1993) introduces little error (Budiansky and Cui,

J. R. Willis (ed.), IUTAM Symposium on Nonlinear Analysis of Fracture, 333–342.

1994). Figure 1 shows long fibers, but a similar picture, and the same notation and assumptions will be used for aligned, but randomly located short fibers having uniform lengths.

In the analysis of the strength of the notched composite, some additional simplifications will be made. We will assume that an isolated matrix crack aligned with the initial flaw will emanate from each flaw tip (Figure 1) when load is applied to the composite. In actuality, fiber composites suffer multiple, closely spaced matrix cracks, with a concentrated density of such cracks near the notch tips. Like crack-tip plasticity in metals, such localized matrix cracks tend to relieve fiber stress concentration at the flaw tips, and so ignoring them makes the calculations of the composite strength conservative. Finally, we will be using an idealized bridging model, wherein the forces applied to the composite by fibers that bridge the matrix crack will be modeled via distributed spring stresses applied to the matrix crack faces, rather than by distributed friction along the fibers in a boundary layer in the interior of the composite. This too tends to concentrate flaw-tip stresses unrealistically, leading to conservative strength estimates, as shown in the "large-scale sliding" study by Xia et al (1994). Nevertheless, comparisons of the relative effectiveness of various kind of fibers may remain approximately valid, and the simplified analyses can be expected to provide lower bounds to the strength of the composite in the presence of a notch.

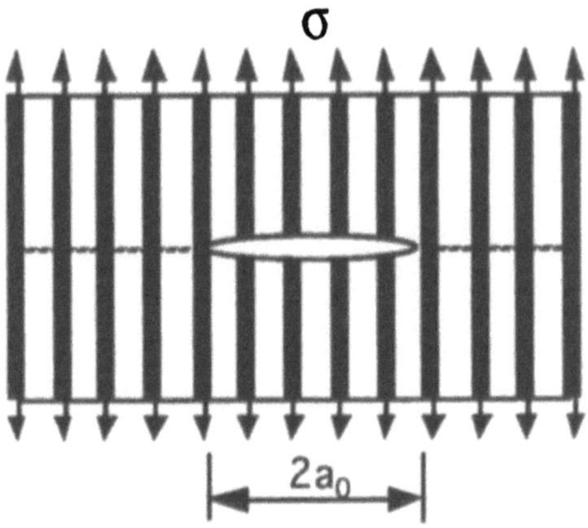

Figure 1. Flawed composite.

We will denote the strength of the composite by σ_s, and let σ_0 be the strength of the unflawed (but matrix-cracked) composite. Then the knockdown factor σ_s/σ_0 is a measure of the notch sensitivity.

2. Long Fibers, Deterministic Fiber Strength

The pertinent results found by Budiansky and Cui (1994) will be summarized. Figure 2 illustrates the bridging model in which the crack-face fiber stresses σ_f were smeared out into bridging stresses $p=c_f\sigma_f$ related by a bridging law to the crack-face opening displacement δ. We define the characteristic length

$$L = RS/\tau \tag{1}$$

where S is the fiber strength, and introduce the parameter

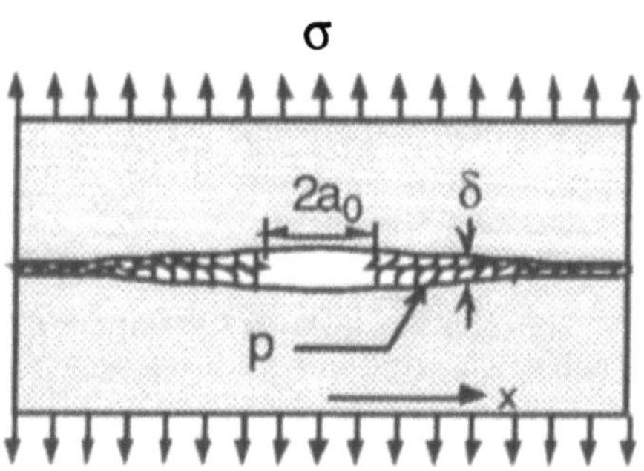

Figure 2. Crack-face spring bridging.

$$\Lambda(S) = \left(\frac{E}{2c_m E_m}\right)\sqrt{\frac{E_f}{S}} \tag{2}$$

where E_f, E_m, and E are the fiber, matrix, and composite moduli, and $c_m=1-c_f$ is the matrix volume fraction. In terms of the non-dimensional crack opening defined by

$$w = 8\Lambda^2(\delta / L) \tag{3}$$

the bridging relation

$$f \equiv \frac{p}{c_f S} = \sqrt{w} \tag{4}$$

was used. (See Xia et al (1994) and Budiansky et al (1995) for discussions of the provenance and limitations of this bridging law.) The criterion for fiber failure is f=1, and when this occurs at the ends of the original flaw, failure of the composite ensues.

Because $K_m=0$, the matrix cracks will extend out to infinity. With $\xi=x/a_0$

$$w(\xi) - w(1) = \left(\frac{16}{\pi}\right)\left(\frac{\omega a_0}{L}\right)\int_1^{\infty} f(t)\log\left|\frac{t^2-\xi^2}{t^2-1}\right|dt. \tag{5}$$

governs the distribution $f(\xi)$ of the non-dimensional bridging stress. Here, via (4), $w=f^2$, and with ν equal to the Poisson's ratio of the composite, the parameter ω is defined as

$$\omega = \frac{(1-\nu^2)c_f E_f E}{2Ac_m^2 E_m^2} = O(1). \tag{6}$$

Here A=O(1) is a parameter that accounts for the orthotropy of the composite (Budiansky and Cui, 1994). Overall equilibrium requires the imposition of the auxiliary condition

$$\int_1^{\infty}\left[f(\xi) - f(\infty)\right]d\xi = f(\infty). \tag{7}$$

With $\sigma_0=c_f S$, the results for the strength ratio σ_s/σ_0 vs. $\omega a_0/L$ shown in Figure 3 were found by setting f(1)=w(1)=1, solving (5) and (7) for $f(\xi)$, and identifying σ_s/σ_0 with $f(\infty)$.

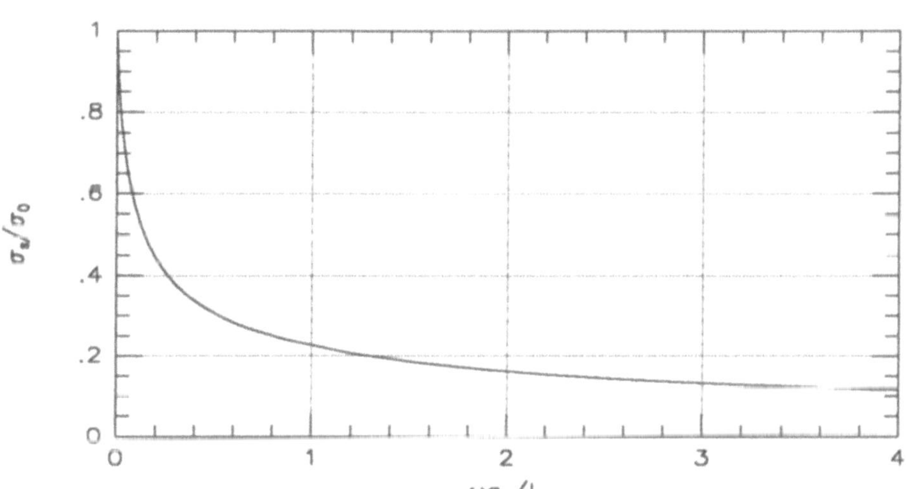

Figure 3. Notch sensitivity, deterministic fiber strength.

(We omit the details of the numerical solution.)

For an SiC/CAS composite, the nominal values R=7 μm, S=2 GPa, and τ=20 MPa give the characteristic length L=700 μm. For c_f=.4, E_f=200 GPa, and E_m=100 GPa, we have $E \approx c_f E_f + c_m E_m$= 140 GPa, and the values ν=.25 and A=.97 (Budiansky and Cui, 1994) give $\omega \approx 1.5$. The consequent theoretical notch sensitivity as a function of the half-crack-size a_0 in millimeters, illustrated in Figure 4, is alarming; but in fact, the experiments by Cady et al (1995) on a notched, 0/90 SiC/CAS composite laminate exhibited essentially no notch sensitivity.

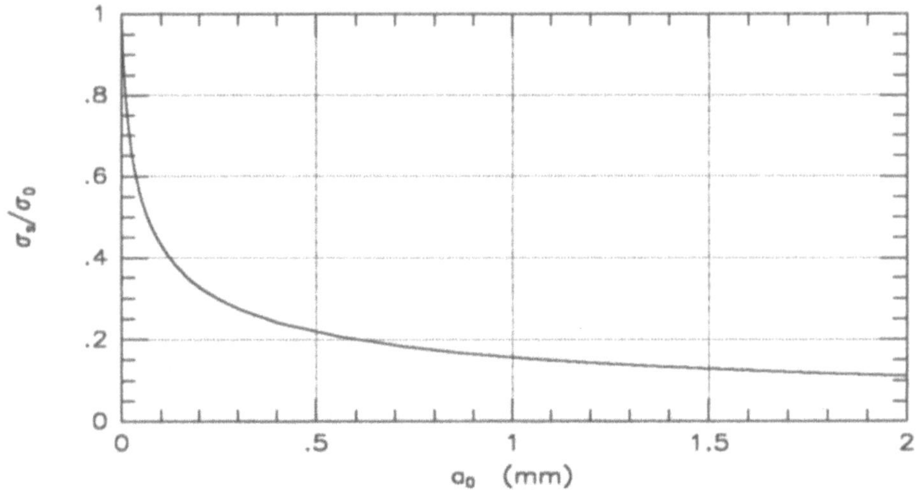

Figure 4. Notch sensitivity, deterministic fibers; example (SiC/CAS).

3. Long Fibers, Stochastic Fiber Strengths

We turn now to consideration of the effect on notch sensitivity of fibers having strengths governed by a Weibull strength distribution. Suppose that the occurrence of inherent flaws in a long fiber obeys a Poisson distribution specified by the probability N[S]dx that the tensile strength of any element dx along the fiber is less than S. The Weibull assumption is that N[S] is proportional to some power S^m. A useful formulation of this law is

$$N[S] = \frac{1}{L}\left[\frac{\Gamma(1+1/m)S}{S_L}\right]^m \tag{8}$$

where S_L is the average strength of fibers of length L, and Γ is the gamma function. We illustrate the significance of the size of the Weibull parameter m in Figure 5, which shows the probability-density function $q(S/S_L)$ for the strength ratio S/S_L of fibers of length L. The values of m shown more-or-less cover the range that has been reported for various fibers. For m=∞ the fiber strength is deterministic, with $S=S_L$; m's less than 2 are unlikely. The validity of Eq. (8) would imply that LS_L^m is invariant, and that the curves in Figure 5 are independent of L. (But experiments by Zok et al (1993) indicate that a simple Weibull distribution does not really account adequately for the presence of the distinct fiber flaw populations associated with pristine fibers and those introduced during composite processing.)

The effective single-matrix-crack bridging law associated with fibers having Weibull strength distributions was studied by Thouless and Evans (1989), and has been

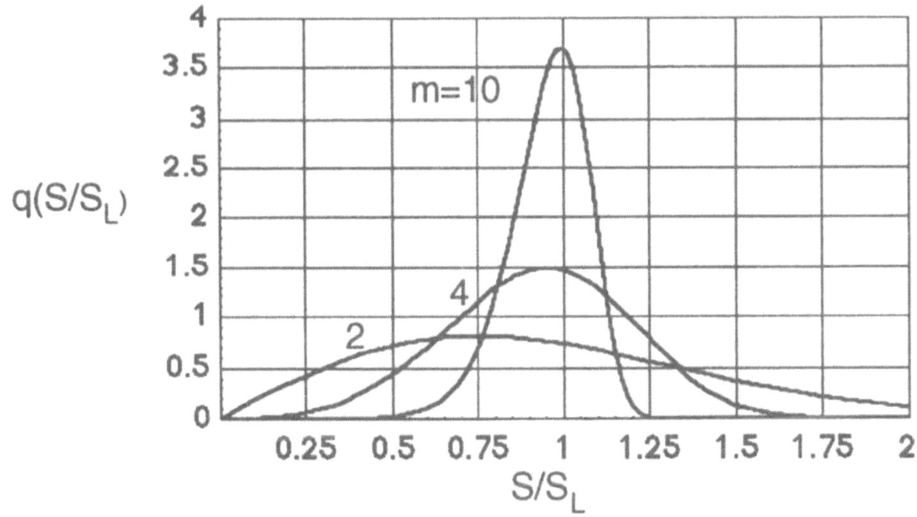

Figure 5. Fiber-strength probability densities for several Weibull exponents.

rederived, with corrections, by Budiansky and Amazigo (to be published). We define the new characteristic length L and the stress S_L by stipulating that the relation

$$L = S_L R / \tau \tag{9}$$

be satisfied by fibers of length L and their average strength S_L; and we keep the definition (2) for the parameter Λ, with S replaced by S_L. Then, with w still defined by Eq. (3), the bridging law (4) generalizes into the form

$$\frac{p}{c_f S_L} \equiv f(w; m, \Lambda) = \sqrt{w} e^{-\alpha_m w^{(m+1)/2}} + \int_{\frac{w}{4\Lambda^2}}^{\sqrt{w}} \alpha_m e^{-\alpha_m s^{m+1}} \left[s - \frac{w}{4\Lambda^2} \right]^{m+1} ds \tag{10}$$

for $w<16\Lambda^4$; for $w>16\Lambda^4$ the integral is dropped. Here α_m is just the constant

$$\alpha_m = \frac{[\Gamma(1+1/m)]^m}{m+1}. \tag{11}$$

(If the average strength S_g is known for a standard gage length L_g of fibers, then L and S_L may be found from the formulas

$$\frac{L}{L_g} = \left(\frac{S_g R}{\tau L_g}\right)^{\frac{m}{m+1}}, \quad \frac{S_L}{S_g} = \left(\frac{S_g R}{\tau L_g}\right)^{-\frac{1}{m+1}} \tag{12}$$

that follow from Eq. (9) and the connection $L S_L^m = L_g S_g^m$.)

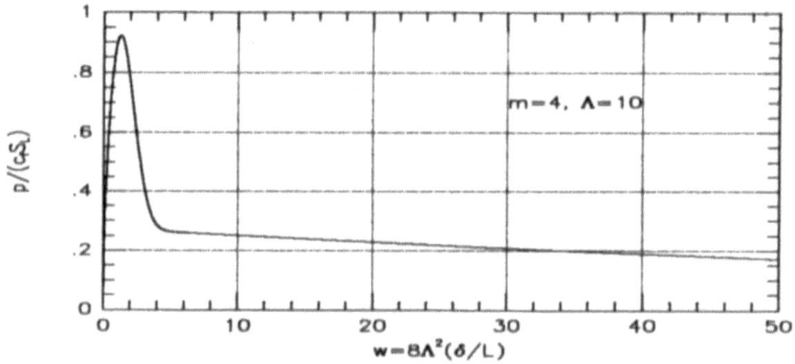

Figure 6. Bridging law, stochastic fiber strength; m=4, $\Lambda(S_L)=10$.

The bridging law (9) can be evaluated by numerical integration, and an example is shown in Figure 6, for m=4 and $\Lambda(S_L)=10$. Typically, in contrast to the deterministic law (4), which confines w to the narrow range (0,1), the curve for stochastic fibers is considerably more broad in its early stage, and then transitions to a very long, low, gradually falling portion.

The integral equation (5) and the auxiliary relation (7) continue to apply to the model problem, with w given in terms of $f=p/(c_f S_L)$ by numerical inversion of the bridging relation. But the strength σ_s is now set by the maximum value achieved by the applied stress σ during opening of the matrix crack. The normalizing value σ_0 for the strength of the unnotched composite is identified with the peak value of p (Figure 6) in the bridging law. The maximum values of $p/(c_f S_L)$ are nearly independent of Λ, and decrease from unity only a little as m decreases. (For m=2 and $\Lambda=10$, $[p/(c_f S_L)]_{max}=.87$.)

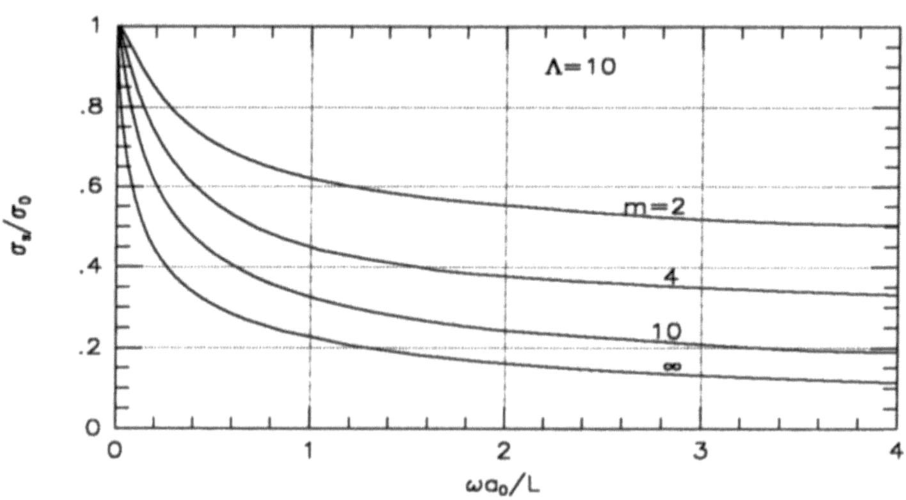

Figure 7. Notch sensitivity, stochastic fiber strengths; $\Lambda(S_L)=10$

We show in Figure 7 the non-dimensional results found for notch sensitivity, for m=2, 4, 10, and the deterministic case m=∞ given in the previous section. For each m, we used $\Lambda(S_L)=10$. (For $m=\infty$ the strength does not depend on Λ, and additional numerical calculations have indicated that the notch strength is not very sensitive to Λ for the small notch sizes that are of interest in the study of notch sensitivity. For each finite m, the strength does becomes nearly proportional to Λ for very long notches.)

The curves of Figure 7 give a clear indication of the diminished notch sensitivity associated with decreasing values of the Weibull exponent. Note, however, that the characteristic length L in the non-dimensional abscissa depends on m, as does the normalizing value σ_0 for the unnotched composite. An explicit presentation of notch sensitivity estimates for SiC/CAS composites is shown in Figure 8, on the following basis. We suppose that for a standard gage length of 25.4 mm (1 inch), the measured average fiber strength S_g is 2 GPa for each m. With the same choices R=7 μm and τ=20 GPa made earlier for deterministic fibers, we used Eqs. (11) and (12) to calculate the characteristic lengths L and associated strengths S_L shown in Table I for the several m's. Then, with ω=1.5 (for the choices c_f=.4, E_f=200 GPa, and E_m=100 GPa used earlier) the results in Figure 7 produced the curves of Figure 8. (This introduces a small error, because Λ=10 was used for each of the curves in Figure 7, and the actual values of $\Lambda(S_L)$ for each m, also given in Table I, are a little different.)

Table I
Stochastic fiber parameters

m	L (mm)	S_L(GPa)	$\Lambda(S_L)$
2	2.3	6.6	6.4
4	1.4	4.1	8.2
10	1.0	2.8	9.9
∞	.7	2.0	11.7

The trends exhibited in Figure 8 confirm the reduced notch sensitivity to be expected from composites containing fibers having a statistical spread of breaking strengths. But some notch sensitivity survives in the millimeter-size notch range even for m=2, and so stochastics alone would not account for the Cady et al (1995) notch-insensitivity observations in SiC/CAS. Perhaps a combination of stochastic fiber strengths, multiple matrix cracking, and the Xia et al (1994) large-scale sliding effect will ultimately be needed to understand their notch-insensitivity data.

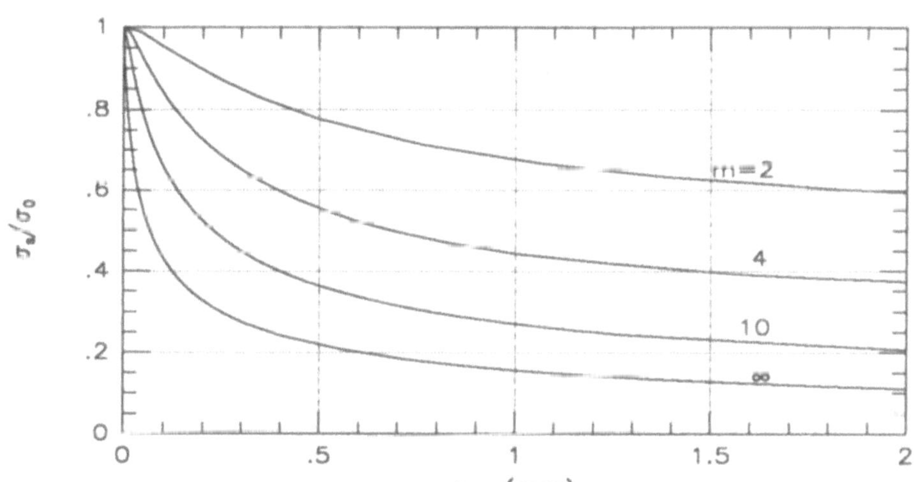

Figure 8. Notch sensitivity, stochastic fiber strengths; example (SiC/CAS).

There are several ways to rationalize the decreased notch sensitivity provided by stochastic fiber strengths. When weaker fibers break in the interior of the matrix they continue to inhibit crack opening as they slide out, allowing the fibers that are stronger than average to sustain their bridging constraint. The "softer" bridging law vis-a-vis the deterministic one tends to reduce the concentration of fiber stress near the notch. And from a related energy viewpoint, there is more dissipation associated with the pullout of broken fibers (Kelly and Macmillan 1985). So short fibers that would always pull out before they break should lead to less notch sensitivity. This was explored by Budiansky and Cui (1995), and is reviewed in the next section.

4. Short Fibers

We illustrate in Figure 9 randomly arrayed, aligned short fibers, or whiskers, in the composite containing an unbridged notch, with matrix cracks emanating from the notch tips. (Sophisticated, detailed analyses of matrix-crack-bridging by randomly inclined short fibers have been conducted by Li and his associates (e.g. Li et al, 1991), mostly in the context of reinforced cementitious materials. The emphasis by Budiansky and Cui was on the notch-strength implications of aligned fibers, and involved much more primitive modeling.) Reverting to the assumption of a deterministic fiber strength S suggests the concept of an "optimal" fiber length $L=RS/\tau$, as the largest length consistent with pullout before fiber fracture, regardless of a bridging fiber location with respect to the matrix crack. Retreating a bit from the modeling of long fibers, we assume that the fibers are rigid. If we let $e \leq L/2$ denote the length of a particular bridging-fiber segment that is being pulled out, then its crack-face fiber stress σ_f is related to the crack opening δ by $\sigma_f(\delta;e)=2\tau(e-\delta)/R$ for $\delta \leq e$, and vanishes for $\delta > e$. Averaging σ_f over all e's between zero and L/2, and writing $p=c_f(\sigma_f)_{ave}$ gives the optimal-fiber bridging law

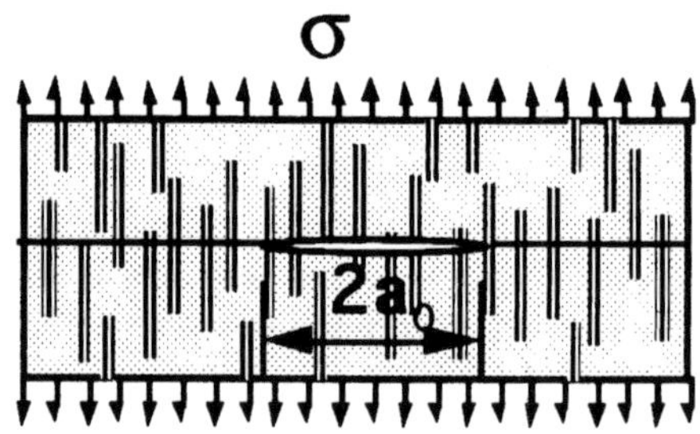

Figure 9. Notched short-fiber composite.

$$\begin{cases} 2p/(c_f S) = [1-2\delta/L]^2 & (0 \leq \delta \leq L/2) \\ \quad = 0 & (\delta > L/2) \end{cases} \tag{13}$$

The maximum bridging stress is $\sigma_0=c_f S/2$, and this is also the reference composite strength for vanishing notch size. Now the matrix crack does not have to extend to infinity, even though the matrix toughness K_m is assumed to vanish; the finite value of bridging stress for $\delta=0^+$ makes it possible for K to vanish at the tip of a matrix crack of finite length $(b - a_0)$. In contrast to Eq. (5), the integral equation

$$1-\sqrt{f(\xi)} = 2\varepsilon_f \eta (a_0/L)\left(\frac{\sigma}{\sigma_0}\sqrt{\beta^2-\xi^2} - \frac{1}{\pi}\int_1^\beta f(t)\log\left|\frac{\sqrt{\beta^2-\xi^2}+\sqrt{\beta^2-t^2}}{\sqrt{\beta^2-\xi^2}-\sqrt{\beta^2-t^2}}\right|dt\right) \tag{14}$$

with $\beta=b/a_0$, governs $f=p/\sigma_0$ in the range $1 \leq \xi \leq \beta$. Here the parameter η is

$$\eta = \frac{2(1-\nu^2)c_f E_f}{AE} = O(1) \tag{15}$$

and $\varepsilon_f = S/E_f$ is the fiber fracture strain. To meet the requirement of zero stress intensity factor at the matrix crack tip, the scalar condition

$$\frac{\sigma}{\sigma_0} = \frac{2}{\pi}\int_1^{\beta} \frac{f(\xi)d\xi}{\sqrt{\beta^2 - \xi^2}} \tag{16}$$

must also be satisfied. For specified values of the notch-size parameter $\varepsilon_f \eta (a_0/L)$, the strength ratio σ_s/σ_0 (Figure 10) was determined as the peak value of σ/σ_0, for increasing values of β, in the numerical solution of (15) and (16) for $f(\xi)$ and σ/σ_0.

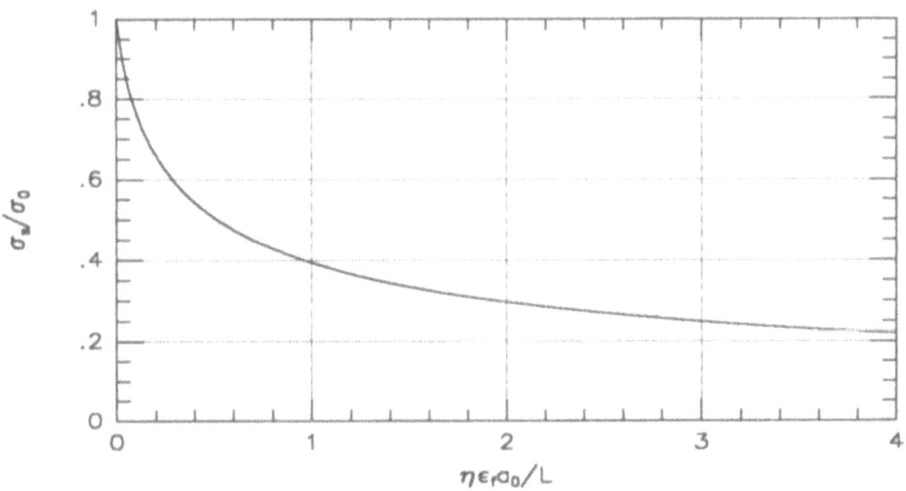

Figure 10. Notch sensitivity, optimal short fibers.

Actually, the abscissa in Figure 10 does not depend on S, but the form shown makes it easy to see that "optimal" short fibers should generally give less notch sensitivity than that given in Figure 3 for deterministic long fibers. The curves look similar, the η and ω parameters are both of order one, and the L parameters for long and short fibers should have comparable sizes. But the presence in the short-fiber abscissa of the factor $\varepsilon_f = S/E_f$, which is around .01, means that for a given notch size a_0, the knockdown factor is much less for short fibers. But we should remember that the normalizing strength parameter σ_0 is only $c_f S/2$ in the short-fiber case, instead of $c_f S$ for long fibers; however, short fibers in whisker form tend to have higher nominal strengths.

We will now finish up by showing an example for a hypothetical CAS matrix containing SiC whiskers. We presume an SiC whisker strength S=4 GPa, whisker radius R=3 μm, and keep τ=20 MPa as the interface friction; this gives an optimal short-fiber length L=600 μm. The value E_f=4 GPa for the whisker modulus gives ε_f=.01, and with c_f=.4, E_m=100 GPa, we get E=220 GPa, and A=.88 (Budiansky and Cui, 1994) for ν=.25. These numbers give η=1.55, and then Figure 10 produces the short-fiber example in Figure 11, shown together with the curves reproduced from Figure 8 for stochastic and deterministic long fibers. Although various properties of the short and long fibers in this comparison differ, they reflect reasonable estimates for whiskers and fabricated fibers.

In this example, the optimal short fibers provide very comforting notch insensitivity, performing much better than even stochastic long fibers with m=2.

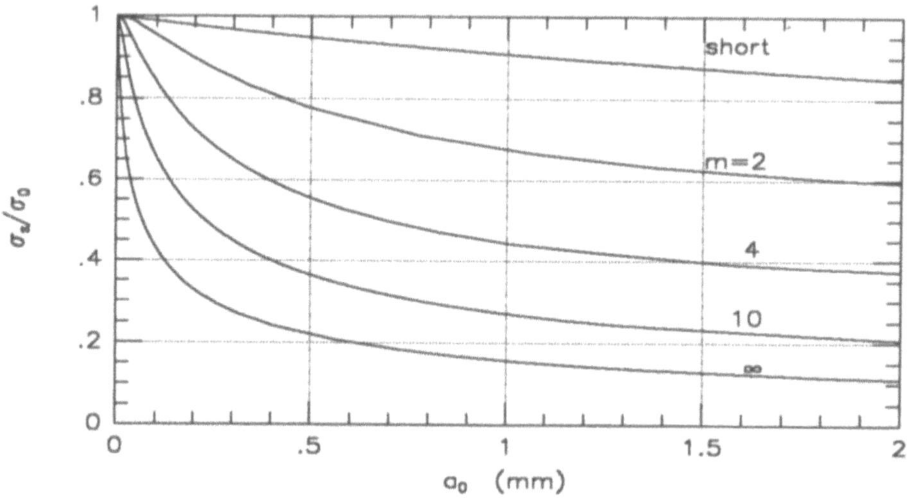

Figure 11. Notch sensitivity examples (SiC/CAS); short fibers, stochastic long fibers.

5. Acknowledgements

This work was partially supported by an ARPA University Research Initiative grant to the University of California at Santa Barbara, and by the Division of Applied Sciences, Harvard University.

6. References

Budiansky, B. and Cui, Y.L. (1994) On the tensile strength of a fiber-reinforced ceramic composite containing a crack-like flaw. J. Mech. Phys. Solids **42**, 1-19.

Budiansky, B., Evans, A.G., and Hutchinson, J.W. (1995) Fiber-matrix debonding effects on cracking in aligned fiber ceramic composites. Int. J. Solids Structures **32**, 315-328.

Budiansky, B. and Cui, Y.L. (1995) Toughening of ceramics by short aligned fibers. Mech. Matls. **21**, 139-146.

Cady, C.M., Mackin, T.J., and Evans, A.G. (1995) Silicon carbide/calcium aluminosilicate: a notch-insensitive ceramic-matrix composite. J. Am. Ceram. Soc. **78**, 77-82

Kelly, A., and Macmillan, N.H. (1985) *Strong Solids*, Clarendon Press, Oxford.

Li, V.C., Wang, Y., and Backer, S. (1991) A micromechanical model of tension-softening and bridging toughening of short random fiber reinforced brittle matrix composites. J. Mech. Phys. Solids **39**, 607-625.

Suo, Z., Ho, S., and Gong X. (1992) Notch ductile-to-brittle transition due to localized inelastic band. ASME J. Engr. Mat. Tech. **115**, 319-326.

Sutcu, M. (1989) Weibull statistics applied to fiber failure in ceramic composites and work of fracture. Acta. Metall. **37**, 651-661.

Thouless, M.D., and Evans, A.G. (1988) Effects of pullout on the mechanical properties of ceramic-matrix composites. Acta. Metall. **36**, 517-522.

Xia, Z.C., Hutchinson, J.W., Evans, A.G., and Budiansky, B. (1994) On large scale sliding in fiber-reinforced composites. J. Mech. Phys. Solids **42**, 1139-1158.

Zok, F.W., X Chen, X., and Weber, C.H. (1995) Tensile strength of SiC fibers. J. Am. Ceram. Soc., **78**, 1965-1968.

NONLINEAR ANALYSIS OF INTERFACIAL CRACKS

N. J.-J. FANG AND J. L. BASSANI
Department of Mechanical Engineering and Applied Mechanics
University of Pennsylvania
Philadelphia, PA 19104
U.S.A.

Abstract

Asymptotic analysis is developed for the problem of a plane strain crack lying on the interface of an elastic-plastic (or elastic-creeping) material and a rigid substrate. The nonlinear plastic (creep) response is taken to follow a power-law hardening relation. In contrast to other recent analyses, we have found asymptotic solutions for a continuous variation of crack-tip mode-mix that agree well with full-field solutions. These crack-tip displacements and stresses are variable-separable in polar coordinates r and θ and exhibit a singularity in stress of $\sigma \propto r^{-1/(n+1)}$ as $r \to 0$, where n is the hardening exponent. The angular variations of these asymptotic fields have been calculated using a finite difference scheme. Unlike the full range of mixed-mode solutions that exist for a homogeneous crack in a power-law hardening material, there appears to be a narrow range around the pure tensile mode for which solutions do not exist. That latter range increases somewhat as the hardening exponent n increases.

1. Introduction

The performance of structural materials is often determined by the response of interfaces between structural components, such as cracking of grain boundaries in polycrystals and fiber/matrix interfaces in composite materials. This paper addresses specifically the mechanics aspect of interfacial fracture. Rice (1988) and Hutchinson and Suo (1992) have provided an extensive review of work dating back to the late 1950's on the subject of fracture along interfaces between dissimilar linearly elastic materials. The crack tip fields of a linear system are found to be oscillatory, except in general when both isotropic materials are incompressible with Poisson's ratio $\nu = 1/2$. In contrast, recent investigations have been inconclusive regarding the structure of the crack tip fields for stationary crack problems at nonlinear bimaterial interfaces.

In recent analyses of interfacial cracks lying between a power-law hardening material and a rigid substrate, Wang (1990), Champion and Atkinson (1991) and Sharma and Aravas (1993) attempted to find crack-tip solutions that are separable in polar coordinates (r,θ) and exhibit the singularity $\sigma \propto r^{-1/(n+1)}$ as $r \to 0$. This stress singularity in the leading order term of the asymptotic expansion can be deduced from J-integral analysis and, consequently, is the same as the Hutchinson-Rice-Rosengren

J. R. Willis (ed.), IUTAM Symposium on Nonlinear Analysis of Fracture, 343–352.

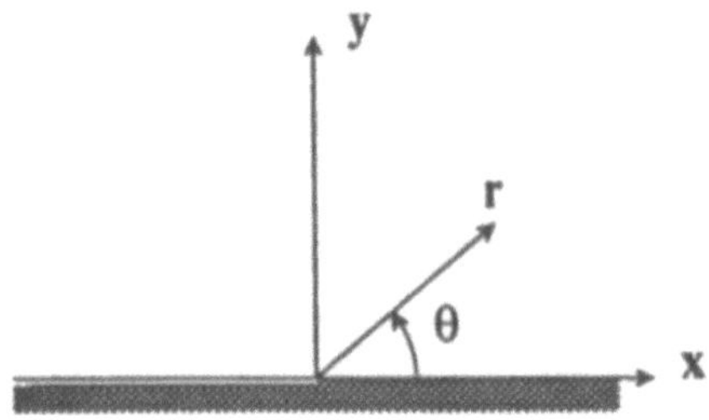

Figure 1. Schematic of the crack tip region.

solutions for cracks in homogeneous materials (Hutchinson, 1968; Rice and Rosengren, 1968). However, as opposed to the stationary homogeneous crack problem which yields mixed-mode solutions varying continuously between pure mode 1 and 2 (Shih, 1974), only a single mode-mix was obtained for each hardening exponent n, in spite of the different numerical schemes used in Wang (1990), Champion and Atkinson (1991) and Sharma and Aravas (1993). The interfacial crack solutions presented in those papers are near mode 1, i.e., tensile, with some (positive) shear.

Several full-field analyses also exist in the literature for elastic/plastic interface cracks under small-scale yielding conditions. Shih and Asaro (1988) and Sharma and Aravas (1993) considered an interfacial crack between a power-law material with $\nu = 0.3$ and a rigid substrate under applied mode I and mixed-mode loading conditions. In a separate study, Shih and Asaro (1989) investigated an interfacial crack between a power-law material with $\nu = 0.3$ and an elastic substrate under mixed-mode loading. Generally, the crack-tip fields from these finite element analyses do not agree with the solutions obtained from the asymptotic analysis discussed previously. However, some components of the near-tip stresses, particularly the shear stresses, appear to attain the singularity $\sigma \propto r^{-1/(n+1)}$.

In this paper, we analyze the interfacial crack problem, with the emphasis on the spectrum of separable mixed-mode asymptotic solutions. The stationary interfacial crack configuration considered, as depicted in Fig. 1, is a semi-infinite crack lying between a power-law hardening material and a rigid substrate and loaded under plane strain conditions. The deformable medium occupies the region $0 \le \theta \le \pi$, and the crack face is taken to be open and, therefore, traction free. Consequently, the boundary conditions satisfied by the asymptotic fields are:

$$\begin{aligned} u_r(r,0) = u_\theta(r,0) = 0 \\ \sigma_{r\theta}(r,\pi) = \sigma_{\theta\theta}(r,\pi) = 0 \end{aligned} \tag{1}$$

We will make use of the mode-mix parameter, M, defined as (Shih, 1974)

$$M = \frac{2}{\pi} \tan^{-1} \left\{ \lim_{r \to 0} \left[\frac{\sigma_{\theta\theta}(r,0)}{\sigma_{r\theta}(r,0)} \right] \right\} \tag{2}$$

as a measure of the relative strengths of tensile to shear stress on the interface near the crack tip. Mode-mix parameter M is $+1/-1$ for mode 1 stressing conditions with tensile/compressive $\sigma_{\theta\theta}$ stress ahead of the crack, and 0 for mode 2 conditions.

The linear elastic, incompressible, asymptotic solution is presented in Section 2 for an interfacial crack configuration in Fig. 1. Section 3 develops the separable mixed-mode asymptotic solutions for interfacial cracks in power-law hardening materials. Angular variations of the mixed-mode displacement and stress fields are solved using finite difference method and a singular value decomposition technique. In a separate paper by Fang and Bassani (1995a), these asymptotic solutions are shown to be in good agreement with small-scale creep finite element calculations, with solutions obtained by Shih and Asaro (1988, 1989) and Sharma and Aravas (1993), and with slip-line solutions in the non-hardening limit.

2. Linear Elastic Asymptotic Solution

In general, singular stress fields at the tip of an interfacial crack with boundary conditions (1) and a linear elastic upper half region are non-variables separable in (r,θ). When the material is incompressible, however, the dominant term in the asymptotic solution reduces to a solution separable in (r,θ) with a stress singularity of the form $\sigma \propto r^{-1/2}$. Then, the leading order displacement and stress fields are:

$$\begin{Bmatrix} u_r(r,\theta) \\ u_\theta(r,\theta) \end{Bmatrix} = \frac{1}{4G}\sqrt{\frac{r}{2\pi}}\left(K_I \begin{Bmatrix} \cos\frac{\theta}{2} - \cos\frac{3\theta}{2} \\ -3\sin\frac{\theta}{2} + \sin\frac{3\theta}{2} \end{Bmatrix} + K_{II} \begin{Bmatrix} -\sin\frac{\theta}{2} + 3\sin\frac{3\theta}{2} \\ -3\cos\frac{\theta}{2} + 3\cos\frac{3\theta}{2} \end{Bmatrix} \right)$$

$$\begin{Bmatrix} \sigma_{r\theta}(r,\theta) \\ \sigma_{rr}(r,\theta) \\ \sigma_{\theta\theta}(r,\theta) \end{Bmatrix} = \frac{1}{4\sqrt{2\pi r}}\left(K_I \begin{Bmatrix} \sin\frac{\theta}{2} + \sin\frac{3\theta}{2} \\ 5\cos\frac{\theta}{2} - \cos\frac{3\theta}{2} \\ 3\cos\frac{\theta}{2} + \cos\frac{3\theta}{2} \end{Bmatrix} + K_{II} \begin{Bmatrix} \cos\frac{\theta}{2} + 3\cos\frac{3\theta}{2} \\ -5\sin\frac{\theta}{2} + 3\sin\frac{3\theta}{2} \\ -3\sin\frac{\theta}{2} - 3\sin\frac{3\theta}{2} \end{Bmatrix} \right) \tag{3}$$

where G is the shear modulus. This asymptotic solution (3) involves two arbitrary constants, the stress intensity factors K_I and K_{II}. As in the case of linear elastic fracture of homogeneous materials, K_I and K_{II} have the conventional interpretations as mode I and II stress intensity factors and are established by the far field conditions such as applied loading and geometry of the cracked material. From (2) and (3), the *elastic* mode-mix parameter, M_e, is then $M_e = (2/\pi)\tan^{-1}(K_I/K_{II})$. M_e takes on values between ± 1 for mode I and 0 for mode II.

3. Asymptotic Solutions of Power-Law Hardening Materials

An asymptotic analysis of the interfacial crack depicted in Fig. 1 is presented for a deformable medium that exhibits power-law hardening behavior characterized by J_2-deformation theory in the region $0 \le \theta \le \pi$, i.e.,

$$\varepsilon_{ij} = \frac{3}{2}\varepsilon_0 \left(\frac{\sigma_e}{\sigma_0}\right)^{n-1} \frac{s_{ij}}{\sigma_0} \tag{4}$$

where ε_{ij} denotes the total strain in the case of J_2-deformation theory or the strain rate in steady state creep, s_{ij} denotes the deviatoric stress components, $\sigma_e = (\frac{3}{2} s_{ij}\, s_{ij})^{1/2}$ is the von Mises effective stress, and n, σ_0, ε_0 are material constants.

If we seek variable-separable solutions in (r,θ) and use the J-integral argument as in the homogeneous stationary crack problem considered by Hutchinson (1968) and Rice and Rosengren (1968), then the power of singularity in stress is determined to be $\sigma \propto r^{-1/(n+1)}$. The dominant terms in the asymptotic expansion can be expressed as:

$$\begin{Bmatrix} u_r(r,\theta) \\ u_\theta(r,\theta) \end{Bmatrix} = \varepsilon_0 \left(\frac{J}{\sigma_0 \varepsilon_0 I(n, M_p)}\right)^{\frac{n}{n+1}} r^{\frac{1}{n+1}} \begin{Bmatrix} \tilde{u}_r(\theta; n, M_p) \\ \tilde{u}_\theta(\theta; n, M_p) \end{Bmatrix}$$

$$\begin{Bmatrix} \sigma_{r\theta}(r,\theta) \\ \sigma_{rr}(r,\theta) \\ \sigma_{\theta\theta}(r,\theta) \end{Bmatrix} = \sigma_0 \left(\frac{J}{\sigma_0 \varepsilon_0 I(n, M_p)}\right)^{\frac{1}{n+1}} \left(\frac{1}{r}\right)^{\frac{1}{n+1}} \begin{Bmatrix} \tilde{\sigma}_{r\theta}(\theta; n, M_p) \\ \tilde{\sigma}_{rr}(\theta; n, M_p) \\ \tilde{\sigma}_{\theta\theta}(\theta; n, M_p) \end{Bmatrix} \tag{5}$$

where the angular functions denoted with a tilde are normalized so that the maximum over θ of $\tilde{\sigma}_e$ is unity, J denotes the J-integral, $I(n, M_p)$ is a number of order unity and has the same definition as in Hutchinson (1968) and Rice and Rosengren (1968) except the lower limit of integration is zero for the interface crack problem, and the *plastic* mode-mix M_p is defined by (2). Using (5) together with equilibrium equations and stress-strain relations, the governing equations of the leading order of asymptotic expansion are simplified to give (Sharma and Aravas, 1993):

$$\begin{aligned} &\frac{d\tilde{\sigma}_{r\theta}}{d\theta} + (s+1)\tilde{\sigma}_{rr} - \tilde{\sigma}_{\theta\theta} = 0 \\ &\frac{d\tilde{\sigma}_{\theta\theta}}{d\theta} + (s+2)\tilde{\sigma}_{r\theta} = 0 \\ &\frac{d\tilde{u}_r}{d\theta} + sn\tilde{u}_\theta - 3\tilde{\sigma}_{r\theta}\tilde{\sigma}_e^{\,n-1} = 0 \\ &\frac{d\tilde{u}_\theta}{d\theta} + (sn+2)\tilde{u}_r = 0 \\ &(sn+1)\tilde{u}_r - \frac{3}{4}(\tilde{\sigma}_{rr} - \tilde{\sigma}_{\theta\theta})\tilde{\sigma}_e^{\,n-1} = 0 \end{aligned} \tag{6}$$

where $s = -1/(n+1)$, the stress exponent, is known from path-independence of the J-integral. This forms a set of four first order nonlinear ordinary differential equations plus one nonlinear algebraic equation. In addition, boundary conditions (1) become:

$$\tilde{u}_r\left(0;n,M_p\right)=\tilde{u}_\theta\left(0;n,M_p\right)=0$$
$$\tilde{\sigma}_{r\theta}\left(\pi;n,M_p\right)=\tilde{\sigma}_{\theta\theta}\left(\pi;n,M_p\right)=0 \tag{7}$$

Equations (6) and (7) constitute a nonlinear eigen-function problem with s as the eigenvalue and $\tilde{u}_i(\theta;n,M_p)$ and $\tilde{\sigma}_{ij}(\theta;n,M_p)$ the corresponding eigenfunctions.

3.1. NUMERICAL ANALYSIS

The boundary value problem defined in (6) and (7) is solved numerically for mixed-mode angular distributions $\tilde{u}_i(\theta;n,M_p)$ and $\tilde{\sigma}_{ij}(\theta;n,M_p)$ using a nonlinear finite difference scheme and a Newton-Raphson iteration. One difficulty in obtaining mixed-mode solutions numerically for the interfacial crack problem is enforcing of the mode-mix condition: $M_p=(2/\pi)\tan^{-1}[\tilde{\sigma}_{\theta\theta}(\theta=0)/\tilde{\sigma}_{r\theta}(\theta=0)]$ in finite differences with Newton-Raphson iterations. We found that careful introduction of the mode-mix condition is necessary for generating solutions other than the ones found by Wang (1990), Champion and Atkinson (1991) and Sharma and Aravas (1993), i.e., the mode-mix condition is an extra constraint that needs to be enforced. In this study, the mode-mix condition is added to the linear Newton-Raphson equations, therefore resulting in an over determined system, i.e. more equations than unknowns. Numerical solutions to this over determined system of equations are then obtained by using a singular value decomposition (SVD) technique (Press et al., 1988).

Specifically, the system of over determined, linearized Newton-Raphson equations is solved using SVD which converges to a solution that is the best approximation in the least-squares sense. Therefore, care must be taken in monitoring the residuals from the full nonlinear discretized finite difference equations during the numerical procedure. As a comparison, we have applied the same technique to obtain the known mixed-mode solutions with $\sigma \propto r^{-1/(n+1)}$ for homogeneous cracks. Mixed-mode solutions obtained using this method are in excellent agreement with the results reported in Shih (1974) and Symington, Shih and Ortiz (1988). However, error analysis for a hardening exponent of $n=5$ shows that relative errors from mixed-mode calculations are of the order $O(10^{-6})$ compared to $O(10^{-12})$ for pure mode 1 and 2 calculations without using SVD. Relative error in the numerical procedure is defined as $\sum f_i/N$ with maximum value of $\tilde{\sigma}_e(\theta)=1$, where f_i are the residuals of nonlinear finite difference equations and N is the number of degrees of freedom. The larger errors from mixed-mode calculations are believed to be the result of the additional numerical approximation, namely SVD, utilized in the solution procedure as well as the inherent difficulty associated with obtaining solutions to these mixed-mode problems.

Mixed-mode solutions to (6) and (7) with SVD for materials with $n=5$ have relative errors of the order $O(10^{-5})$ compared to $O(10^{-12})$ for the solution $M_p=0.937$ reported in Wang (1990), Champion and Atkinson (1991) and Sharma and Aravas (1993), which converged without SVD. The accuracy of the procedures with and without SVD technique for the interface crack is consistent with that for the crack in a homogeneous material. Thus, it is believed the mixed-mode numerical solutions presented in this paper for interfacial cracks are good approximations of the *exact* solutions. Several of these solutions are shown in Figs. 2 and 3 for $n=5$ and 10,

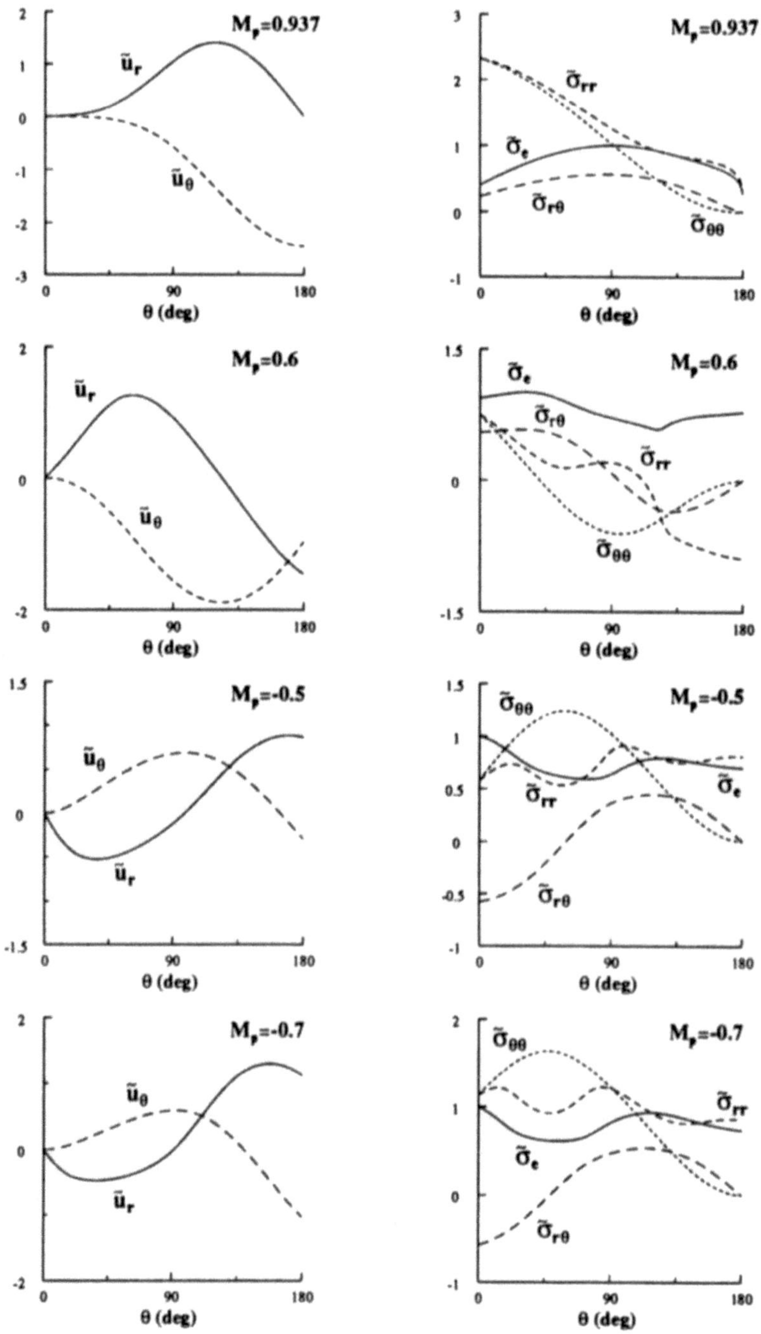

Figure 2. Angular variation of displacements and stresses as in Equation (5) for n=5.

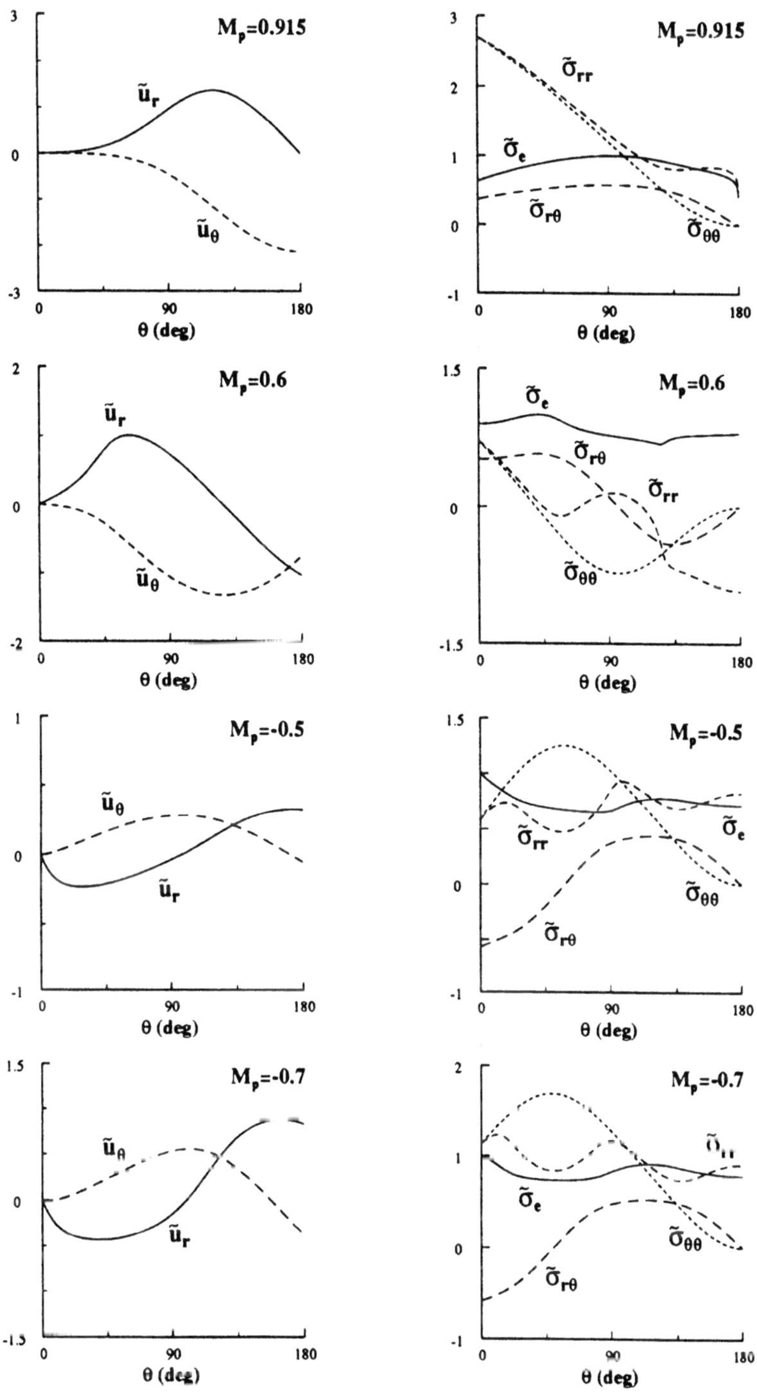

Figure 3. Angular variation of displacements and stresses as in Equation (5) for n=10

TABLE 1. Selected values of $I(n, M_p)$ for interfacial cracks with $n = 5$ and 10.

M_p	$n = 5$	$n = 10$
0.937	2.789	----
0.915	----	2.650
0.600	0.847	0.517
-0.500	0.372	0.140
-0.700	0.899	0.467

respectively, for both positive and negative values of M_p. The solutions obtained by Wang (1990), Champion and Atkinson (1991) and Sharma and Aravas (1993) correspond to $M_p = 0.937$ for $n = 5$ and $M_p = 0.915$ for $n = 10$. Displacements and stresses in both figures are normalized such that the maximum value of effective stress $\tilde{\sigma}_e$ is unity. In addition, Table 1 lists values of $I(n, M_p)$ calculated for solutions in Figs. 2 and 3.

The numerical solutions presented in this section show that these separable asymptotic solutions with $\sigma \propto r^{-1/(n+1)}$ for an elastic-plastic or elastic-creep interface crack admit a continuous variation of the mode-mix value at the crack tip. It should be noted that, however, the range of mode-mix existing at the crack tip changes with the hardening exponent n. It has been found that the upper/lower limit of M_p values decreases/increases as the hardening exponent n of the material increases. In spite of numerous attempts, there is no solution found to converge with values of M_p approaching ± 1. For instance, the upper limits of mode-mix parameter decrease from the value of +1 for $n = 1$ to 0.98 for $n = 5$, to 0.95 for $n = 10$, approximately. On the other hand, the lower limits of mode-mix at the crack tip increases from -1 for $n = 1$, to -0.98 for $n = 5$, to -0.84 for $n = 10$. This finding is consistent with the extreme case of perfect plasticity $(n \to \infty)$ where slip-line fields within the range of $-0.7638 \le M_p(n \to \infty) \le 0.8897$ are found to exist near the crack tip (Fang and Bassani, 1995a). Certain admissible stress discontinuities in these slip-line solutions also provide insight into why mixed-mode solutions for finite n are difficult to obtain, particularly using shooting methods.

4. Full-Field Solutions and Discussion

Fang and Bassani (1995a) have considered full-field mixed-mode problems under small-scale creep conditions to verify interfacial crack solutions developed here. Finite element solutions for various mixed-mode loading conditions display crack tip fields that agree well with predicted asymptotics. Within the region surrounding the crack tip and dominated by the leading order asymptotic solutions, stress states are found both to have spatial and temporal behavior associated the short-time similarity solution of Riedel and Rice (1980) and Bassani and McClintock (1981), i.e. $\sigma \propto (K^2 / rt)^{1/(n+1)}$ as

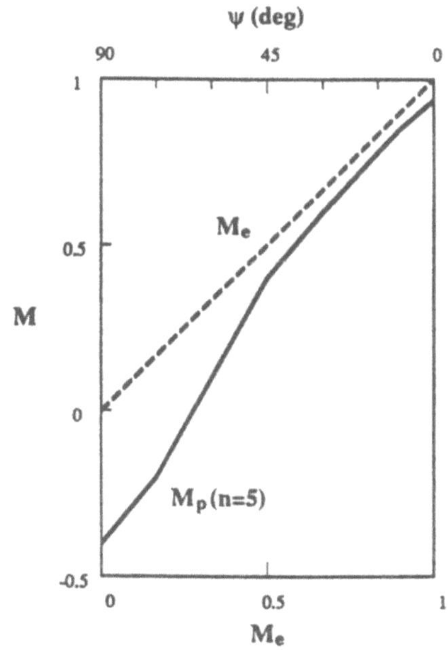

Figure 4. Crack tip mode-mix M_p *vs.* applied mode-mix M_e for small-scale creep.

$r \to 0$ for small time t. The temporal behavior associated with the similarity solution is another verification for the existence of these variable-separable fields. Furthermore, from the small-scale creep finite element analysis, the relationship between crack tip mode-mix and applied elastic loading can be found and is shown in Fig. 4.

In summary, the asymptotic analysis of the crack tip fields near an interface crack between power-law hardening materials and a rigid substrate presented in this paper demonstrates the existence of a continuous spectrum of mixed-mode variable-separable solutions. A scheme combining both finite difference and singular value decomposition has proven effective, and possibly the best scheme utilized to date, for determining mixed-mode variable-separable solutions for both homogeneous as well as interfacial crack problems. The same procedure has been utilized to solve for mixed-mode crack-tip fields for a propagating crack (Fang and Bassani, 1995b).

The solutions presented in this paper admit a nearly full range of mode-mix values at the crack tip with nearly pure tensile modes inadmissible, at least in terms of variable-separable solutions. This latter finding is consistent with the perfect plasticity limit ($n \to \infty$) reported in Fang and Bassani (1995a). Finally we note that for any solution $\tilde{u}_i(\theta; n, M_p)$ and $\tilde{\sigma}_{ij}(\theta; n, M_p)$ to homogeneous system (6) with (1), -1 times that solution also satisfies (1) and (6). The negative M_p asymptotic fields in Figs. 2 and 3 are solutions where $\sigma_{\theta\theta} > 0$ on $\theta = 0$ and $u_\theta = -u_y < 0$ on $\theta = \pi$, i.e. an open crack solution. Nevertheless, from small-scale-creep solutions we found that for pure mode II loading with positive shear ($M_e = 0$) and for small positive values of M_e (positive shear with some tensile loading) the crack tip solutions have M_p negative with $\sigma_{\theta\theta} < 0$ and $u_\theta > 0$ on $\theta = 0$ as $r \to 0$. We also found asymptotic solutions with M_p slightly negative, e.g. $\sigma_{r\theta} > 0$ and $\sigma_{\theta\theta} < 0$ on $\theta = 0$, that correspond to an open crack, i.e. $u_\theta = -u_y < 0$ on $\theta = \pi$; this situation cannot exist in the linear case which can be easily seen from (3).

5. Acknowledgment

This research was supported by the National Science Foundation MRL program, Grant no. DMR-91-20668, and the Laboratory for Research on the Structure of Matter at the University of Pennsylvania.

6. References

Bassani, J. L. and McClintock, F. A. (1981) Creep relaxation of stress around a crack tip, *Int. J. Solids Struct.* **17**, 479-492.

Champion, C. R. and Atkinson, C. (1991) A crack at the interface between two power-law materials under plane strain loading, *Proc. R. Soc. Lond.* **A432**, 547-553.

Fang, N. J.-J. and Bassani, J. L. (1995a) Nonlinear near-tip fields of interfacial cracks, *submitted for publication.*

Fang, N. J.-J. and Bassani, J. L. (1995b) Crack propagation under mixed-mode creep conditions, *submitted for publication.*

Hutchinson, J. W. (1968) Singular behavior at the end of a tensile crack in a hardening material, *J. Mech. Phys. Solids* **16**, 13-31.

Hutchinson, J. W. and Suo, Z. (1992) Mixed mode cracking in layered materials, *Advances in Applied Mechanics* **29**, Academic Press, New York.

Press, W. H., Flannery, B. P., Teukolsky, S. A., and Vetterling, W. T. (1988) *Numerical Recipes,* Cambridge University Press, Cambridge.

Rice, J. R. (1988) Elastic fracture mechanics concepts for interfacial cracks, *J. Appl. Mech.* **55**, 98-103.

Rice, J. R. and Rosengren, G. F. (1968) Plane strain deformation near a crack tip in a power-law hardening material, *J. Mech. Phys. Solids* **16**, 1-12.

Riedel, H. and Rice, J. R. (1980) Tensile cracks in creeping solids, *Fracture Mechanics: Twelfth Conference, ASTM STP 700,* American Society for Testing and Materials, Philadelphia, pp. 112-130.

Sharma, S. and Aravas, N. (1993) On the development of variable-separable asymptotic elastoplastic solutions for interfacial cracks, *Int. J. Solids Struct.* **30**, 695-723.

Shih, C. F. (1974) Small-scale yielding analysis of mixed mode plane-strain crack problems, *Fracture Analysis, ASTM STP 560,* American Society for Testing and Materials, Philadelphia, pp. 187-210.

Shih, C. F. and Asaro, R. J. (1988) Elastic-plastic analysis of cracks on bimaterial interfaces: Part I-Small scale yielding, *J. Appl. Mech.* **55**, 299-316.

Shih, C. F. and Asaro, R. J. (1989) Elastic-plastic analysis of cracks on bimaterial interfaces: Part II-Structure of small-scale yielding fields, *J. Appl. Mech.* **56**, 763-779.

Symington, M., Shih, C. F., and Ortiz, M. (1988) *Tables of plane strain mixed-mode plastic crack tip fields,* Brown University Report MRG/DMR-8714665/1.

Wang, T. C. (1990) Elastic-plastic asymptotic fields for cracks on bimaterial interfaces, *Engng. Fract. Mech.* **37**, 527-538.

MATRIX CRACKING IN CERAMIC MATRIX COMPOSITES WITH CREEPING FIBERS

B. N. COX and D. B. MARSHALL
Rockwell Science Center
1049 Camino Dos Rios
Thousand Oaks, CA 91360

R. M. McMEEKING and M. R. BEGLEY
Dept. Mechanical and Environmental Engineering
University of California, Santa Barbara
Santa Barbara, CA 93106

1. Introduction

High strength can be achieved in ceramic fibers by manufacturing them as fine-grained polycrystals. Unfortunately, fine-grained fibers are prone to creep, because grain boundaries provide easy paths for diffusion. The matrix in many composites, whether formed by chemical vapor infiltration or by some other route, has a much coarser grain structure. Therefore, service life of CMCs reinforced by fine-grained fibers can be limited by fiber creep at temperatures where the matrix remains elastic. This paper is concerned with how fiber creep affects the failure by matrix cracking of a CMC under monotonic loading.

Of primary interest is the passage to failure of 0/90° laminates, which offer at least a partial solution to the requirement of strength under transverse loads. The first damage observed in these materials under loads aligned with the 0° fibers is matrix cracking, which initiates in the 90° plies [1,2]. In many applications, failure may be considered to occur when these matrix cracks grow into through cracks by crossing the entire composite. 1) *Stiffness critical applications*. As long as matrix cracks remain comparable in size to the ply thickness, they cause a relatively minor reduction in the composite stiffness, because crack openings are limited by the crack length [3]. However, when matrix cracks traverse the entire composite, the separated pieces of matrix sustain average strains much lower than those in the fibers. The contribution of the matrix to the composite modulus is then severely degraded. 2) *Hermeticity*. As long as matrix cracks remain small and unconnected, they do not compromise the ability of the composite to act as a seal or liner, impermeable to gases. However, gas permeability obviously rises dramatically in the presence of through cracks. 3) *Hostile environments*. Matrix cracks admit

J. R. Willis (ed.), IUTAM Symposium on Nonlinear Analysis of Fracture, 353–365.

corrosive gases into the interior of the composite, where they attack fibers and the fiber/matrix interfaces. Composite strength and toughness can then fall away very quickly. Even small matrix cracks will introduce some corrosive gases. However, degradation may be limited to near-surface regions and relatively harmless until through cracks form, at which point failure will usually be imminent [4].

At room temperature, a lower bound exists to the stress at which cracks initiating in neighboring 90° plies link up to form through cracks. The bound obtains regardless of the presence of matrix flaws or flaws in clusters of fibers. The lower bound is a certain critical matrix cracking stress, σ_{crit}, to be introduced below. While σ_{crit} varies to some extent with laminar dimensions and residual stresses, it is easily characterized for a single material and remains a fundamental, robust engineering parameter for failure at room temperature.

At high temperatures where fibers creep, this useful simplicity is lost. The critical matrix cracking stress, which now takes a temperature dependent value $\sigma_{crit}(T)$, is still a bound for *rate independent* cracking, i.e., the immediate response to load, but subcritical crack growth mediated by fiber creep will eventually lead to through cracks at stresses much lower than σ_{crit} [5,6]. Lifetime becomes a function of temperature and stress, dictated by the rate of subcritical crack growth. This is the subject considered here.

Fiber creep will also affect the residual ultimate strength of the composite. In applications where matrix cracking is not fatal per se, changes in residual strength must also be modeled. Some remarks will be made on the role of subcritical matrix crack growth on residual strength, but this issue will otherwise be left for future research.

2. Initial, Rate-Independent Matrix Cracking

In 0/90° laminates loaded in the 0° fiber direction, the first manifestation of matrix cracking is tunneling cracks in the 90° plies [1,2]. The tunneling cracks initiate at intrinsic flaws in the 90° plies, which, being loaded transversely, tend to be weaker than the 0° plies. If any stable phase of matrix cracking occurs, it consists of the subsequent spreading of the cracks produced by tunneling into the adjacent 0° plies. The stable phase of growth occurs in the plane strain configuration indicated in Fig. 1a. Before discussing details of the initiation of matrix cracks by tunneling, it is helpful to examine the factors governing the stability and stress levels of the plane strain phase.

Within the plane strain crack, the 90° ply acts effectively as a sharp notch or unbridged interval of width $2h_{90}$, while fiber bridging in the 0° plies acts in the intervals $h_{90} < |x_1| < a$ to shield the crack tip, where x_1 and the crack length a are defined in Fig. 1a. A useful model of the relation between the bridging tractions p and the crack opening displacement $2u$ is [7,8]

$$p = \frac{2fE}{(1-f)E_m}\left[\frac{\tau E_f u}{R}\right]^{1/2} \quad , \tag{1}$$

where τ is the interfacial friction stress (a function of temperature); f is the fiber volume fraction; R is the fiber radius; E_f and E_m are the fiber and matrix moduli; and $E = fE_f + (1-f)E_m$. While this relation is based on several assumptions that may not always be valid in a given CMC [9], it is likely to indicate trends quite well.

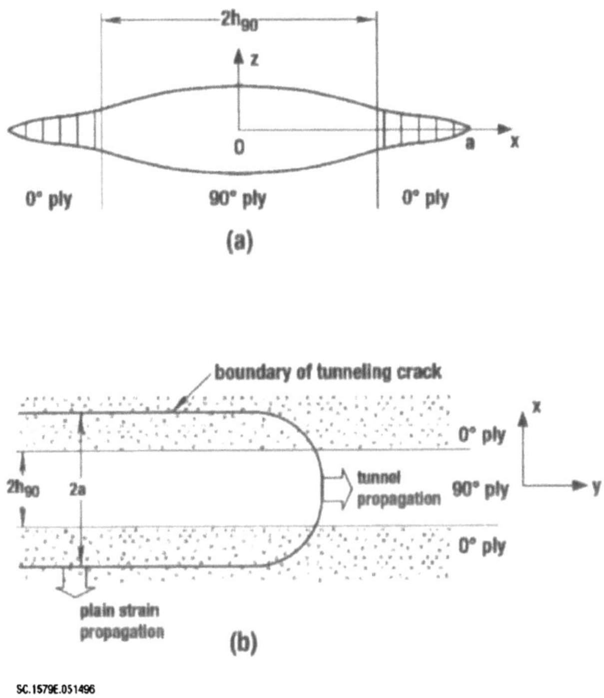

Figure 1. (a) Schematic of a plane strain crack spreading from a 90° ply. *(b)* Schematic of a tunneling crack propagating along a 90° ply and lapping into the adjacent 0° plies.

Crack propagation is assumed here to occur when the crack tip stress intensity factor, K_{tip}, takes a critical value $K_c = \sqrt{E'\Gamma_0}$, where E' is an effective modulus for the laminate [3] and Γ_0 is the effective fracture energy for the matrix crack in the 0° ply. This criterion implies that the matrix is an elastic, brittle material.

Because p is an increasing function of u and crack propagation occurs at a nonzero value of K_{tip}, the steady state crack configuration can be attained for long bridged cracks where the critical stress for propagation, σ_p, becomes independent of the crack length [10]. This limiting stress is the matrix cracking stress, σ_{mc}. It can be readily deduced from $p(u)$ by evaluating the complementary energy density [11]:

$$\sigma_{mc} = \left[\frac{6\tau f^2 E_f E^2 \Gamma_0}{R(1-f)^2 E_m^2} \right]^{1/3} \quad . \tag{2}$$

A characteristic length, the "bridging length scale," a_m, is associated with the traction law, $p(u)$. For the model of fiber bridging underlying Eq. (1), it has the form [7,10]

$$a_m = \frac{\pi E'}{4} \left(\frac{3\Gamma_0}{2} \right)^{1/3} \left[\frac{4\tau f^2 E_f E^2}{R(1-f)^2 E_m^2} \right]^{-2/3} \quad . \tag{3}$$

The stability of the plane strain phase of crack growth is determined entirely by the ratio, h_{90}/a_m, of the 90° ply half-width to the bridging length scale (Fig. 2). If $h_{90}/a_m < 8/3\pi^2$, $\sigma_p(a)$ is a decreasing function; crack growth is unstable and always requires an applied stress $\sigma_a > \sigma_{mc}$. If $h_{90}/a_m > 8/3\pi^2$, $\sigma_p(a)$ is an increasing function; crack growth is stable, commencing at some $\sigma_a < \sigma_{mc}$. (The critical ratio $8/3\pi^2$ is found by equating the critical stress for a Griffith crack of size h_{90}/a_m to the steady state matrix cracking stress, σ_{mc}.)

Whatever the value of h_{90}, σ_p eventually reaches the matrix cracking stress, σ_{mc}, approaching close to it for crack lengths large compared to a_m. It is for this reason that σ_{mc} serves as a design limit for forming long matrix cracks for *any* distribution of flaws. In current CMCs, a_m is typically ~ 0.1 mm at room temperature [10]. Thus, $h_{90}/a_m > 8/3\pi^2$ in most cases, so that crack growth is stable; while $\sigma_p \approx \sigma_{mc}$ for cracks ~ 1 mm or more.

Now consider the events leading up to the propagation of a plane strain crack. The critical stress for tunnel crack propagation can be computed from energy considerations [12,13]. When the tunneling crack is long, the critical stress approaches a constant asymptote, σ_t, which is a function $\sigma_t(d)$ of the distance, $d = a - h_{90}$, to which the tunneling crack laps into the adjacent 0° plies as it propagates along the 90° ply. This function depends on the 90° ply width and the ratio $\eta = \Gamma_{90}/\Gamma_0$ of the effective fracture toughnesses of the 90° and 0° plies. The minimum, $\sigma_t^{(min)}$, of $\sigma_t(a)$ is generally a lower bound to the initiation stress for matrix cracking. The initiation stress will be significantly higher than $\sigma_t^{(min)}$ only if intrinsic flaws are much less than the 90° ply width.

Estimates for existing CMCs suggest this is not usually the case. The sum $h_{90} + d_{min}$, where d_{min} is the value of d corresponding to $\sigma_t = \sigma_t^{(min)}$, will be the half-width of the plane strain crack created by the tunneling event. For $\eta < 1/2$, $d_{min} = 0$; the tunneling crack is confined entirely to the 90° ply. Some nonzero increment in the applied load will be required before propagation of the newly created plane strain crack into the 0° plies occurs. For $\eta > 1/2$, $d_{min} > 0$; and in this case it can be shown that the tunneling crack simultaneously satisfies the criterion for propagation as a plane strain crack into the 0° plies [14,15]. Any increase in load after the tunneling crack has formed will be accompanied immediately by its broadening in the plane strain crack configuration [15].

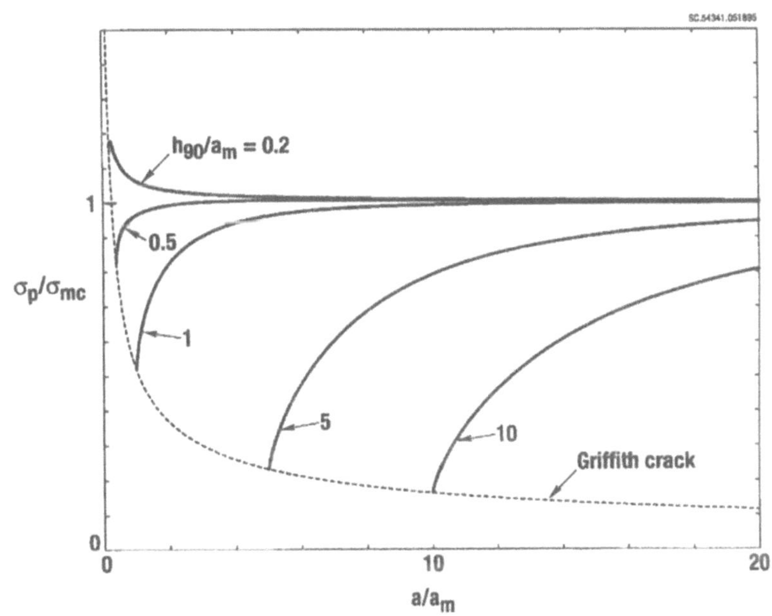

Figure 2. The critical stress for plane strain crack growth.

The correct value of the toughness ratio, η, is difficult to specify for some particular CMC from microstructural considerations alone. It will probably always have to be measured. However, the range $0 \le \eta \le 1$ seems reasonable for composites with identical materials in the 0° and 90° plies [15].

One further condition must be satisfied for a stable phase of plane strain cracking to occur. The tunneling crack must not penetrate beyond the adjacent 0° ply during its formation. If it does, it will enter new unbridged zones, i.e., the next 90° plies, the critical stress for its formation will drop, and the initiating crack will radiate unstably out in all directions without limit. This consideration imposes bounds on the minimum ratio, h_0/h_{90}, of the 0° and 90° ply half-widths for stable cracking, given a value of the toughness ratio, η. A stability map can thus be created from solutions to the tunneling problem (Fig.

3). Figure 3 shows the approximate location of a typical SiC-SiC composite on the map, as well as the repositioning that would be expected if values of the micro-structural parameters f (volume fraction), R (fiber radius), τ (friction stress), or η (toughness ratio) were halved.

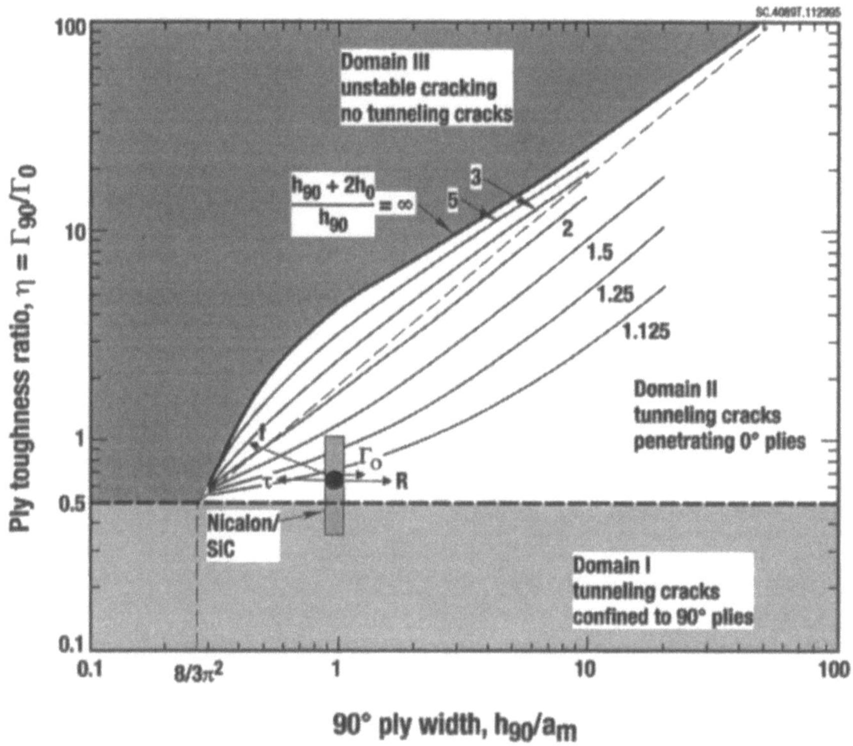

Figure 3. Map showing the regimes where a phase of stable plane strain crack growth can follow crack initiation by tunneling.

Figure 4 shows an alternative view of the role of ply widths in the cracking process. Here the boundaries of stable and unstable cracking are plotted out against the 90° ply half-width, h_{90}, for the fixed ratio $h_{90}/h_0 = 1$. One solid curve shows the critical stress for a plane strain crack (no bridging) of width $2h_{90}$. The higher solid curve shows the stress at which the plane strain crack curve intersects the far boundary of the first 0° plies, i.e., $\sigma_p(h_{90} + 2h_0)$. All stresses are normalized against σ_{mc}, the limiting critical stress for an infinite crack wholly contained in an infinite (hypothetical) 0° ply. The horizontal dashed lines show the stress values at which crack initiation will occur by tunneling for several values of the toughness ratio, η. Here it is assumed that an initial flaw comparable to or slightly greater than the 90° ply widths pre-exists, so that the steady state tunneling stress, σ_t, is appropriately taken as the initiation stress. If only smaller flaws exist, the dashed line should be raised. Whether a stable phase of plane strain crack growth exists is determined by where the horizontal dashed line meets the vertical line drawn at the given value of 90° ply width, h_{90}/a_m. This point is denoted P.

If $\eta < 1/2$, P will fall below the Griffith curve. In other words, after initiation by tunneling, a further load increment to σ_G will be required before the crack extends stably in plane strain. This stable growth will terminate when the applied load rises to $\sigma_p(h_{90} + 2h_0)$.

If $1/2 < \eta < \eta_{crit}$, P will fall between the curve $\sigma_p(h_{90} + 2h_0)$ and the Griffith curve (the case shown in Fig. 4). Here η_{crit} is the critical value of η for the ply width ratio unity in Fig 3. Plane strain crack growth will follow immediately after crack initiation upon further load increase. Once again, stable growth will terminate when the applied load reaches $\sigma_p(h_{90} + 2h_0)$.

If $\eta > \eta_{crit}$, P will fall above the curve $\sigma_p(h_{90} + 2h_0)$ and unlimited unstable cracking will accompany crack initiation.

Thus $\sigma_p(h_{90} + 2h_0)$ should be identified with the engineering design limit, σ_{crit}. Whatever the ply toughness ratio, any matrix cracks will arrest without exceeding the first 0° plies at all stresses below $\sigma_{crit} = \sigma_p(h_{90} + 2h_0)$.

3. Subcritical Crack Growth Mediated by Fiber Creep

Whenever crack arrest depends on fiber bridging, it will give way to subcritical crack growth at high temperatures because of fiber creep. Thus, subcritical crack growth will fix a finite lifetime for all applied loads between σ_G and σ_{crit} in Fig. 4.

A bridging law for creeping fibers coupled to the matrix by friction takes the relatively simple form

$$\dot{u} = 2\lambda p\left[\dot{p} + \beta p\right] \tag{4}$$

provided: there are no rate effects in the interfacial friction; fiber creep is confined to the slip zone adjacent to the matrix crack by the stress concentration due to the crack; the matrix remains elastic; and fiber creep follows a linear law

$$\dot{\varepsilon}_f = \dot{\sigma}_f / E_f + \beta\sigma \quad , \tag{5}$$

where ε_f and σ_f are the fiber strain and stress and β is a creep coefficient [5]. For general fiber creep laws, an explicit relation between u and p and their time derivatives cannot always be derived; but this is unlikely to have any qualitative effect on the subcritical crack growth problem.

The condition for matrix crack extension is

$$K_{tip} = K_a - K_b = K_c \tag{6}$$

where K_a is the applied stress intensity factor and K_b its reduction due to bridging. Consider the differential

$$dK_{tip} = \frac{\partial K_{tip}}{\partial a} da + \frac{\partial K_{tip}}{\partial t} dt \quad . \tag{7}$$

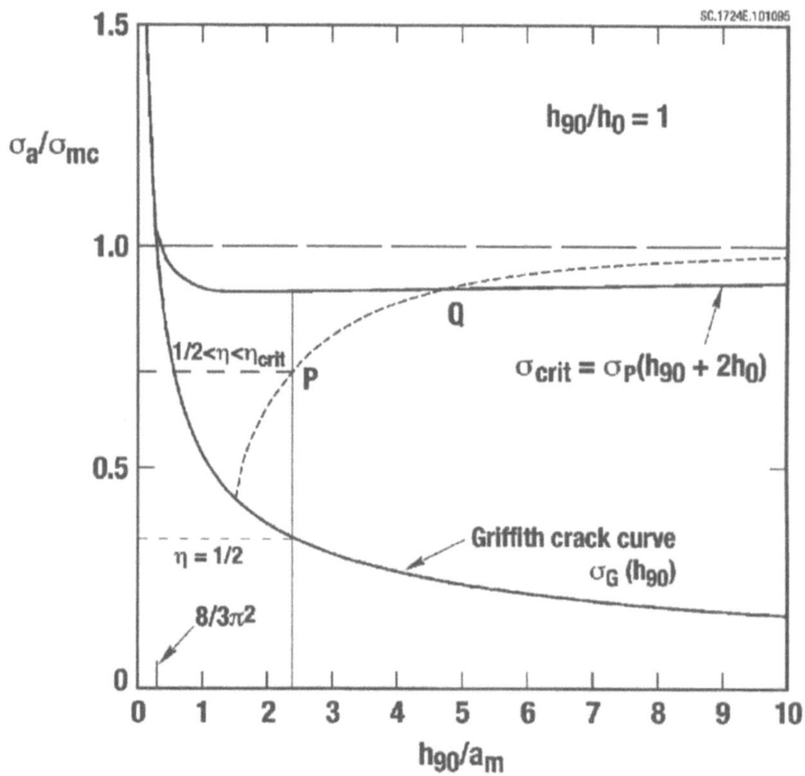

Figure 4. Diagram defining σ_G and σ_{crit}, which bound the domain where life at high temperature will be determined by subcritical plane strain crack growth; and showing schematically the boundary between stresses for which fiber creep rupture follows subcritical crack propagation and vice versa.

For stable rate-independent growth, which is the regime of interest, the first term must be negative. The second term is positive, reflecting the decay of bridging. Maintenance of the condition Eq. (6) implies $dK_{tip} = 0$ and via Eq. (7) there follows a governing differential equation for the crack velocity, da/dt. Numerical methods for solving this equation may be found in Refs. 6 and 16.

Typical solutions for the crack velocity are shown in Fig. 5. Essential characteristics are as follows. 1) The velocity history can be divided into an initial deceleration transient; and a quasi-steady state at long crack lengths, defined by increasing insensitivity of da/dt to the initial conditions [6]. ("Quasi-steady state" is preferred to "steady state" because there is no invariant of the motion as $a \to \infty$ [16].) The deceleration transient is dominated by creep relaxation of fibers right next to the 90° ply, while the quasi-steady state is dominated by the loading history of fibers in a small, propagating zone just behind the crack tip [6]. 2) Unless the 0° ply width is much greater than the 90° ply width, crack propagation across the first 0° ply (whose limits are not marked in Fig. 5), which is followed by catastrophic cracking and therefore defines lifetime in the sense of this paper, is spent entirely in the deceleration transient. 3) The velocity and therefore the time to failure are very strong functions of the applied stress, with da/d$t \to \infty$ and lifetime $t_c \to 0$ as $\sigma_a \to \sigma_{mc}$. 4) At long crack lengths, the crack continues to accelerate, but increasingly slowly.

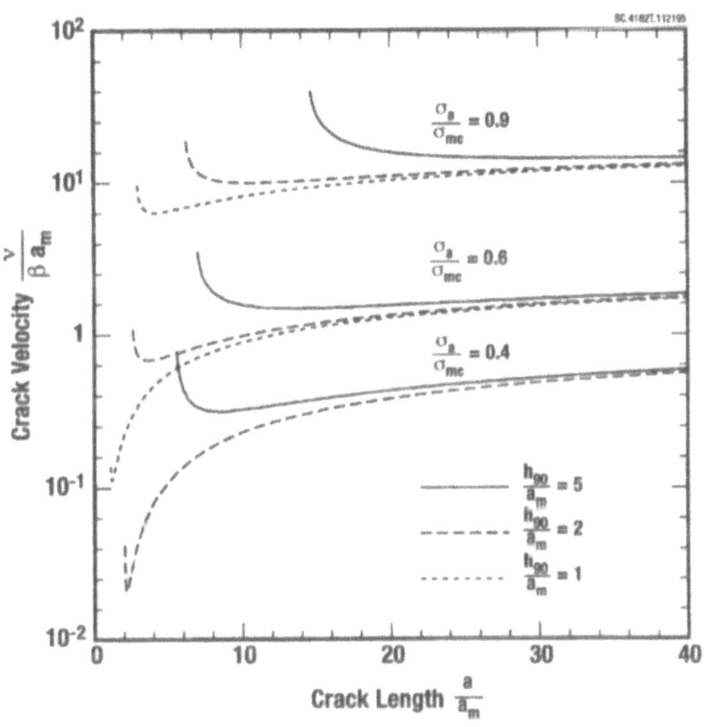

Figure 5. Crack growth rate histories for cracks bridged by creeping fibers (from [6]). Curves are shown for three different 90° ply widths and three different applied stress levels. The initial crack length, which exceeds the 90° ply width for the cases shown, is defined by arrest of the rate-independent plane strain crack that grows from the initiating tunneling crack. The calculated crack growth is pursued well into hypothetically semi-infinite 0° plies.

4. Other Aspects of Composite Failure

4.1 FIBER FAILURE BY CREEP RUPTURE

In the calculations of Fig. 5, fibers were assumed never to fail. In reality, they will fail by creep rupture if the fiber strain exceeds some critical value, ε_{crit}. Fiber failure may occur either after through cracks have formed or during the phase of subcritical crack growth. The former will always be the case for applied stresses, σ_a, sufficiently close to or above σ_{mc}, since crack propagation will then be either very fast or dynamic (depending or whether σ_a is above or below σ_{crit}). Because subcritical matrix crack propagation rates are such strong functions of applied stress, there will be a transition to fiber failure during subcritical propagation as σ_a is reduced. The velocity becomes vanishingly small as the applied load approaches the minimum, σ_G, required for penetration of the initiating tunneling crack into the 0° plies. Lower velocities imply an increasing time for the accumulation of fiber creep strain; and the available time increases much faster than the fiber's creep rate declines [6]. New experiments are required to confirm this transition.

4.2 CRACK PROPAGATION FOLLOWING FIBER FAILURE IN THE SMALL SCALE BRIDGING LIMIT

In the domain where fiber failure occurs during subcritical crack growth, the surviving bridging zone will often be much smaller than other crack and specimen dimensions. Small scale bridging conditions will then apply. After some growth, the crack configuration will approximate a steady state configuration governed solely by the applied stress intensity factor, K_a. The velocity will approach a steady state value, V_{ss}, which will be a function of K_a rather than directly of crack length. Complete solutions for this steady state have been presented in Ref. [17]. The solutions show many similarities to those for equilibrium craze zones in polymers, e.g., Refs. [18-21], but also some distinct features peculiar to the mechanics of brittle matrix composites [17]. In particular, crack growth occurs only when K_a exceeds K_C. At the lowest stresses, i.e., small K_a-K_c

$$V_{ss} \sim (K_a - K_c)^{3/2} \varepsilon_{crit}^{1/2} \quad , \qquad (8)$$

where ε_{crit} is the critical fiber strain. Small K_a-K_c is probably the most important regime for the present context, since fiber failure during subcritical crack growth occurs only for stresses just above σ_G, which means that K_a is near K_C.

As illustrated by the asymptotic result Eq. (8), the crack velocity following fiber failure increases much faster than when the fibers remain intact (cf. Fig. 5). For practical purposes, the onset of fiber failure in the wake of a subcritical crack could be considered equivalent to structural failure.

4.3 THE LIMITS OF SINGLE CRACK MODELS

The constitutive model of Eq. (4) was derived for a single matrix crack and the resulting crack growth curves were all calculated in the absence of any stress redistribution effects that might arise from similar, nearby cracks. At room temperature, multiple matrix cracks are the common case; and rather than being noninteracting, their separation is in fact determined by marginal overlap between their interfacial slip zones. In a body with multiple matrix cracks, the overlap ought to be factored into the bridging law for creeping fibers. Multiple cracks should indeed be expected in a smooth specimen. But at high temperatures, recent experiments on *notched* SiC/SiC composites have shown that while multiple cracks do initiate, only one propagates very far, so that the problem reverts to that of a single dominant matrix crack [22]. Whether multiple or single cracks will occur at high temperatures in the presence of blunt stress concentrators remains a topic of current research.

When arrays of cracks arise in the 90° plies in smooth laminates, crack interaction effects ought to be computed, e.g., by employing weight functions for arrays of cracks instead of those for single cracks [13]. Crack interactions will accelerate the formation of through cracks via crack coalescence. For an isolated matrix crack growing away from a 90° ply through neighboring 0° plies, unstable growth to a through crack will commence when it has reached the far side of the 0° plies. If two cracks grow towards one another into the same 0° ply from successive 90° plies, they each need cross only half the 0° ply before coalescence leads to catastrophic cracking; and their mutual interaction will accelerate even this diminished phase of subcritical crack growth. The boundary marked σ_{crit} in Fig. 4 will move down, because of the stricter limits that must be satisfied for the initiation (tunneling) phase to terminate in crack arrest. How far down the boundary will move will depend on how the 0° ply half-width, h_0, compares to the bridging length scale, a_m. If $h_0/a_m >> 1$, the change will be slight, because the rate-independent plane strain cracking stress, $\sigma_p(a)$, will approach close to the limit σ_{mc} before crack interaction effects become significant. If $h_0 \sim a_m$, the fall in σ_{crit} will be considerable, but readily calculated [13]. The failure map retains the same appearance.

Even when multiple cracks exist, the trends and failure maps presented here will remain qualitatively true. The overlap of slip zones does not change the essential features of the bridging constitutive law, namely that p is an increasing function of u and decays with time. Furthermore, the law already contains implicit assumptions about micromechanics which are doubtful in detail. The constitutive law of Eq. (4) serves as a good guide to possible fracture behavior, with its details *always* understood to require empirical calibration, preferably using fracture data rather than micromechanical tests [9]. Whatever the exact form chosen for a bridging law like Eq. (4), once it is calibrated against experiments, it will probably yield accurate predictions for a wide range of other stress levels.

Acknowledgments BNC was supported under AFOSR Contract No. F49620-94-C-0030; DBM under Rockwell Independent R&D Funding; and RMM and MRB under the ARPA University Research Initiative (ONR Contract No. N-0014-92-J-1808).

References

1. D.S. Beyerle, S.M. Spearing, and A.G. Evans, "Damage Mechanisms and the Mechanical Properties of a Laminated 0/90 Ceramic/Matrix Composite," *J. Amer. Ceram. Soc.* **75**, 3321-30 (1992)

2. F. Lamouroux, M. Steen, and J.L. Valles, "Uniaxial Tensile and Creep Behaviour of an Alumina Fibre-Reinforced Ceramic Matrix Composite: I-Experimental Study," *J. European Ceram. Soc.*, in press.

3. Z.C. Xia, R.R. Carr, and J.W. Hutchinson, "Transverse Cracking in Fiber-Reinforced Brittle Matrix, Cross-Ply Laminates," *Acta Metall. Mater.* **41**, 2365-75 (1993).

4. F. Heredia, J.C. McNulty, F.W. Zok, and A.G. Evans, "Oxidation Embrittlement Probe for Ceramic Matrix Composites," *J. Amer. Ceram. Soc.***78**, 2097-2100 (1995).

5. M.R. Begley, A.G. Evans, and R.M. McMeeking, "Creep Rupture in Ceramic Matrix Composites with Creeping Fibers, " *J. Mech. Phys. Solids* **43**, 727-40 (1995).

6. M.R. Begley, B.N. Cox, and R.M. McMeeking, "Time Dependent Crack Growth in Ceramic Matrix Composites with Creeping Fibers," *Acta Metall. Mater.* **43**, 3927-36 (1995).

7. D.B. Marshall, B.N. Cox, and A.G. Evans, "The Mechanics of Matrix Cracking in Brittle-Matrix Fiber Composites," *Acta Metall. Mater.* **33**, 2013-21 (1985).

8. L.N. McCartney, "Mechanics of Matrix Cracking in Brittle-Matrix Fibre-Reinforced Composites," *Proc. Roy. Soc. London* **A409**, 329-50 (1987).

9. B.N. Cox, "Life Prediction for Bridged Cracks," in *Life Prediction for Titanium Matrix Composites,*" ed. W.S. Johnson, J. Larsen, and B.N. Cox, ASTM STP (ASTM, Philadelphia, 1995), in press.

10. B.N. Cox and D.B. Marshall, "Concepts for Bridged Cracks in Fracture and Fatigue," *Acta Metall. Mater.* **42**, 341-63 (1994).

11. D.B. Marshall and B.N. Cox, "A J Integral Method for Calculating Steady State Matrix Cracking Stresses in Composites," *Mechanics of Materials* **7**, 127-33 (1988).

12. S. Ho and Z. Suo, "Tunneling Cracks in Constrained Layers," *J. Appl. Mech.* **60**, 890-4 (1993).

13. Z.C. Xia and J.W. Hutchinson, "Matrix Cracking of Cross-Ply Ceramic Composites," *Acta Metall. Mater.* **42**, 1933-45 (1994).

14. J.L. Beuth, "Cracking in Thin Bonded Films in Residual Tension," *Int. J. Solids Structures* **29**, 1657-75 (1992).

15. B.N. Cox and D.B. Marshall, "Crack Initiation in Brittle Fiber Reinforced Laminates," *J. Amer. Ceram. Soc.*, in press.

16. B.N. Cox and L.R.F. Rose, "Time or Cycle Dependent Crack Bridging," *Mechanics of Materials* **19**, 39-57 (1994).

17. M.R. Begley, B.N. Cox, R.M. McMeeking, "Creep Crack Growth with Small Scale Bridging in Ceramic Matrix Composites," submitted to *Acta Metall. Mater.*

18. W.G. Knauss, "Time Dependent Fracture and Cohesive Zones," *Journal of Engineering Materials and Technology* **115**, 262-7 (1993).

19. L.-O. Fager, J.L. Bassani, C.-Y. Hui, and D.-B. Xu, "Aspects of Cohesive Zone Models and Crack Growth in Rate-Dependent Materials", *International Journal of Fracture* **52**, 119-44 (1991).

20. E.J. Kramer and E.W. Hart, "Theory of Slow, Steady Crack Growth in Polymer Glasses," *Polymer* **25**, 1667-78 (1984).

21. R.A. Schapery, "A Theory of Crack Initiation and Growth in Viscoelastic Media," *International Journal of Fracture* **11**, 141-59 (1975).

22. D.R. Mumm, W.L. Morris, M.S. Dadkhah, and B.N. Cox, "High Temperature Crack Growth in Ceramic Composites," *Thermal and Mechanical Test Methods and Behaviour of Continuous-Fiber Ceramic Composites, ASTM STP 1309*, ed. M.G. Jenkins, S.T. Gonczy, E. Lara-Curzio, N.E. Ashbaugh, and L.P. Zawada (ASTM, Philadelphia, 1996).

FIBER FRAGMENTATION IN A SINGLE FILAMENT COMPOSITE AND THE STRENGTH OF FIBER-REINFORCED CERAMICS

S. L. PHOENIX[†], M. IBNABDELJALIL[††] and C.-Y. HUI[†]
[†]*Theoretical and Applied Mechanics*
Cornell University, Ithaca, NY 14853, USA
[††]*Engineering Science and Mechanics, Virginia Polytechnic Institute and State University, Blacksburg, VA 24061, USA*

Abstract. Recent advances in understanding the statistical fragmentation process of a fiber in a single filament composite have also been a key in understanding the failure process in fiber-reinforced ceramics, especially the strength distribution and size effects. Comparisons with Monte Carlo simulations suggest that accurate reliability estimates are possible for very long composites. This paper reviews the key theoretical developments.

1. Introduction and Overview

For the past 25 years, going back to Aveston *et al.* (1971), the single filament composite has been an important tool in the study of the failure of fibrous composites. The model involves a single brittle fiber embedded along the center-line of a 'dog-bone', matrix specimen, where the matrix has considerable plasticity or ductility. As the overall strain is increased, the fiber fails progressively at random flaws producing an increasing number of shorter and shorter fragments. Around a fiber break the tensile stress decays from the far field value to zero at the break over a certain fiber length, which grows with strain level. In these unloading zones no further breaks occur as the flaws there tend to become 'shielded' from further stress increases. Eventually the fragmentation ceases leaving a limiting distribution of fiber fragments along the dog-bone as unloading zones saturate the fiber.

In interpreting experiments, it is typically assumed that the magnitude of the shear traction in the unloading zone is a constant, τ_s, as though the matrix is plastic, or, Coulomb sliding friction prevails; elsewhere the shear tractions are zero. The actual mechanics of load interaction near a break is complicated (Nairn (1992)), but the practical realism of the constant τ_s assumption is surprising. Often the fibers are assumed to follow a Weibull, weakest-link model for strength with shape parameter ρ and scale parameter, σ_0 measured at some gage length l_0. An even simpler assumption is that the fiber strength is just a unique constant σ_0, (i.e., $\rho = \infty$). Regardless, the experimental fragment length distribution reflects large variability.

Most frequently the single filament composite experiment has been used to determine the interfacial shear stress τ_s, say for a new matrix or fiber surface treatment (e.g. Netravali *et al.* (1989), Henstenburg and Phoenix (1989), and Wagner and Eitan (1990). It has also been been used to

J. R. Willis (ed.), IUTAM Symposium on Nonlinear Analysis of Fracture, 367–378.

determine the fiber strength distribution at the mean fragment length (Henstenburg and Phoenix (1989) and Gulino and Phoenix (1991)).

In modelling the failure of ceramic fiber-ceramic matrix composites, the single filament composite concept has become an important element. Curtin (1991b) developed the key ideas, pointing out that the ceramic matrix allows the fibers to engage in 'global load sharing' wherein the load of a failed fiber is transferred laterally, by frictional shear coupling through matrix slabs, equally onto all surviving fibers. Thus an individual fiber behaves like a single filament composite in a large effective medium of the average response of all the other fibers and matrix. Literature now exists on determining the strength distribution of the composite, and various approximations for the composite mean strength, failure strain, coefficients of variation, size effects and fiber pullout distributions. See for example Phoenix and Raj (1992), Phoenix (1993), Neumeister (1993), Curtin (1993) and Ibnabdeljalil and Phoenix (1995). These theories are based on approximations, which all incorrectly model the subtleties of unloading zones near fiber breaks. They work well for $\rho > 7$, but exhibit large inaccuracies for $\rho < 2$ which has importance in modelling fibers heavily damaged during processing of the composite.

Developing accurate statistical theory for fiber fragmentation has proved challenging. Monte Carlo simulation models were developed by Fraser *et al.* (1983), and by Henstenburg and Phoenix (1989) who gave basic scalings. Curtin (1991a) developed a theory for $\rho > 0$ using results of Widom (1966) for a problem equivalent to the 'car parking problem' of Rényi (1958). Kimber and Keer (1982) considered the same problem in the context of matrix cracking. The problems of Widom (1966), Rényi (1958), and Kimber and Keer (1982) all correspond to unique fiber strength ($\rho \to \infty$).

With respect to the mean composite strength, Neumeister (1993) developed what appeared to be an accurate approximation as compared to a numerical implementation of the theory of Curtin (1991a). A puzzling feature, however, was the appearance of an anomaly for $0 < \rho < 2$ wherein the strain to failure suddenly jumped to infinity. In fact, the failure strain is finite for all $\rho > 0$. This result turned out to be an artifact of the solution of Curtin (1991a), which has turned out not to be exact.

Recently, Hui *et al.* (1995) developed an exact, closed form solution for fragmentation in the single filament composite, and some generalizations are in Hui *et al.* (1996). The key was to derive exact equations governing the evolution of fiber fragments for all $\rho \geq 0$ (including random initial breaks with spacing $1/\alpha_0$). The final distribution for normalized fragment length was also obtained. The special case of unique fiber strength ($\rho = \infty$), was a limiting case with the normalized fragment lengths lying between 1/2 and 1. The convergence was slow, being $O(\rho^{-1/2})$. As $\rho \to 0$, the normalized fragment length was found to be exponentially distributed.

Hui et al. (1995) also applied this fiber fragmentation solution to the composite model of Curtin (1991b) to obtain the ultimate strength μ^* for all $\rho > 0$. Expressions were given in terms of elementary functions that allowed computation of μ^* and the associated dimensionless fiber stress s^* to any degree of accuracy for any $\rho > 0$.

Recently, Phoenix, Ibnabdeljalil and Hui (1996) used the solution of Hui *et al.* (1995) to determine approximations to the distribution for strength of a long composite with a finite number of fibers n. The results were favorably compared to numerical results from a Monte Carlo simulation model. The comparison suggests that predictions are possible for composite lengths orders of magnitude beyond what a Monte Carlo simulation can handle. We now review key features of these new developments.

2. Basic Model and Scalings for Strength and Length

Prior to any loading, flaws and discontinuities (flaws of strength zero) are assumed to occur along the fiber according to a compound Poisson process in distance where the average number of flaws per unit length with strength less than σ is given by

$$\Lambda(\sigma) = (1/l_0)(\sigma/\sigma_0)^\rho + \alpha_0. \tag{1}$$

The first term leads to the usual Poisson-Weibull model for fiber strength where $\sigma_0 > 0$ is the Weibull scale parameter relative to a continuous fiber test length l_0 in simple tension, and the exponent $\rho > 0$ is the usual Weibull shape parameter. The second term adds random discontinuities along the fiber with mean rate α_0 per unit length.

Around a fiber break, the length required to build up to the far field fiber stress σ from zero is the slip length $l_f(\sigma) = r\sigma/(2\tau_s)$ where r is the fiber radius and τ_s is the interfacial shear stress. Within this length, the fiber is 'shielded' from further breaks. The normalizing scales for strength and length are respectively

$$\sigma_c = \sigma_0\{\tau_s l_0/(\sigma_0 r)\}^{1/(\rho+1)} \quad \text{and} \quad \delta_c = l_0\{\sigma_0 r/(l_0\tau_s)\}^{\rho/(\rho+1)}. \tag{2}$$

A simple interpretation is that at the fiber stress level σ_c the mean number of Weibull flaws ($\alpha_0 = 0$) over the length δ_c, which have strength less than or equal to σ_c, is exactly one. Also at the stress level σ_c, the length δ_c is exactly double the slip length $l_f(\sigma_c)$, that is, the total shielded length around a fiber break. Thus, we work henceforth in terms of a normalized stress $s = \sigma/\sigma_c$, and distance is always actual distance divided by δ_c. Then the mean number of flaws per unit dimensionless length at the dimensionless stress s is

$$\Lambda(s) = \alpha + s^\rho \tag{3}$$

where $\alpha = \alpha_0\delta_c = \alpha_0 l_0/(\sigma_c/\sigma_0)^\rho$ is the mean number of initial breaks in δ_c. Furthermore, the slip length becomes $s/2$ at fiber stress s. The main complications are that new breaks cannot occur in slip zones around old breaks, which may overlap, and the stress carried by a fiber in a slip zone is reduced depending on the distance to the nearest break.

3. Distribution of Fragment Lengths versus Stress Level

A key quantity is $p(s, x)$, the density of the number per unit length of inter-break spacings or fragments of length x at stress s. The normalizing condition is

$$\int_0^\infty xp(s,x)dx = 1, \quad \text{for all} \quad s \geq 0. \tag{4}$$

Hui *et al.* (1995) derived the governing differential equations for $p(s,x)$ in terms of fragments of length x both being created and disappearing as the stress s is increased. Here we quote only their solution for the important special case $\rho > 0$ and $\alpha = 0$. We let

$$A_0(s) = s^{2\rho} \exp(\lambda s^{\rho+1}) \Psi(s^{\rho+1}/2), \tag{5}$$

$$\Psi(s^{\rho+1}/2) = \exp[-2\lambda \int_0^{s^{\rho+1}/2} t^{-1}(1-e^{-t})dt], \tag{6}$$

and $\lambda = \rho/(\rho+1)$. Then

$$p(s,x) = A_0(s) \exp(-s^\rho x), \qquad s < x \tag{7a}$$

$$p(s,x) = 2\rho \int_x^s t^{-1} A_0(t) \exp[-t^\rho(x+t/2)]\, dt + p(x,x), \quad s/2 < x < s \tag{7b}$$

$$p(s,x) = 2\rho \int_x^{2x} t^{-1} A_0(t) \exp[-t^\rho(x+t/2)]\, dt + p(x,x), \quad x < s/2 \tag{7c}$$

where

$$p(x,x) = x^{2\rho} \exp[-x^{\rho+1}/(1+\rho)\, \Psi(x^{\rho+1}/2). \tag{8}$$

Hui *et al.* (1995) reduced this to Widom's (1966) results in the limit $\alpha = 0$ and $\rho \to \infty$ with $s \to \infty$. They also obtained $F(x)$, the fraction of fiber segments in an infinitely long fiber with length less than x at saturation ($s \to \infty$), and showed that $F(x)$ has a lower tail for small $x > 0$ that behaves as $c(\rho)x^{2\rho+1}$ for an explicit constant $c(\rho)$. Fig. 1 illustrates the behavior of $F(x)$ on Weibull probability coordinates showing the slow convergence in ρ.

4. Statistics for Composite Strength at a Cross-Section

For a composite with n fibers we consider an applied far field fiber stress s, and consider the statistics of the load carried on a cross-sectional plane in terms of nearby breaks, which reduce the fiber loads at that plane. The first quantity of interest is the asymptotic mean normalized stress function $\mu(s)$ for a large composite ($n \to \infty$) for $s > 0$. We ignore the tensile load carried by the matrix so the 'effective' composite stress is the calculated stress times the fiber volume fraction, f. This is understood in all calculations.

To determine $\mu(s)$, one method has been based on estimating $g(s,y)$, $0 \leq y < \infty$, the probability density function for the absolute distance y to the nearest break in a fiber from a plane. An alternative and equivalent method for calculating $\mu(s)$, is to travel along a given fiber, and calculate the average

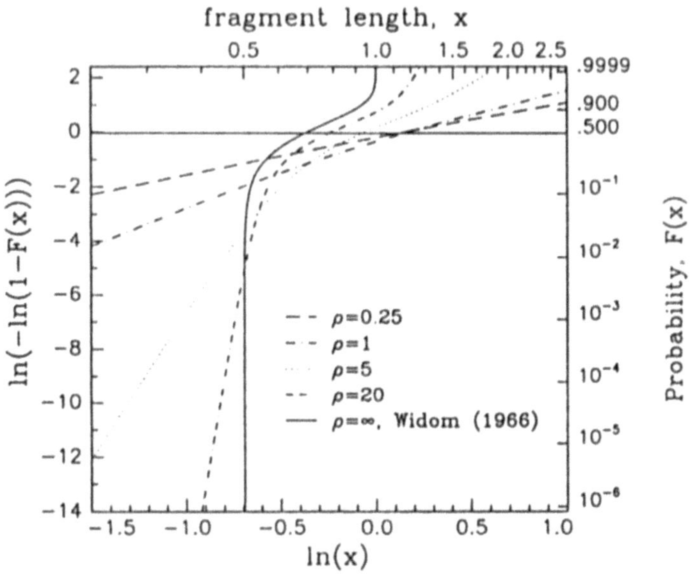

Figure 1. Limiting fragment length distribution, $F(x)$, on Weibull coordinates for $\alpha = 0$ and various ρ values. Note the power form of the lower tail with exponent $2\rho+1$. (Taken from Hui *et al.* (1995).)

fiber stress. It is quickly seen that

$$\mu(s) = \int_0^s (x^2/2)p(s,x)dx + s\int_s^\infty \{(x-s)+s/2\}p(s,x)dx, \tag{9}$$

which uses the exact closed form solution to $p(s,x)$ given by (5) to (8).

The next key quantity is the variance function $\Gamma_0(s)/n$ for a composite cross-section, where $\Gamma_0(s)$ is the variance in the stress of an arbitrarily selected fiber. A calculation travelling along a fiber leads to

$$\Gamma_0(s) = (1/3)\int_0^s x^3 p(s,x)dx + s^2\int_s^\infty xp(s,x)dx$$

$$- (2/3)s^3\int_s^\infty p(s,x)dx - \mu(s)^2. \tag{10}$$

Also of interest is the covariance function $\Gamma_\delta(s)/n$ for the stress at two cross-sections separated by a normalized distance $\delta > 0$. Actually its local behavior for small δ is what is needed in using the asymptotic theory for Gaussian processes from Leadbetter, Lindgren and Rootzén (1983). For larger ρ, Phoenix *et al.* (1996) have shown

$$\Gamma_\delta(s) \approx \Gamma_0(s)\{1 - 6(\delta/s)^2 + o(\delta/s)^2\}. \tag{11}$$

The next quantity of interest is the maximum of the asymptotic mean stress $\mu(s)$ called μ^*, being the composite stress where the load-strain curve begins to decrease. The corresponding value of s is called s^*, so $\mu^* = \mu(s^*)$. From an asymptotic analysis Hui *et al.* (1995) obtained extremely accurate approximations good for $\rho \geq 1$ and for $0 < \rho < 0.2$. Based on these approximations, an accurate approximation for all $\rho > 0$ is simply

$$s^* \approx \{(2/\rho)(4\rho+2)/(4\rho+1)\}^{1/(\rho+1)}, \tag{12}$$

being a slight improvement over the version in Hui *et al.* (1995). Earlier approximations in the literature have reflected incorrectly the unbounded behavior as $\rho \to 0$.

Apart from being able to calculate the true value of μ^* numerically for all $\rho \geq 0$ using (7) and (9), Hui *et al.* (1995) obtained the three term asymptotic expansion

$$\mu^* \approx s^*[1 + s^{*\rho+1}/2 + \theta s^{*2\rho+2}]\exp\{- s^{*\rho+1}(1-\lambda s^{*\rho+1}/8)\} \tag{13}$$

where $\lambda = \rho/(\rho+1)$ and $\theta = \{(7\rho+12)/(2\rho+3)\}/24$. This approximation has less than 1% error for $\rho > 3$ and at $\rho = 1$ the error is only about 8% low. Empirically, it seems that

$$\mu^* \approx 1 - c\rho + d\rho^2 \tag{14}$$

for small $\rho > 0$, where $c \approx 0.5$ and $d \approx 0.2$ and it is consistent with the fact $\mu^* = 1$ at $\rho = 0$, $\alpha = 0$. Figure 2 plots these approximations against the true μ^* obtained numerically. Phoenix *et al.* (1996) pursue this issue analytically.

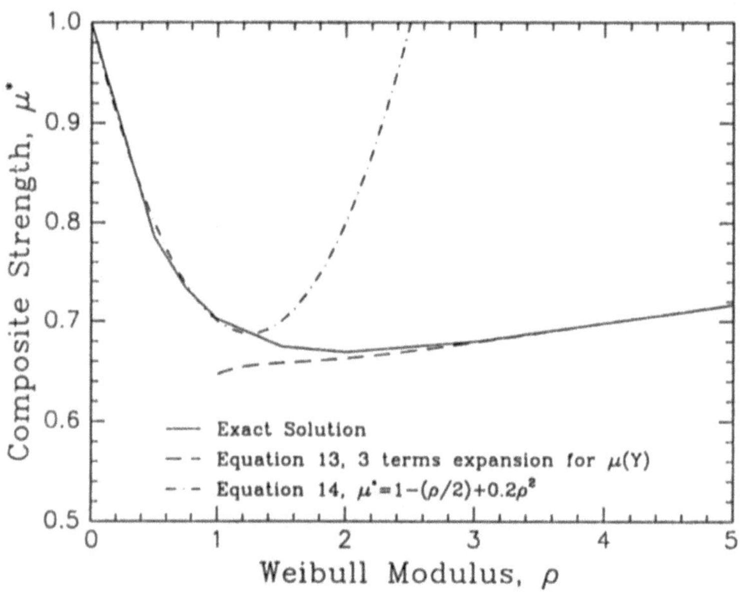

Figure 2. Plot of normalized asymptotic mean composite strength μ^* versus shape parameter ρ for $\alpha = 0$. Shown are exact solution from Hui *et al.* (1995) and approximations based on large and small ρ.

Next we evaluate the asymptotic, normalized, standard deviation for the strength of a composite cross-section. This is denoted γ^*_n and it is given by

$$\gamma^*_n = \{\Gamma_0(s^*)/n\}^{1/2}. \tag{15}$$

Phoenix *et al.* (1996) have done this for $\rho \geq 0$ and $\alpha = 0$ by numerically integrating in (10) with $s = s^*$. Figure 3 shows a plot of $\gamma^*_n n^{1/2} = \{\Gamma_0(s^*)\}^{1/2}$. Also shown is an approximation for $\{\Gamma_0(s^*)\}^{1/2}$ based on one in Phoenix and Raj (1992), and is given by

$$\Gamma_0(s^*) \approx s^{*2}\{[1 - \exp\{-s^{*\rho+1}\}]/3 - [1 - \exp\{-s^{*\rho+1}\}]^2/4\}, \tag{16}$$

and we take s^* as given by (12). This approximation works well for $\rho > 0.8$.

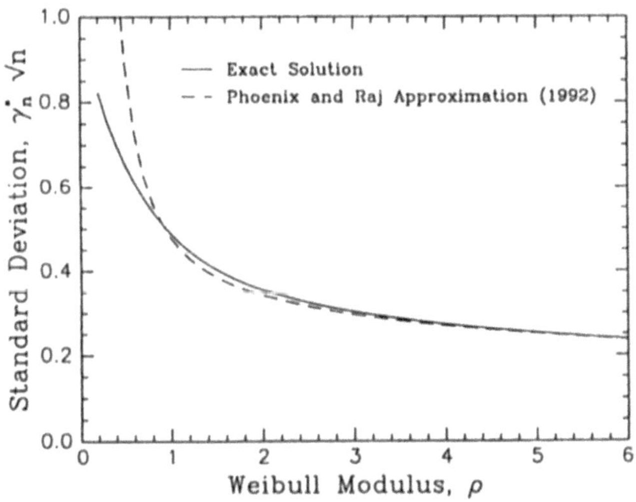

Figure 3. Plot of normalized asymptotic standard deviation $n^{1/2}\gamma^*_n$ for composite strength versus shape parameter ρ for $\alpha = 0$. Shown are exact solution and an approximation based on Phoenix and Raj (1992).

Lastly we determine a refined mean μ^*_n and standard deviation γ^*_n, respectively, when n is small. A refinement, from Phoenix and Raj (1992), is

$$\mu^*_n = \mu^* + \Delta^*_n, \tag{17}$$

$$\Delta^*_n \approx n^{-2/3}\{(\rho/3)s^{*\rho+2} + [(1-\rho)/3]s^{*2\rho+3}\}^{2/3}\{(\rho+1)^2 s^{*\rho}[1-2s^{*\rho+1}]\}^{-1/3} \tag{18}$$

For $\rho > 4$, Δ^*_n is roughly $0.25n^{-2/3}$. To obtain a corrected standard deviation γ^*_n Phoenix *et al.* (1996) suggest an adaptation of a result for classical bundles, which is

$$\gamma^{**}_n = \gamma^*_n[1 - (0.317)(\mu^*/\gamma^*_n)^2\rho^{-2/3}e^{4/(3\rho)}n^{-4/3}]^{1/2}. \tag{19}$$

The main result (Phoenix and Raj (1992)) is that the strength at a composite cross-section is approximately normally distributed with mean μ^*_n and standard deviation γ^{**}_n. The covariance, $\Gamma_\delta(s^*)/n$, for two planes δ apart is roughly $\Gamma_0(s^*)\{1 - 6(\delta/s^*)^2\}/n$.

5. Asymptotic Distributions for the Strength of Long Composites

To estimate $H_{m,n}(\sigma/\sigma_c)$, the distribution function for the strength of a long composite of length $L = m\delta_c$, there are two approaches (Phoenix *et al.* (1996)): The first is to view the strength of planes along the composite as being represented by a continuous Gaussian process with mean μ^*_n and covariance function, $\Gamma_\delta(s^*)/n$, and then to use asymptotic results for minima in Gaussian processes. The second is to view the composite as a discrete chain of m' short bundles or 'slabs' each of length $\beta\delta_c$, so that $L = m'\beta\delta_c$, and where $0 < \beta \leq 1$ is an adjustable parameter. Note that $m' = m/\beta$. Then each bundle has independent strength following a normal (Gaussian) distribution with mean μ^*_n and standard deviation γ^{**}_n. The composite fails when the weakest plane or bundle fails.

In the Gaussian process version (Leadbetter, Lindgren and Rootzén (1983)) we let

$$a_{m,n} = \gamma^{**}_n/[2\log_e(m)]^{1/2} \tag{20a}$$

and

$$b_{m,n} = \mu^*_n - \gamma^{**}_n\{[2\log_e(m)]^{1/2} - \log_e(s^*\pi/2)/[2\log_e(m)]^{1/2}\}. \tag{20b}$$

Then the approximation to $H_{m,n}(\sigma/\sigma_c)$ is the double exponential form

$$H_{m,n}(\sigma/\sigma_c) \approx 1 - \exp\{-\exp\{[(\sigma/\sigma_c) - b_{m,n}]/a_{m,n}\}\}, \quad \sigma \geq 0, \tag{20c}$$

and its median is approximately

$$\sigma^*_{m,n}/\sigma_c \approx b_{m,n} + \gamma^{**}_n\log_e(\log_e(2))/[2\log_e(m)]^{1/2}. \tag{20d}$$

In the chain-of-bundles version let $z = [(\sigma/\sigma_c) - \mu^*_n]/\gamma^{**}_n$ be the standardized stress. From Feller (1968) the standard normal distribution function has a lower tail described by

$$\Phi^*(z) = (2\pi)^{-1/2}\exp(-z^2/2)/|z|, \quad z \ll 0. \tag{21}$$

Using the fact that $(1-\Phi^*)^{m'} \approx \exp(-m'\Phi^*)$ for large m' we may write the approximation

$$H_{m,n}(\sigma/\sigma_c) \approx 1 - \exp\{-(m/\beta)\Phi^*([(\sigma/\sigma_c) - \mu^*_n]/\gamma^{**}_n)\}, \quad \sigma \geq 0. \tag{22}$$

In essence we are assuming that the composite strength is given by the weakest of $m' = m/\beta$ cross-sections spaced $\beta\delta_c$ apart, where β is chosen so that the strengths are effectively independent and identically distributed. A double exponential version of (22) is less accurate. Using results from Cramér (1946), the median, $\sigma^*_{m,n}$, is approximately

$$\sigma^*_{m,n}/\sigma_c \approx \mu^*_n - \gamma^{**}_n\{[2\log_e(m/\beta)]^{1/2} - \{\log_e(\log_e(m/\beta)) + \log_e(4\pi) + 2\log_e(\log_e(2))\}/\{2[2\log_e(m/\beta)]^{1/2}\}. \tag{23}$$

In comparing the medians for the two versions, (20) and (23), there is no single choice of β that renders these equivalent. Phoenix *et al.* (1996) show that the effective link length $\beta\delta_c$ decreases slowly with increasing length $L = m\delta_c$ or lower tail probability of interest.

In our analysis of Monte Carlo data we consider the 'reverse weakest link transform'

$$\Phi_{m'}([(\sigma/\sigma_c) - \mu^*_n]/\gamma^{**}_n) = 1 - \{1 - H_{m,n}(\sigma/\sigma_c)\}^{1/m'} \qquad (24)$$

where $H_{m,n}(\sigma/\sigma_c)$ is the empirical distribution function calculated from the Monte Carlo data at $L = m\delta_c = (m'\beta)\delta_c$. The idea is adjust β to superimpose $\Phi_{m'}([(\sigma/\sigma_c) - \mu^*_n]/\gamma^{**}_n)$ onto $\Phi([(\sigma/\sigma_c) - \mu^*_n]/\gamma^{**}_n)$, the asymptotic normal distribution for cross-section strength.

6. Comparison of Theory and Monte Carlo Simulations

The Monte Carlo simulation program for composite failure is described in Ibnabdeljalil and Phoenix (1995). The comparison between theory and simulation is carried out for the typical vale $\rho = 5$, for $n = 50$ fibers in the composite and for composite lengths of $m = L/\delta_c = 2, 5, 10, 20$, and 50. There were 500 simulations for each case taking in total several hundred hours of CPU time on a Sun (Sparc 2) workstation. Fig. 4 shows reverse weakest-link scaling of the data, using (24), for the case $\beta\delta_c \approx 0.7\delta_c$. Choices such as $\beta = 0.4$ or 0.8 give noticeably poorer agreement in this range of m. The uncertainty in β, calculated by comparing the formulas for the medians, is barely noticeable as the numerical effects are about 0.009 (as compared to $\mu^*_n \approx 0.74$). For $n = 50$ the finite size corrections to the mean (18) and standard deviation (19) turn out to be much more important at +2.5 % and −16%, respectively.

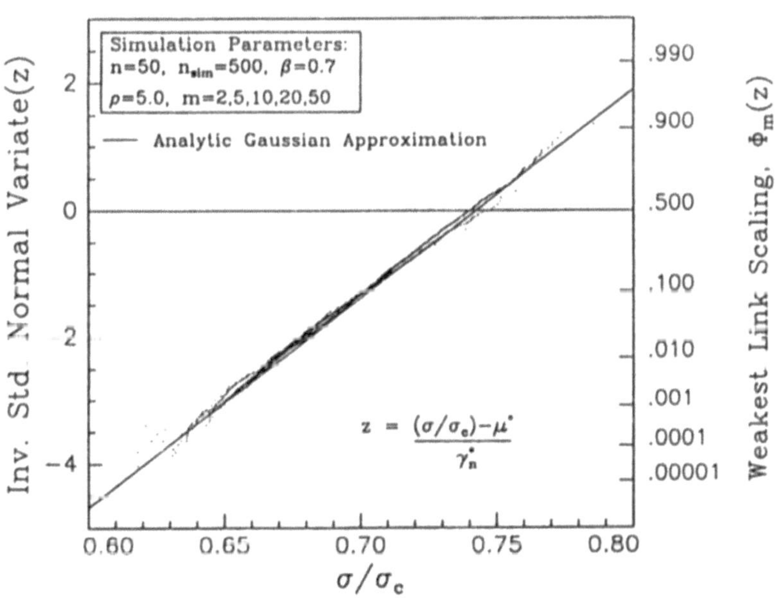

Figure 4. Comparison of 'reverse' weakest link transform of simulated composite strength distribution (n = 50, $m = L/\delta_c = 2, 5, 10, 20, 50$ and $\beta = 0.7$) with theoretical normal distribution for cross-section.

Figure 5 shows the relationship between median lifetime and composite length $m = m'\beta = L/\delta_c$. Also plotted are the theoretical medians from the two versions where the Gaussian process version appears less accurate. Agreement is good to values of m up to 10^4, and one expects that an extrapolation would be valid to at least $m = 10^6$, far beyond the present capability of simulation. Note that at $m = 50$, the median strength is down by about 8%, and at $m = 100{,}000$ it is projected to be down by 18%.

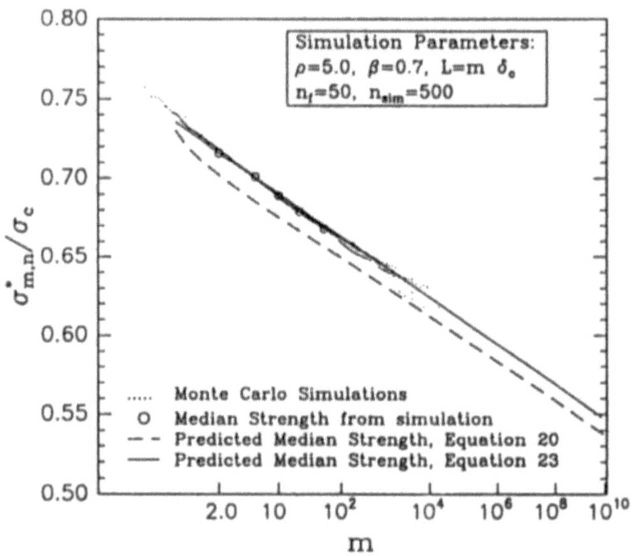

Figure 5. Plot of median composite strength $\sigma^*_{m,n}/\sigma_c$ for two approximations (20d) and (23) together with results from Monte Carlo simulations for $n = 50$, $m = L/\delta_c = 2$, 5, 10, 20, 50 and $\beta = 0.7$.

Figure 6 shows plots of the composite strength distribution $H_{m,n}(\sigma/\sigma_c)$ given by (22) using $\beta = 0.4$ against the Monte Carlo data on double exponential coordinates where (20c) would plot as a straight line. Clearly (22) does a better job of fitting the data than the double exponential form (20c) (not shown) would do since the plots are quite nonlinear. The value $\beta = 0.4$ was used since it does a better job of fitting the lower tails. Actually the simulation requirements in determining lower tail probabilities for a given length, are about the same as those to dermine the median at longer lengths. As mentioned earlier, longer lengths seem to require slightly smaller effective values for β.

7. Discussion and Conclusions

Compared to Monte Carlo simulations, the Gaussian process version does not fare quite as well as the chain-of-bundles version, which has the adjustable parameter β. The value of $\beta\delta_c = 0.7\delta_c$ seems reasonable in view of the fact that the 'effective' exclusion zone length around a break is about $s^*\delta_c = 0.88\delta_c$. In the Gaussian process version, as $m \to \infty$ only the local decay of $\Gamma_\delta(s)$ near $\delta = 0$ is used but the quadratic approximation (11) substantially overestimates this decay (underestimates the true covariance) for larger δ, say $\delta = 0.7\delta_c$, and this may be the cause of disagreement for moderate values of m. In any case the agreement between theory and Monte Carlo simulation is excellent lending confidence to extrapolations of median strength and high-reliability calculations (lower tails of the composite strength distribution) to larger composites.

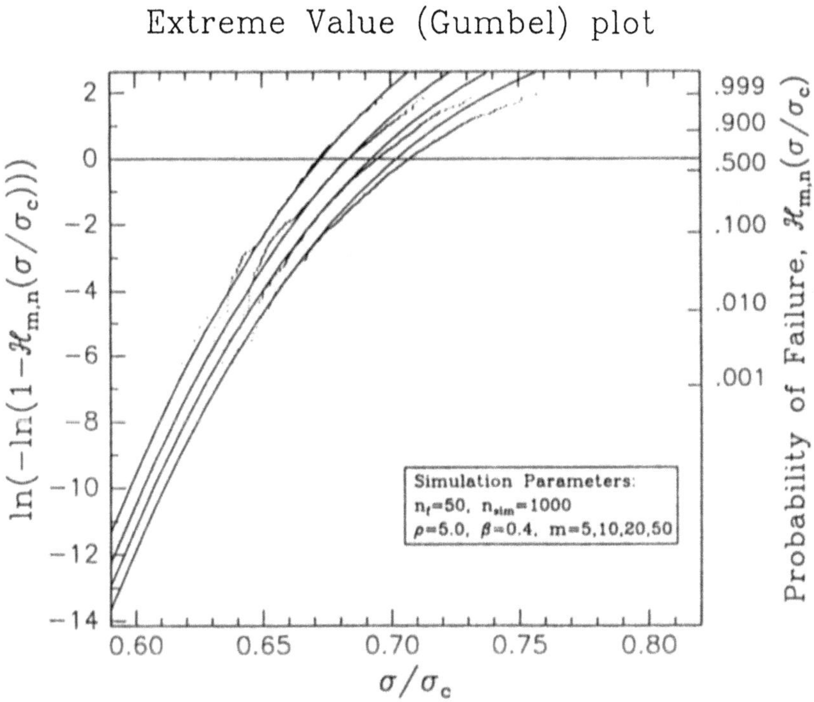

Figure 6. Comparison of weakest-link version (22) of $H_{m,n}(\sigma/\sigma_C)$, the composite strength distribution, with Monte Carlo simulation results for $n = 50$, $m = L/\delta_C = 2, 5, 10, 20$, and 50 and $\beta = 0.4$.

8. Acknowledgement

SLP and CYH acknowledge the support of the AFOSR-URI through grant F49620-93-1-0235.

References

Aveston, J., Cooper, G. A. and Kelly, A. (1971) Single and multiple fracture: The properties of fibre composites. Conference Proc., National Physical Laboratory, (Guildford, U.K.) *The Properties of Fibre Composites*, pp. 15-26, IPC Science and Technology Press, Ltd. Teddington, U.K.

Cramér, H. (1946) *Mathematical Methods of Statistics*, pp. 374-377. Princeton University Press. Princeton.

Curtin, W. A. (1991) Exact theory of fibre fragmentation in a single-filament composite. *J. Mater. Sci.* **26**, 5239-5253.

Curtin, W. A. (1991) Theory of mechanical properties of ceramic matrix composites. *J. Am. Ceram. Soc.* **74,** 2837-2845.

Curtin, W. A. (1993) The tough to brittle transition in brittle matrix composites. *J. Mech. Phys. Solids* **41,** 217-245.

Feller, W. (1968) *An Introduction to Probability Theory and its Applications (third edition)*, pp. 174-179. John Wiley & Sons, New York.

Fraser, W. A., Ancker, A. T., DiBenedetto and Elbirli, B. (1983) Evaluation of surface treatments for fibers in composite materials. *Polym. Compos.* **4,** 238-249.

Gulino, R. and Phoenix, S. L. (1991) Weibull strength statistics for graphite fibers measured from the break progression in a model graphite/glass/epoxy microcomposite, *J. Mater. Sci.* **26,** 3107-3118.

Henstenburg, R. B. and Phoenix, S. L. (1989) Interfacial shear strength studies using the single-filament-composite test. Part II: A probability model and Monte Carlo simulation. *Polym. Compos.* **10,** 389-408.

Hui, C.-Y., Phoenix, S. L., Ibnabdeljalil, M. and Smith, R. L. (1995) An exact closed form solution for fragmentation of Weibull fibers in a single filament composite with applications to fiber reinforced ceramics. *J. Mech. Phys. Solids* **43,** 1551-1585.

Hui, C.-Y., Phoenix, S. L., and Kogan, L. (1996) Exact closed form solutions for fragmentation of fibers in a single filament composite. *J. Mech. Phys. Solids*, to appear.

Ibnabdeljalil, M. and Phoenix, S. L. (1995) Scalings in the statistical failure of brittle matrix composites with discontinuous fibers; Analysis and Monte Carlo simulations. *Acta Metall. Mater.* **43,** 2975-2983.

Kimber, A. C. and Keer, J. G. (1982) On the theoretical average crack spacing in brittle matrix composites containing continuous aligned fibres. *J. Mater. Sci. Lett.* **1,** 353-354.

Leadbetter, M. R. Lindgren, G. and Rootzén, H. (1983) *Extremes and Related Properties of Random Sequences*, pp. 171-172, 205-206. Springer-Verlag, New York.

Nairn, J. A. (1992) A variational mechanics analysis of the stresses around breaks in embedded fibers. *Mechanics of Materials* **13,** 131-154.

Netravali, A. N., Henstenburg, R. B., Phoenix, S. L. and Schwartz, P. (1989) Interfacial shear strength studies using the single-filament-composite test. Part I: Experiments on graphite fibers in epoxy. *Polym. Compos.* **10,** 226-241.

Neumeister, J. M. (1993) A constitutive law for continuous fiber reinforced brittle matrix composites with fiber fragmentation and stress recovery. *J. Mech. Phys. Solids* **41,** 1383-1404.

Phoenix, S. L. and Raj, R. (1992) Scaling in fracture probabilities for a brittle matrix fiber composite. *Acta Metall. et Mater.* **40,** 2813-2828.

Phoenix, S. L., Ibnabdeljalil, M. and Hui, C.-Y. (1996) Size effects in the distribution for strength of brittle matrix composites: Analysis and Monte Carlo simulations. *Int. J. Sol. Struc.* to appear.

Phoenix, S. L. (1993) Statistical issues in the fracture of brittle matrix fibrous composites. *Composites Sci. Tech.* **48,** 65-80.

Rényi, A. (1958) On a one-dimensional problem concerning random space filling. *Magy. Tud. Akad. Matematikai Kut. Intéz. Kozl. (Budapest)* **3,** 109-127; see also *Math. Rev.* **21,** (1960) 577.

Wagner, H. D. and Eitan, A. (1990) Interpretation of fragmentation phenomenon in single-filament composite experiments, *Appl. Phys. Lett.* **56,** 1965-1967.

Widom, B. (1966) Random sequential addition of hard spheres to a volume, *J. Chem. Phys.* **44,** 3888-3894.

PREDICTING NON-LINEAR BEHAVIOUR IN MULTIPLE-PLY CROSS-PLY LAMINATES RESULTING FROM MICRO-CRACKING

L. N. McCARTNEY
Centre for Materials Measurement & Technology
National Physical Laboratory
Teddington
Middx. U.K. TW11 0LW

Abstract. Laminated composites exhibit non-linear stress-strain behaviour because of the initiation and growth of arrays of cracks in the plies of laminates leading to stress transfer between cracked and uncracked plies and to consequential changes in the effective thermoelastic constants of laminates. Recent progress with the development of stress-transfer models for transverse cracking in simple cross-ply and multiple-ply laminated cross-ply composites is described. The equilibrium equations, the out-of-plane stress-strain relations and the interface conditions (assuming perfect bonding) together with boundary conditions involving stresses, are satisfied exactly. The remaining in-plane stress-strain relations and boundary conditions involving displacements are satisfied on averaging through the thickness of each ply. The model is based upon a single assumption and the predicted stress and displacement fields lead to stationary values of the Reissner energy functional. The generalised in-plane strain model of stress transfer, when coupled with ply refinement methods, offers a powerful technique for predicting accurately both stress and displacement distributions, and the effective thermoelastic constants of cracked laminates.

It is also shown how solutions may be used with energy balance methods to predict simple relations governing the initiation and growth of ply cracking, and non-linear stress-strain behaviour. While the model is formulated at the ply level the criteria governing the formation of cracks in transverse plies are such that they involve only parameters that can be defined at the macroscopic laminate level, e.g. the effective thermoelastic constants of the cracked laminate, crack closure stresses and the energy absorbed in the laminate by ply crack formation. Thus the use of a high quality stress-transfer model leads to the construction of a bridge between a model formulated at the ply level, and the macroscopic behaviour of multiple-ply laminates containing arrays of transverse cracks.

J. R. Willis (ed.), IUTAM Symposium on Nonlinear Analysis of Fracture, 379–390.

1. Introduction

It is well known that the considerable experimental data reported in the literature indicate that the stress/strain behaviour of laminated composites exhibits non-linear behaviour following the initiation of microstructural damage in the form of transverse cracks normal to the direction of loading. In this case, fracture events are distributed throughout the laminate and lead to non-linear behaviour at the macroscopic level, although the mechanics governing the localised fracture events are linear.

Some attention has been devoted to developing models of stress transfer in laminates that are based on variational techniques. The normal method of analysis assumes a stress representation that satisfies the equations of equilibrium and stress boundary conditions. The representation involves functions of the axial coordinate which are determined by minimising the complementary energy of the system using variational techniques. For $[0/90]_s$ laminates, Hashin [1-4] and Nairn [5] have developed such solutions, while Nairn [6] has derived corresponding solutions for other types of cross-ply laminate. These solutions, when used to predict the dependence of axial Young's modulus on crack density lead to lower bound estimates. Nairn [7] has recently also developed a displacement-based variational model that leads to an upper-bound prediction for the axial modulus of the cracked laminate. A characteristic of stress based variational methods is that the corresponding displacement distribution, related by the stress-strain equations, is not specified. However, consistent solutions for both stress **and** displacement distributions in multiple-ply cross-ply laminates can be derived using analytical methods (see McCartney [8, 9]). Furthermore, the solution developed, automatically leads to stationary values of the Reissner energy functional so that a corresponding solution could have been derived by variational methods. The objective of this paper is to show how a recently developed analytical method [9] of calculating the stress and displacement distributions in a multiple ply cracked laminate may be used to predict the formation of transverse cracking as a function of biaxial loading, taking full account of the effects of thermal residual stresses.

2. Representation for stress and displacement

A set of 2N+2 plies are perfectly bonded together to form a laminate. At this stage each ply is regarded as being made of a different transverse isotropic material such that the laminate is symmetric about the mid-plane. When subject to biaxial loading only one half of the laminate needs to be considered. The N+1 plies in this half are regarded as having thicknesses h_i, i = 1...N+1, where the suffix i is also attached to the thermoelastic constants. In the sequel E, ν, μ, and α denote respectively Young's modulus, Poisson's ratio, shear modulus, and the thermal expansion coefficient. The temperature difference between stress-free temperature of the laminate and the current temperature is denoted by ΔT.

Rectangular Cartesian coordinates (x, y, z) are introduced with the origin lying on the centre plane of the laminate, such that x is in the through-thickness direction, y is in the axial direction and z is in the width direction. The interfaces between the plies are defined to be at $x = x_i$, i = 1...N, with $x_0 = 0$ and $x_{N+1} = h$. The 90° plies in the laminate may have cracks (but not necessarily) only on the planes $y = \pm L$. The laminate is assumed to be subject to generalised plane strain deformation such that

$$u_i = u_i(x, y) \ , \quad v_i = v_i(x, y) \ , \quad w_i = \varepsilon_c^* z \ , \quad i = 1...N+1 \ , \tag{1}$$

where ε_c^* is a uniform transverse strain whose value is to be determined by demanding that the effective transverse stress applied to the laminate has a specified value. The assumption (1) leads to a transverse stress distribution $\sigma_{zz}(x,y)$ that is non-uniform from which the effective stress is calculated by averaging this stress over the laminate edge. The resulting stress distribution is representative of that found in laminates well away from their edges where transverse stress transfer effects also occur.

The following representation for the stress and displacement field has been derived from the equilibrium equations, strain/displacement and stress/strain relations using the methods described in [9].

2.1 STRESS FIELD

The fundamental assumption is that the shear stress component σ_{xy} has the following form, linear in the through-thickness coordinate x, for i = 1...N+1

$$\sigma_{xy}^{i} = \frac{1}{h_i}\left[C_{i-1}'(y)\,(x_i - x) + C_i'(y)\,(x - x_{i-1}) \right] . \tag{2}$$

Functions $C_i(y)$, i = 1...N+1, depend only on the axial coordinate y, and are for the moment regarded as being arbitrary. It is to be understood that $C_0(y)$ and $C_{N+1}(y)$ are identically zero. On substituting (2) into one of the equilibrium equations, it can be shown that, for i = 1...N+1

$$\sigma_{xx}^{i}(x,y) = \frac{x-x_i}{2h_i}\Big[(x-x_i)C_{i-1}''(y) - (x-x_i+2h_i)C_i''(y)\Big] + S_i(y), \qquad (3)$$

where $S_i(y)$ is determined later, while the second equilibrium equation leads to

$$\sigma_{yy}^{i}(y) = \frac{1}{h_i}\big[C_{i-1}(y) - C_i(y)\big] + \sigma_i\,, \quad i = 1...N+1\,, \qquad (4)$$

indicating that the axial stress in each ply is independent of x. The parameter σ_i denotes the uniform axial stress that would arise in the ith ply if the laminate were undamaged and subject to the same applied tractions and temperature.

2.2 DISPLACEMENT FIELD

On using (2-4) in conjunction with all but one of the stress/strain equations it can be shown [9], for i = 1...N+1, that the displacement components are

$$\begin{aligned} u_i(x,y) = {} & \frac{(x-x_i)^2}{6\tilde{E}_t^{\,i} h_i}\Big[(x-x_i)\,C_{i-1}''(y) - (x-x_i+3h_i)\,C_i''(y)\Big] \\ & - \frac{x-x_i}{h_i}\frac{\tilde{\nu}_a^i}{\tilde{E}_A^{\,i}}\big[C_{i-1}(y) - C_i(y)\big] + (x-x_i)\left[\frac{1}{\tilde{E}_t^{\,i}}S_i(y) + \varepsilon_i\right] + U_i(y), \end{aligned} \qquad (5)$$

$$\text{where} \qquad \varepsilon_i = -\frac{\tilde{\nu}_a^i}{\tilde{E}_A^{\,i}}\sigma_i + \tilde{\alpha}_t^i\,\Delta T - \nu_t^i\,\varepsilon_c^*\,, \qquad (6)$$

$$v_i(x, y) = \frac{x - x_i}{2h_i}\left[(x - x_i)\left(\frac{\tilde{\nu}_a^i}{\tilde{E}_A^{\,i}} - \frac{1}{\mu_a^i}\right)C'_{i-1}(y) - \left(\frac{\tilde{\nu}_a^i}{\tilde{E}_A^{\,i}}(x - x_i) - \frac{1}{\mu_a^i}(x - x_i + 2h_i)\right)C_i'(y)\right]$$
$$- \frac{(x - x_i)^3}{24\tilde{E}_t^{\,i}h_i}\left[(x - x_i)C_{i-1}'''(y) - (x - x_i + 4h_i)C_i'''(y)\right]$$
$$- \frac{(x - x_i)^2}{2\tilde{E}_t^{\,i}}S_i'(y) - (x - x_i)U_i'(y) + V_i(y)\,, \quad i = 1...N+1\,. \qquad (7)$$

In (5-7), elastic constants with a tilde are those defined in [9] for an uncracked laminate subject to generalised plane strain conditions. Thermoelastic constants having an upper case suffix relate to in-plane properties while those with a lower case suffix imply association with out-of-plane deformation.

3. Interfacial and boundary conditions

As all tractions and displacement components must be continuous in the laminate,

$$\sigma_{xx}^{i} = \sigma_{xx}^{i+1}\,, \sigma_{xy}^{i} = \sigma_{xy}^{i+1}\,, u_i = u_{i+1}\,, v_i = v_{i+1} \text{ on } x = x_i\,,\ i = 1..N. \qquad (8)$$

Symmetry about $x = 0$ implies that

$$\sigma_{xy}^{1} \equiv u_1 \equiv 0 \quad \text{on } x = 0\,, \qquad (9)$$

and, assuming that the outer surface of the laminate is stress free,

$$\sigma_{xx}^{N+1} \equiv \sigma_{xy}^{N+1} \equiv 0 \quad \text{on } x = h\,. \qquad (10)$$

For any plies that are **uncracked** on $y = \pm L$

$$v_i(x, \pm L) = \pm \varepsilon_c L \,, \quad \sigma^i_{xy}(x, \pm L) = 0 \,, \tag{11}$$

where ε_c is the effective axial strain applied to the cracked laminate. Any plies that are **cracked** on $y = \pm L$ will be stress free, i.e.

$$\sigma^i_{yy}(x, \pm L) = 0 \,, \quad \sigma^i_{xy}(x, \pm L) = 0 \,. \tag{12}$$

For generalised plane strain conditions and uniaxial loading conditions the effective transverse applied stress σ_t is given by

$$\sigma_t = \frac{1}{2hL}\sum_{i=1}^{N+1}\int_{-L}^{L}\int_{x_{i-1}}^{x_i}\sigma^i_{zz}\,dx\,dy \,, \tag{13}$$

where $\sigma_{zz}{}^i$ is calculated using the transverse stress/strain relation

$$\sigma^i_{zz} = \nu^i_t\sigma^i_{xx} + \nu^i_A\frac{E^i_T}{E^i_A}\sigma^i_{yy} - E^i_T\alpha^i_T\Delta T + E^i_T\varepsilon^*_c \,. \tag{14}$$

4. Recurrence relations

By imposing the continuity of σ_{xx} , u_i and v_i at the interfaces $x = x_i$, $i = 1...N$, using (3), (5), (7) and (8-10) it can be shown that

$$S_{i-1} = S_i + \frac{h_i}{2}\left[C''_{i-1} + C''_i\right], \quad S_{N+1} = 0, \tag{15}$$

$$U_i = U_{i-1} + \frac{h_i^2}{6\tilde{E}^i_t}\left[C''_{i-1} + 2C''_i\right] - \frac{\tilde{\nu}^i_a}{\tilde{E}^i_A}\left[C_{i-1} - C_i\right] + h_i\left[\frac{1}{\tilde{E}^i_t}S_i + \varepsilon_i\right], \quad U_0 = 0, \tag{16}$$

$$V_i = V_{i-1} + \frac{h_i^3}{24\tilde{E}_t^{\,i}}\left[C_{i-1}^{\prime\prime\prime} + 3C_i^{\prime\prime\prime}\right] - \frac{h_i}{2}\left[\left(\frac{\tilde{\nu}_a^i}{\tilde{E}_A^{\,i}} - \frac{1}{\mu_a^i}\right)C_{i-1}^{\prime} - \left(\frac{\tilde{\nu}_a^i}{\tilde{E}_A^{\,i}} + \frac{1}{\mu_a^i}\right)C_i^{\prime}\right]$$

$$+ \frac{h_i^2}{2\tilde{E}_t^{\,i}} S_i^{\prime} - h_i U_i^{\prime}\,, \qquad i = 1...N+1\,, \tag{17}$$

where $V_1(y)$ is regarded as being specified for the operation of the recurrence relation (17), but is eliminated when developing the solution of the problem [9].

5. Governing differential equations

The representation defined by (2-7) and (15-17) automatically satisfies the equilibrium equations and all but one of the relevant stress/strain relations in each ply, together with the interface and boundary conditions (8-10) for any functions $C_i(y)$, i = 1...N. The axial stress/strain relations are not satisfied but are replaced by their averages (through the thickness of each ply) as follows

$$\bar{\varepsilon}_{yy}^{i} = -\frac{\tilde{\nu}_a^i}{\tilde{E}_A^{\,i}}\bar{\sigma}_{xx}^{i} + \frac{1}{\tilde{E}_A^{\,i}}\bar{\sigma}_{yy}^{i} + \alpha_A^i\,\Delta T - \nu_A^i\frac{E_T^i}{E_A^i}\varepsilon_c^{*}\,, \qquad i = 1...N+1\,. \tag{18}$$

It is worth noting that the Reissner energy functional has a stationary value only when (18) is satisfied. It can be shown [9] that the unknown functions $C_i(y)$, i = 1...N, of the representation must then satisfy

$$\sum_{i=1}^{N} F_{ij}\,C_i^{\prime\prime\prime\prime}(y) + \sum_{i=1}^{N} G_{ij}\,C_i^{\prime\prime}(y) + \sum_{i=1}^{N} H_{ij}\,C_i(y) \equiv 0\,, \quad j = 1..N, \tag{19}$$

defining N homogeneous simultaneous differential equations. The coefficients F_{ij} , G_{ij} and H_{ij} are constants that are readily calculated numerically. The equations (19) when subject to appropriate boundary conditions (specified in [9, 10]) derived from (11-13) may be solved using standard techniques (see [11] for specific details regarding their application to cracked laminates).

6. Thermoelastic constants of cracked laminates

By making use of the analysis presented in sections 2-5 the effective in-plane thermoelastic constants may be calculated using the method described in [9, 10]. For an effective axial stress σ and effective transverse stress σ_t applied to a laminate containing n cracks per unit length, the stress/strain relations for a cracked multiple-ply laminate are of the form

$$\varepsilon(n) = \frac{\sigma}{E_A(n)} - \frac{\nu_A(n)}{E_A(n)}\sigma_t + \alpha_A(n)\Delta T \,, \tag{20}$$

$$\varepsilon^*(n) = -\frac{\nu_A(n)}{E_A(n)}\sigma + \frac{\sigma_t}{E_T(n)} + \alpha_T(n)\Delta T \,, \tag{21}$$

where $\varepsilon(n) = \varepsilon_c$ and $\varepsilon^*(n) = \varepsilon_c^*$ are respectively the effective axial and transverse strains of the laminate containing n cracks per unit length, and where $E_A(n)$, $E_T(n)$, $\nu_A(n)$, $\alpha_A(n)$ and $\alpha_T(n)$ denote the effective thermoelastic constants of the cracked laminate. It must be emphasised that the multiple-ply stress transfer model described in this paper automatically leads to stress/strain relations of the form (20) and (21) where the coefficient of σ_t in (20) is identical to the coefficient of σ in (21), i.e. the tensor for the elastic constants of a cracked laminate is symmetric; an essential property if a cracked laminate having non-uniform stress and displacement distributions is to behave as the corresponding effective uncracked laminate having reduced properties that take account of the presence of ply cracks. Most stress transfer models used in the literature, if extended for biaxial loading, would not lead to this symmetry.

By applying the multiple-ply analysis to the basic problem of just two plies representing the 0° and 90° plies in a $[0/90]_s$ cross-ply laminate, results identical to those based on existing theory [8, 12, 13] have been obtained. For such solutions the following inter-relationships between the thermoelastic constants have been derived [12, 13]

$$\frac{\dfrac{\nu_A}{E_A} - \dfrac{\nu_A(n)}{E_A(n)}}{\dfrac{1}{E_A(n)} - \dfrac{1}{E_A}} = \frac{\dfrac{1}{E_T(n)} - \dfrac{1}{E_T}}{\dfrac{\nu_A}{E_A} - \dfrac{\nu_A(n)}{E_A(n)}} = \sqrt{\frac{\dfrac{1}{E_T(n)} - \dfrac{1}{E_T}}{\dfrac{1}{E_A(n)} - \dfrac{1}{E_A}}} = \frac{\alpha_T(n) - \alpha_T}{\alpha_A(n) - \alpha_A} = k, \tag{22}$$

where $$k = \frac{1}{1 - \nu_A \nu_A^{(90)}} \left(\frac{E_A}{E_T} \nu_A^{(90)} - \nu_A \right), \tag{23}$$

and E_A and ν_A are respectively the axial Young's modulus and Poisson's ratio of the laminate in an uncracked state. The parameter $\nu_A^{(90)}$ is the axial Poisson's ratio of the material in the 90° ply. More accurate stress and displacement solutions can, however, be obtained by representing the 0° ply by a series of refined layers all having the properties of the 0° ply, and similarly for the 90° ply. The layers used are refined near the interfacial region in order to resolve the stress singularity that exists at the transverse crack tips. The accurate solutions resulting from such an application of the analysis presented above have been used to establish numerically that the inter-relationships (22) are satisfied also by multiple-ply solutions.

7. Prediction of transverse cracking

The mechanics of the formation of fully developed transverse cracks in cross-ply laminates has been studied in some detail [12, 13] for the case when the laminate is modelled by just two plies, one representing the 0° ply and the other the 90° ply. Let $\gamma = \gamma_t a/h$ denote the effective fracture energy for transverse cracking in the laminate; h denoting the laminate half-thickness, and γ_t the fracture energy for transverse cracking in the 90° plies. The fracture energy is defined so that 2γ would be equated to an energy release rate. Let the parameters σ_o and σ_o^* be the effective axial and transverse stresses that must be applied so that any cracks in the 0° and 90° plies of an orthogonally cracked laminate are **just** closed. It has been shown [13, 14] that the criterion for the formation of p new cracks in **unit** length of laminate already having n existing cracks, is

$$2\gamma p < \frac{1}{2}(\sigma-\sigma_o)^2\left(\frac{1}{E_A(n+p)}-\frac{1}{E_A(n)}\right)+\frac{1}{2}(\sigma_t-\sigma_o^*)^2\left(\frac{1}{E_T(n+p)}-\frac{1}{E_T(n)}\right)$$

(24)

$$-\,(\sigma-\sigma_o)(\sigma_t-\sigma_o^*)\left(\frac{\nu_A(n+p)}{E_A(n+p)}-\frac{\nu_A(n)}{E_A(n)}\right).$$

It is worth noting that the result (24) is independent of the thermal expansion coefficients of the cracked and uncracked laminate; a property that arises from the introduction of the crack closure stresses σ_o and σ_o^* and the use of (22). On using (22) the result (24) may also be written

$$4\gamma p\left(\frac{1}{E_A(n+p)}-\frac{1}{E_A(n)}\right)=\left[\varepsilon(n+p)-\varepsilon(n)\right]^2, \tag{25}$$

indicating that the use of an accurate stress transfer model of a cracked cross-ply laminate has led to an exceedingly simple result that involves only macroscopic properties defined at the laminate level, and which takes full account of the effects of biaxial loading and of the thermal residual stresses. By making use of (22) the relation (25) is also valid if the axial Young's modulus and strain are replaced by the corresponding transverse values.

8. Discussion

It is beyond the scope of this paper to show how the analysis described above can successfully solve a wide range of stress transfer problems for multiple-ply cross-ply laminates. The first principal result is that the stress transfer model predicts effective stress/strain relations (20) and (21), that are of the same form as those for uncracked laminates. The method of analysis leads to reliable predictions for the values of the thermoelastic constants appearing in (20) and (21). In addition it can be shown that the thermoelastic constants satisfy the inter-relationships (22) ensuring that at the point of crack closure the cracked laminate behaves in exactly the same way as an uncracked laminate. These relationships are required to derive the second principal result (24), or equivalently (25), that governs crack formation in the laminate. The transverse cracking relation (24) involves the applied biaxial stresses, whereas the much simpler result (25) involves the axial applied strain. It is worth noting that (24) and (25) were first derived for a two ply model of the laminate [13] for the case when n = 0.

It is emphasised that the result (24) involves only macroscopic parameters even though the stress transfer model was formulated at the ply level involving many parameters associated with the individual plies. The macroscopic parameters are the effective thermoelastic constants of a cracked laminate whose values have been calculated numerically. For a two ply model of the laminate, analytical expressions for the thermoelastic constants may be derived [13]. The stresses σ_0 and σ_0^* required to **just** close any cracks that are present in the 0° and 90° plies of the laminate are the other parameters appearing on the R.H.S. of (24). The physical interpretation of these parameters as crack closure stresses means that they too can be defined at the macroscopic level even though expressions for them are given (see [13], Appendix C) in terms of individual ply properties. The stresses σ_0 and σ_0^* and the corresponding strains ε_0 and ε_0^* define the point (usually in compression) at which unloading stress-strain curves would converge from any stages of prior loading, and any stages of transverse crack formation. The inter-relationships (22) ensure that solutions have this essential property. The result (24) has been used [13] to investigate simultaneous cracking by setting n = 0, and progressive cracking by setting n = p for the case of regular cracking leading to uniform crack densities. By setting p = 1, for the case of irregular cracking arising from statistically distributed fracture energies at sites along the laminate length at which cracks can potentially form, predictions have been made [13] of the non-linear stress/strain behaviour that arises from transverse crack formation. Remarkably, the simple result (25) that is equivalent to (24) is of general validity (i.e. it applies to biaxial loading and takes account of thermal residual stresses) and yet, in addition to the effective fracture energy γ, it involves only the effective axial applied strain and the axial Young's modulus for a cracked laminate. This relationship could, therefore, form the basis of an experimental technique for determining the appropriate value of the effective fracture energy γ.

To conclude, the analysis presented in this paper demonstrates that detailed and accurate micromechanical modelling at the ply level in multiple-ply laminates leads automatically to relatively simple modelling at the macroscopic laminate level. The results indicate that detailed micromechanical modelling at the ply level is required only to determine the dependence of the thermoelastic constants on transverse crack density.

Acknowledgement

Research in this paper arises from:- i) NPL Strategic Research Programme, and ii) collaboration with AEA Technology, Harwell, within the 'Materials Measurement Programme' of underpinning research financed by the UK Dept. of Trade and Industry.

References

1 Hashin, Z., Mech. of Mater. **4**, (1985), pp. 121-136.
2 Hashin, Z., Eng. Fract. Mech. **25**, (1986), pp. 771-778.
3 Hashin, Z., J. Appl. Mech. **54**, (1987), pp. 872-879.
4 Hashin, Z., Comp. Sci. Tech. **31**, (1988) pp. 247-260.
5 Nairn, J.A., J. Comp. Mater. **23**, (1989), pp. 1106-1129.
6 Nairn, J.A., Eng. Fract. Mech. **41**, (1992), pp. 203-221.
7 Nairn, J.A., 'Some new variational mechanics results on composite microcracking', Proc. 10th Int. Conf. on Composite Materials, Whistler, B.C. Canada, Aug. pp. 14-18, 1995.
8 McCartney, L.N., J. Mech. Phys. Solids, **40**, (1992), pp. 27-68.
9 McCartney, L.N., 'A recursive method of calculating stress transfer in multiple-ply cross-ply laminates subject to biaxial loading', NPL Report DMM(A)150, January 1995.
10 McCartney, L.N., Hannaby, S. and Cooper, P.M., Proc. of 3rd Int. Conf. on Deformation and Fracture of Composites, University of Surrey, 27-29 March 1995, pp.56-65.
11 Hannaby, S.A., 'The solution of ordinary differential equations arising from stress transfer mechanics' NPL Report DITC 223/93, November 1993.
12 McCartney, L.N., Composites, **24**, (1993), pp. 84-92.
13 McCartney, L.N., 'The prediction of non-uniform cracking in biaxially loaded cross-ply laminates', NPL Report DMM(A)142, July 1994.
14 McCartney, L.N., "Prediction of microcracking in composite materials", in "FRACTURE: A Topical Encyclopaedia of Current Knowledge Dedicated to Alan Arnold Griffith", pp. 905-916, (ed. G.P.Cherepanov), Krieger Publishing Company, Melbourne, USA, 1995.

APPENDIX: The Scientific Programme

Session 1
Opening of Symposium
Z.P. Bažant: Scaling in nonlinear fracture mechanics
B. Karihaloo: Scaling effects in the fracture of quasi-brittle materials and structures

Session 2
A.D. Karstensen, A. Nekkal & *J.W. Hancock*: Constraint estimation
N.P. O'Dowd & P.J. Budden: Finite element analysis of a bimaterial
G.X. Shan, *O. Kolednik* & F.D. Fischer: Geometry and size effects in ductile fracture – FEM studies and theoretical considerations
C.F. Shih, L. Xia & L. Cheng: Ductile/brittle fracture – modeling fracture process region by cell elements

Session 3
J.F. Knott: Effects of size-scale on fracture processes in engineering materials
J.W. Hutchinson: Crack growth in ductile solids from the fracture process
V. Tvergaard: Relations between crack growth resistance and fracture process parameters under large scale yielding

Session 4
B. Cotterell, K.Y. Lam, Z. Chen & A.G. Atkins: The effect of geometry and size on the growth of ductile fracture
K.P. Herrmann & T. Hauck: Interfacial cracking in thermomechanically loaded elastoplastic bimaterials
S. Suresh, A.S Kim & Y. Sugimura: The effect of plasticity on the growth of a crack approaching an interface: theory, experiments and applications
J. Lemaitre: Conditions of crack arrest by interfaces

Session 5
H. Horii, Y. Okui & J. Inoue: Micromechanics-based continuum theory and thermomechanical formulation for localization phenomena
S. Nemat-Nasser: Dynamic crack growth in inelastic solids with application to adiabatic shear band
A.J. Rosakis, G. Ravichandran & M. Zhou: Dynamically growing shear bands in metals: high speed optical and infrared measurements of temperature and deformation fields

J. R. Willis (ed.), IUTAM Symposium on Nonlinear Analysis of Fracture, 391–394.

M. Ortiz & G. Gioia: The two-dimensional structure of dynamic boundary layers and dynamic shear bands in thermoviscoplastic solids

Session 6

L. Knopoff: A mesoscale model of friction and the final stress on crack surfaces

R. Madariaga & A. Cochard: Dynamic friction and the origin of the complexity of earthquake sources

D.L. Turcotte: Earthquakes, fracture, complexity

Short "advertisements" for posters: G.E. Beltz, F.M. Borodich, M.M. Davydova, N.A. Fleck, K.-S. Kim, M.J. Korteoja, V. Mishakin, A.B. Movchan, N.V. Movchan, T. Sadowski, E. Visse

Session 7

E. Sharon, S.P. Gross & *J. Fineberg*: Local crack branching as a mechanism for instability in dynamic fracture

J.S. Langer: Dynamic stability of fracture

L.I. Slepyan & J.P. Dempsey: Radial crack dynamics with closure

Session 8

T. Siegmund & *A. Needleman*: Numerical studies of fast crack growth in elastic-plastic solids

J.R. Rice, J. Morrissey, Y. Ben-Zion, P.H. Geubelle & G. Perrin: Statistical elastodynamics of cracking through solids of heterogeneous toughness

D. Gross: Fast crack growth in viscoplastic solids

K.C. Hwang, T.F. Guo & Y. Huang: Near-tip fields for cracks in non-classical materials

Poster Session I

G.E. Beltz, J.R. Rice, C.F. Shih & L. Xia: A self-consistent model for cleavage in the presence of plastic flow

F.M. Borodich: Some fractal models of fracture

M.M. Davydova: Nonlinear kinetics of microcrack accumulation and topological characteristics of fracture

A.R. Akisanya & *N.A. Fleck*: Interfacial cracking from the free edge of a bimaterial strip

A.F. Bastawros & *K.-S. Kim*: Laser-moire analysis of the near-tip incremental-plastic-deformation fields in FCC copper single crystals

M.J. Korteoja, A. Lukkarinen, K.J. Niskanen & K. Kaski: Computational modelling of plastic strains in paper

V. Mishakin: The influence of the process of plastic deformation and damage of polycrystalline materials on the effect of acoustoelasticity
A.B. Movchan & J.R. Willis: Nonlinear problems of modelling dislocations and cracks bridged by fibres
N.V. Movchan & J.R. Willis: Fibre reinforcement effect in composites with frictional interfaces
T. Sadowski: Dynamic damage modelling in semi-brittle ceramics under compressive impulse loading
P. Mialon & *E. Visse*: Large displacement energy release rate

Session 9
W.A. Curtin: Non-linear damage evolution and failure in materials
P.M. Duxbury: Mathematical modelling of the structure-sensitive properties of materials
P.L. Leath & N.N. Chen: The statistics of failure probability in heterogeneous materials with crossover from tough to brittle behavior
K. Kaski, M.J. Korteoja, A. Lukkarinen & T.T. Rautiainen: Computer modelling of disordered plastic and viscoelastic systems

Session 10
O.B. Naimark: Kinetic transitions in ensembles of defects (microcracks) and some nonlinear aspects of fracture
J.-B. Leblond, M. Gologanu & J. Devaux: Void growth in ductile metals subjected to cyclic loadings
L.R. Botvina: Phase transitions in fracture and crystallization processes
Short "advertisements" for posters:
M.J. Alava, J.P. Dempsey, S.A. Elsoufiev, J.-F. Ganghoffer, H. Gao, G. Lvov, Yu.K. Maksimenkov, T.T. Rautiainen, A. Roytman, A. Yuse, A. Zolochevsky, V.I. Erofeyev

Session 11
L.R.F. Rose: Nonlinear crack-bridging models
B. Budiansky & J.C. Amazigo: Fracture-strength effects of randomly flawed long and short fibers in ceramic composites

Poster Session II
V.I. Räisänen, *M.J. Alava* & R.M. Nieminen: Fracture in layered fibrous structures
J.P. Dempsey, R.M. Adamson & S.V. Mulmule: Scale effects on the fracture of sea ice

S.A. Elsoufiev: Nonlinear methods of strength computation
J.-F. Ganghoffer & J. Schultz: Expansion of a cavity in a rubber block under stress: application of the asymptotic expansion method to the analysis of the stability and bifurcation conditions
H. Gao: Nonlinear elastic effects in dynamic fracture instabilities
G. Lvov & E. Kostenko: Creep and brittle fracture of shells
Yu.K. Maksimenkov & I.A. Tsurpal: Nonlinear solutions of crack theory problems
T.T. Rautiainen, M.J. Alava & K. Kaski: Fracture in dissipative disordered systems
S. Shishkanova, *A. Roytman* & F. Neves: A discrete model of estimating the moment of annular crack growth under bending
A. Yuse & M. Sano: Instabilities and pattern formation in directional crack propagation
A. Zolochevsky: Identification of damage variables in anisotropic materials with different behaviour in tension and compression
V.I. Erofeyev: Elastic wave propagation in a damaged medium with microstructure

Session 12
A.N. Guz': The study and analysis of non-classical problems of fracture and failure mechanics
N.J.-J. Fang & *J.L. Bassani*: Mixed-mode creep cracking
B.N. Cox & M.R. Begley: Creeping cracks in ceramic matrix composites

Session 13
S.L. Phoenix: An exact closed form solution for fragmentation of fibers in a single filament composite with applications to fiber reinforced ceramics
R.M. McMeeking & M.A. McGlockton: Numerical models for rupture of uniaxial composites with brittle fibers
L.N. McCartney: Predicting non-linear behaviour in composites resulting from micro-cracking
H.M. Jensen: Compressive failure of fiber reinforced ceramics

Mechanics

FLUID MECHANICS AND ITS APPLICATIONS

Series Editor: R. Moreau

Aims and Scope of the Series

The purpose of this series is to focus on subjects in which fluid mechanics plays a fundamental role. As well as the more traditional applications of aeronautics, hydraulics, heat and mass transfer etc., books will be published dealing with topics which are currently in a state of rapid development, such as turbulence, suspensions and multiphase fluids, super and hypersonic flows and numerical modelling techniques. It is a widely held view that it is the interdisciplinary subjects that will receive intense scientific attention, bringing them to the forefront of technological advancement. Fluids have the ability to transport matter and its properties as well as transmit force, therefore fluid mechanics is a subject that is particularly open to cross fertilisation with other sciences and disciplines of engineering. The subject of fluid mechanics will be highly relevant in domains such as chemical, metallurgical, biological and ecological engineering. This series is particularly open to such new multidisciplinary domains.

1. M. Lesieur: *Turbulence in Fluids.* 2nd rev. ed., 1990 — ISBN 0-7923-0645-7
2. O. Métais and M. Lesieur (eds.): *Turbulence and Coherent Structures.* 1991 — ISBN 0-7923-0646-5
3. R. Moreau: *Magnetohydrodynamics.* 1990 — ISBN 0-7923-0937-5
4. E. Coustols (ed.): *Turbulence Control by Passive Means.* 1990 — ISBN 0-7923-1020-9
5. A.A. Borissov (ed.): *Dynamic Structure of Detonation in Gaseous and Dispersed Media.* 1991 — ISBN 0-7923-1340-2
6. K.-S. Choi (ed.): *Recent Developments in Turbulence Management.* 1991 — ISBN 0-7923-1477-8
7. E.P. Evans and B. Coulbeck (eds.): *Pipeline Systems.* 1992 — ISBN 0-7923-1668-1
8. B. Nau (ed.): *Fluid Sealing.* 1992 — ISBN 0-7923-1669-X
9. T.K.S. Murthy (ed.): *Computational Methods in Hypersonic Aerodynamics.* 1992 — ISBN 0-7923-1673-8
10. R. King (ed.): *Fluid Mechanics of Mixing.* Modelling, Operations and Experimental Techniques. 1992 — ISBN 0-7923-1720-3
11. Z. Han and X. Yin: *Shock Dynamics.* 1993 — ISBN 0-7923-1746-7
12. L. Svarovsky and M.T. Thew (eds.): *Hydroclones.* Analysis and Applications. 1992 — ISBN 0-7923-1876-5
13. A. Lichtarowicz (ed.): *Jet Cutting Technology.* 1992 — ISBN 0-7923-1979-6
14. F.T.M. Nieuwstadt (ed.): *Flow Visualization and Image Analysis.* 1993 — ISBN 0-7923-1994-X
15. A.J. Saul (ed.): *Floods and Flood Management.* 1992 — ISBN 0-7923-2078-6
16. D.E. Ashpis, T.B. Gatski and R. Hirsh (eds.): *Instabilities and Turbulence in Engineering Flows.* 1993 — ISBN 0-7923-2161-8
17. R.S. Azad: *The Atmospheric Boundary Layer for Engineers.* 1993 — ISBN 0-7923-2187-1
18. F.T.M. Nieuwstadt (ed.): *Advances in Turbulence IV.* 1993 — ISBN 0-7923-2282-7
19. K.K. Prasad (ed.): *Further Developments in Turbulence Management.* 1993 — ISBN 0-7923-2291-6
20. Y.A. Tatarchenko: *Shaped Crystal Growth.* 1993 — ISBN 0-7923-2419-6

Kluwer Academic Publishers – Dordrecht / Boston / London

Mechanics

FLUID MECHANICS AND ITS APPLICATIONS

Series Editor: R. Moreau

21. J.P. Bonnet and M.N. Glauser (eds.): *Eddy Structure Identification in Free Turbulent Shear Flows.* 1993 ISBN 0-7923-2449-8
22. R.S. Srivastava: *Interaction of Shock Waves.* 1994 ISBN 0-7923-2920-1
23. J.R. Blake, J.M. Boulton-Stone and N.H. Thomas (eds.): *Bubble Dynamics and Interface Phenomena.* 1994 ISBN 0-7923-3008-0
24. R. Benzi (ed.): *Advances in Turbulence V.* 1995 ISBN 0-7923-3032-3
25. B.I. Rabinovich, V.G. Lebedev and A.I. Mytarev: *Vortex Processes and Solid Body Dynamics.* The Dynamic Problems of Spacecrafts and Magnetic Levitation Systems. 1994 ISBN 0-7923-3092-7
26. P.R. Voke, L. Kleiser and J.-P. Chollet (eds.): *Direct and Large-Eddy Simulation I.* Selected papers from the First ERCOFTAC Workshop on Direct and Large-Eddy Simulation. 1994 ISBN 0-7923-3106-0
27. J.A. Sparenberg: *Hydrodynamic Propulsion and its Optimization.* Analytic Theory. 1995 ISBN 0-7923-3201-6
28. J.F. Dijksman and G.D.C. Kuiken (eds.): *IUTAM Symposium on Numerical Simulation of Non-Isothermal Flow of Viscoelastic Liquids.* Proceedings of an IUTAM Symposium held in Kerkrade, The Netherlands. 1995 ISBN 0-7923-3262-8
29. B.M. Boubnov and G.S. Golitsyn: *Convection in Rotating Fluids.* 1995 ISBN 0-7923-3371-3
30. S.I. Green (ed.): *Fluid Vortices.* 1995 ISBN 0-7923-3376-4
31. S. Morioka and L. van Wijngaarden (eds.): *IUTAM Symposium on Waves in Liquid/Gas and Liquid/Vapour Two-Phase Systems.* 1995 ISBN 0-7923-3424-8
32. A. Gyr and H.-W. Bewersdorff: *Drag Reduction of Turbulent Flows by Additives.* 1995 ISBN 0-7923-3485-X
33. Y.P. Golovachov: *Numerical Simulation of Viscous Shock Layer Flows.* 1995 ISBN 0-7923-3626-7
34. J. Grue, B. Gjevik and J.E. Weber (eds.): *Waves and Nonlinear Processes in Hydrodynamics.* 1996 ISBN 0-7923-4031-0
35. P.W. Duck and P. Hall (eds.): *IUTAM Symposium on Nonlinear Instability and Transition in Three-Dimensional Boundary Layers.* 1996 ISBN 0-7923-4079-5
36. S. Gavrilakis, L. Machiels and P.A. Monkewitz (eds.): *Advances in Turbulence VI.* Proceedings of the 6th European Turbulence Conference. 1996 ISBN 0-7923-4132-5
37. K. Gersten (ed.): *IUTAM Symposium on Asymptotic Methods for Turbulent Shear Flows at High Reynolds Numbers.* Proceedings of the IUTAM Symposium held in Bochum, Germany. 1996 ISBN 0-7923-4138-4
38. J. Verhás: *Thermodynamics and Rheology.* 1997 ISBN 0-7923-4251-8
39. M. Champion and B. Deshaies (eds.): *IUTAM Symposium on Combustion in Supersonic Flows.* Proceedings of the IUTAM Symposium held in Poitiers, France. 1997 ISBN 0-7923-4313-1

Kluwer Academic Publishers – Dordrecht / Boston / London

Mechanics

SOLID MECHANICS AND ITS APPLICATIONS

Series Editor: G.M.L. Gladwell

Aims and Scope of the Series

The fundamental questions arising in mechanics are: *Why?, How?,* and *How much?* The aim of this series is to provide lucid accounts written by authoritative researchers giving vision and insight in answering these questions on the subject of mechanics as it relates to solids. The scope of the series covers the entire spectrum of solid mechanics. Thus it includes the foundation of mechanics; variational formulations; computational mechanics; statics, kinematics and dynamics of rigid and elastic bodies; vibrations of solids and structures; dynamical systems and chaos; the theories of elasticity, plasticity and viscoelasticity; composite materials; rods, beams, shells and membranes; structural control and stability; soils, rocks and geomechanics; fracture; tribology; experimental mechanics; biomechanics and machine design.

1. R.T. Haftka, Z. Gürdal and M.P. Kamat: *Elements of Structural Optimization.* 2nd rev.ed., 1990 ISBN 0-7923-0608-2
2. J.J. Kalker: *Three-Dimensional Elastic Bodies in Rolling Contact.* 1990 ISBN 0-7923-0712-7
3. P. Karasudhi: *Foundations of Solid Mechanics.* 1991 ISBN 0-7923-0772-0
4. *Not published*
5. *Not published.*
6. J.F. Doyle: *Static and Dynamic Analysis of Structures.* With an Emphasis on Mechanics and Computer Matrix Methods. 1991 ISBN 0-7923-1124-8; Pb 0-7923-1208-2
7. O.O. Ochoa and J.N. Reddy: *Finite Element Analysis of Composite Laminates.* ISBN 0-7923-1125-6
8. M.H. Aliabadi and D.P. Rooke: *Numerical Fracture Mechanics.* ISBN 0-7923-1175-2
9. J. Angeles and C.S. López-Cajún: *Optimization of Cam Mechanisms.* 1991 ISBN 0-7923-1355-0
10. D.E. Grierson, A. Franchi and P. Riva (eds.): *Progress in Structural Engineering.* 1991 ISBN 0-7923-1396-8
11. R.T. Haftka and Z. Gürdal: *Elements of Structural Optimization.* 3rd rev. and exp. ed. 1992 ISBN 0-7923-1504-9; Pb 0-7923-1505-7
12. J.R. Barber: *Elasticity.* 1992 ISBN 0-7923-1609-6; Pb 0-7923-1610-X
13. H.S. Tzou and G.L. Anderson (eds.): *Intelligent Structural Systems.* 1992 ISBN 0-7923-1920-6
14. E.E. Gdoutos: *Fracture Mechanics.* An Introduction. 1993 ISBN 0-7923-1932-X
15. J.P. Ward: *Solid Mechanics.* An Introduction. 1992 ISBN 0-7923-1949-4
16. M. Farshad: *Design and Analysis of Shell Structures.* 1992 ISBN 0-7923-1950-8
17. H.S. Tzou and T. Fukuda (eds.): *Precision Sensors, Actuators and Systems.* 1992 ISBN 0-7923-2015-8
18. J.R. Vinson: *The Behavior of Shells Composed of Isotropic and Composite Materials.* 1993 ISBN 0-7923-2113-8
19. H.S. Tzou: *Piezoelectric Shells.* Distributed Sensing and Control of Continua. 1993 ISBN 0-7923-2186-3

Kluwer Academic Publishers – Dordrecht / Boston / London

ERCOFTAC SERIES

1. A. Gyr and F.-S. Rys (eds.): *Diffusion and Transport of Pollutants in Atmospheric Mesoscale Flow Fields.* 1995 ISBN 0-7923-3260-1
2. M. Hallbäck, D.S. Henningson, A.V. Johansson and P.H. Alfredsson (eds.): *Turbulence and Transition Modelling.* Lecture Notes from the ERCOFTAC/IUTAM Summerschool held in Stockholm. 1996 ISBN 0-7923-4060-4
3. P. Wesseling (ed.): *High Performance Computing in Fluid Dynamics.* Proceedings of the Summerschool held in Delft, The Netherlands. 1996 ISBN 0-7923-4063-9
4. Th. Dracos (ed.): *Three-Dimensional Velocity and Vorticity Measuring and Image Analysis Techniques.* Lecture Notes from the Short Course held in Zürich, Switzerland. 1996 ISBN 0-7923-4256-9

KLUWER ACADEMIC PUBLISHERS – DORDRECHT / BOSTON / LONDON

MIX
Papier aus verantwortungsvollen Quellen
Paper from responsible sources
FSC® C105338

If you have any concerns about our products,
you can contact us on
ProductSafety@springernature.com

In case Publisher is established outside the EU,
the EU authorized representative is:
Springer Nature Customer Service Center GmbH
Europaplatz 3, 69115 Heidelberg, Germany

Printed by Libri Plureos GmbH
in Hamburg, Germany